D0847722

G. Evelyn Hutchinson and the Invention of Modern Ecology

G. Evelyn Hutchinson. Courtesy of Yemaiel Aris.

G. Evelyn Hutchinson and the Invention of Modern Ecology

Nancy G. Slack

Foreword by Edward O. Wilson

Yale

UNIVERSITY PRESS

New Haven & London

Published with assistance from the Fourth Century Fund.

Yale University Press books may be purchased in quantity for educational, business, or
promotional use. For information, please e-mail sales.press@yale.edu (U.S. office) or
sales@yaleup.co.uk (U.K. office).

Designed by James J. Johnson and set in Fairfield Medium type by
Tseng Information Systems, Inc.
Printed in the United States of America.

Library of Congress Cataloging-in-Publication Data
Slack, Nancy G.
G. Evelyn Hutchinson and the invention of modern ecology /
Nancy G. Slack; foreword by Edward O. Wilson.
p. cm.
Includes bibliographical references and index.
ISBN 978-0-300-16138-0 (cloth : alk. paper)
1. Hutchinson, G. Evelyn (George Evelyn), 1903–1991. 2. Hutchinson, G. Evelyn
(George Evelyn), 1903–1991—Influence. 3. Hutchinson, G. Evelyn (George Evelyn),
1903–1991—Friends and associates. 4. Ecologists—England—Biography.
5. Naturalists—England—Biography. 6. Ecology—History—20th century.
7. Scientists—England—Biography. 8. Scholars—England—Biography. I. Title.
QH31.H87S58 2010
570.92—dc22
[B]
2010019069

A catalogue record for this book is available from the British Library.

This paper meets the requirements of ANSI/NISO Z39.48-1992 (Permanence of Paper).

10 9 8 7 6 5 4 3 2 1

THIS BOOK IS DEDICATED, WITH THANKS, TO

W. T. (Tommy) Edmondson

AND TO MY HUSBAND,

Glen A. Slack

Both Forbes and Darwin realize struggle but see that it has produced harmony. Today perhaps we can see just a little more. The harmony clearly involves great diversity, and we now know . . . that every level is surprisingly diverse. We cannot say whether this is a significant property of the universe; without the model of a less diverse universe, a legitimate but fortunately unrealized alternative, we cannot understand the problem. We can, however, feel the possibility of something important here, appreciate the diversity, and learn to treat it properly.

G. Evelyn Hutchinson,
The Ecological Theater and the Evolutionary Play, 1965

Contents

Foreword by Edward O. Wilson   ix

Preface   xi

Acknowledgments   xvii

CHAPTER 1. A Man So Various   1

CHAPTER 2. "The Circumstances of My Upbringing"   15

CHAPTER 3. Becoming a Zoologist: The Hunter and Gatherer   42

CHAPTER 4. From Naples to South Africa: A Failed Career?   63

CHAPTER 5. First Years at Yale   89

CHAPTER 6. Chief Biologist: The Yale North India Expedition   100

CHAPTER 7. The First Crop: W. T. Edmondson, Gordon Riley,
Ed Deevey, and Max Dunbar   115

CHAPTER 8. The Old and the New Limnology   138

CHAPTER 9. Radioisotopes: Radiation Ecology and
Systems Ecology   159

CHAPTER 10. Biogeochemistry: From Linsley Pond to the
   Guano Islands   171

CHAPTER 11. The Three Wives and Yemaiel   199

CHAPTER 12. Good Friends: Margaret Mead and
   Gregory Bateson   235

CHAPTER 13. Fond Correspondents: Rebecca West and
   Evelyn Hutchinson   250

CHAPTER 14. From the N-dimensional Niche to Santa Rosalia   275

CHAPTER 15. Hutchinson the Environmentalist   294

CHAPTER 16. The Polymath: Art History and Many Other Fields   320

CHAPTER 17. The Last Years—From Yale to England   334

CHAPTER 18. Concluding Remarks: Hutchinson's Legacy
   in Ecology   371

Notes   395
Bibliography   431
Index   443

Foreword

G. Evelyn Hutchinson was the last great Victorian naturalist, a pioneer of modern ecology and justifiably called its founder, the one who brought the discipline into the Modern Synthesis of evolutionary theory. He was a polymath, master expositor, and teacher, and one of the few scientists who could unabashedly be called a genius. Because of Hutchinson's extraordinary scholarship and breadth of his influence, historians of science and general readers will be grateful for Nancy Slack's well-written, reliable, and insightful biography.

Hutchinson lived in the golden age of ecology, when useful discoveries in the field were easy to make and fit into concepts just in the process of forming. There were still relatively few scientists who addressed ecology effectively as a fundamental science, and many fewer still who had the capacity to formulate and test the new important ideas effectively. As in any golden age, there were heroes. They led entire schools of thought, and students tried to emulate them.

The golden age lasted, very roughly, five decades, from the 1930s into the 1970s. Hutchinson was ideally suited for his major role in its forefront. First of all, he seemed interested in almost everything. It was hardly an exaggeration to say that Hutchinson never met a fact or logical theoretical advance he didn't love, providing it was intelligently presented to him. He was, as Nancy Slack describes him, open and welcoming to any information relevant to his sphere of scholarly interest, which was so wide as to defy any simple description of its content. Of equal importance for the magnitude of his importance, he treated everyone, senior scientists and students alike, with equal respect.

Hutchinson was shy, but the social reticence he displayed only added to his reputation. His students—and others like myself who only experienced him peripherally—were inspired by the combination of his intellectual warmth with the sense that here before us was a man of deep knowledge and understanding. We all knew that, large or small, our discoveries and best ideas would receive his sincere encouragement. (My own formulation of character displacement coauthored with W. L. Brown, for example, was one influence that led to Hutchinson's analysis of the ecological niche; and I've boasted about that throughout my career.) To the next generation, many who became leaders of their own, he was like Linnaeus and his seventeen "apostles," who were told as they set off for distant lands from the Levant and Japan to South and North America: Go there, collect and learn all you can, and return and report what you find of importance.

As this biography documents, Evelyn Hutchinson will be remembered concretely as the founder of the synthetic disciplines of limnology and biogeochemistry and, in fact, much of basic ecology itself in its present form. But his life means much more than that.

He will be remembered for his spirit and quality of mind. If an angel of the Lord were to purge the Temple of Science of those who are drawn by a joyful sense of their superior intellect, and those there for utilitarian purposes, only a few would be left behind, and he would be one of those—and that is why we love him. Einstein said that of Max Planck in 1932, and we can now appropriately say it of Evelyn Hutchinson.

EDWARD O. WILSON

Preface

G. Evelyn Hutchinson was, according to Stephen Jay Gould, the most important ecologist of the twentieth century and, according to newspaper accounts, the "Father of Ecology." He himself insisted that that title belonged to Darwin, but he did not deny his paternity of a large number of outstanding young ecologists.

Hutchinson was instrumental in the birth of several new fields, including systems ecology and radiation ecology, and in new advances in population ecology, biogeochemistry, limnology, and, together with his students, even behavioral ecology. As E. O. Wilson wrote in his book *Naturalist,* Hutchinson deserves to be called the father of evolutionary ecology.

He initiated the "colonization" of ecology by mathematical theoreticians, as one historian wrote. Hutchinson's niche theory propelled the "pioneer existence on the intellectual frontier of ecology," in the words of two ecologists. These are striking images of Hutchinson and his students as successful colonists of the then-open field of modern ecology. Although a number of biologists and historians of science have written about his work, this is the first biography of Hutchinson. His life, from 1903 to 1991, spanned almost the entire twentieth century; he published books and papers for seventy-five of those years. Although he was born and educated in Cambridge, England, and spent his early professional years in Italy and South Africa, nearly his whole professional life was spent at Yale

University. There he taught more than four decades of graduate students, a role that he felt was as important as his own scientific work.

This book will be of special interest to ecologists, limnologists, and other biologists, as well as to environmentalists, historians, and many others. Hutchinson was a fascinating person as well as a renowned and innovative scientist. He and his graduate students worked in a great variety of fields, in most of which Hutchinson also published papers or books. His fields spread out from his fascination with the ecology of water bugs to so many aspects of ecology and other related—and not so re-lated—fields, as is shown in the diagram that appears in chapter 1.

As one now well-known ecologist wrote, "I swallowed hard and dared to ask the great polymath to be the advisor for my thesis. That began a rich adventure which so many shared." Hutchinson knew and wrote about not only ecology but also a great many other subjects from the biosphere to medieval animal art, stone circles in Britain, homosexuality, and women in science. The "Marginalia" column he wrote for *American Scientist* over several decades was read by a great variety of scientists and by others as well. Dame Rebecca West, the English novelist, was so impressed by one of these "Marginalia" essays that she wrote to Hutchinson and asked to meet him. That meeting led to a close friendship and a thirty-five-year correspondence. Hutchinson had a particular talent for friendship. He kept up with his graduate students, both men and women, long after they left Yale. In addition to Rebecca West, he was a good friend of both Mar-garet Mead and Gregory Bateson.

Hutchinson had three very different and interesting wives. The first was a well-known biologist; the second shared his deep interests in art, music, and literature; and the third (the wife of his old age), of Caribbean origin, provided him with insights into the black community and traveled with him to Japan to receive the Kyoto Award (the Japanese equivalent of the Nobel Prize, which does not exist in the field of ecology).

Hutchinson received many awards, including the Franklin Medal, which put him in the company not only of Ben Franklin, but also of Albert Einstein, Niels Bohr, Enrico Fermi, and Thomas Edison. Although this was an award for inventors, not for ecology, he warned in his acceptance speech, long before the current concerns about climate change: "I hope the things we are doing to the atmosphere will cancel each other out." He also received several major awards for his important environmental work, much of which is not well known. The issues he worked on include the eutrophication of lakes, the use of Agent Orange during the Vietnam War, and the preservation of the island of Aldabra (see chapter 15).

As the director of graduate students in ecology and other fields at Yale (including animal behavior and the history of science), Hutchinson had no peer. His innovative teaching methods, partly developed from his own interesting career at Cambridge University and probably in part from his Yale mentor, Ross G. Harrison, are attested to in interviews with many of his graduate students and are important in terms of training graduate students today.

In addition to his direct influence as a teacher, many other people who were not his students learned much from him through his writings. Not many scientists write superbly; Hutchinson was one of the few. His very first book, *The Clear Mirror*, an account of his travels in Ladakh on the Yale University North India Expedition when he was still in his twenties, was a literary success.

I was one of many graduate students in ecology who did not study at Yale but were much influenced by Hutchinson's writings. My research concerned community and evolutionary ecology, and I read many of Hutchinson's papers and books, including my favorite one, *The Ecological Theater and the Evolutionary Play*. I dedicated my dissertation, and its subsequent publication, to Hutchinson. I had never met him while a graduate student, but shortly afterward, when I was an assistant professor, he was invited to give a talk at my university. It was 1971, and he had many invitations, but he accepted this one. He delivered his lecture, "The Natural History of Lakes," to both the biology faculty and students, and also held a "rap session" for the fifteen graduate students and me. He kept us enthralled for a very long time answering our questions about eutrophication and a great many environmental and other ecological topics. He took each question seriously and gave a short dissertation on each topic, citing references in true Hutchinsonian fashion. Afterward he asked if any of us had recently done any research on aquatic plants; I had. He listened carefully to my account of *Sphagnum* ecology in Michigan bogs and bog lakes, my postdoctoral research of the previous summer (with Dale Vitt). He asked pertinent questions and treated me as a scientist whose ideas were to be taken seriously. Hutchinson's account of this research, not yet published at the time of our discussion, appeared in the third volume of his *Treatise of Limnology*. Similar stories from young scientists appear throughout this book.

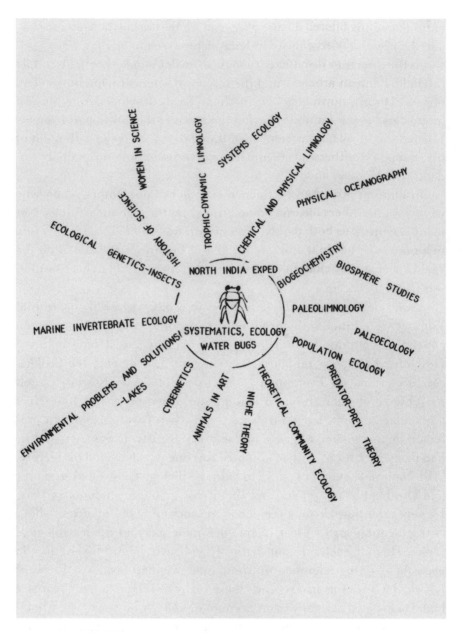

P-1. Fields in which G. Evelyn Hutchinson and his graduate students worked. Hutchinson's earliest work was on systematics and ecology of water bugs (1918 ff.) and in limnology, South Africa (1927–28) and the North India Expedition (1932). He himself did research in nearly all these fields except physical oceanography; additional fields such as behavioral ecology, taken up by the Peter Klopfer branch (see figure 1), could also be included.

When I was offered a sabbatical year at Yale working in the Section of the History of Medicine and Life Sciences, I wrote to Hutchinson and then visited him to discuss writing an article about his life and work. I did write such an article, in *Endeavour*, together with Larry Slobodkin, one of his well-known graduate students. I spent the fall of 1990, before he returned to his family in England in 1991, with Hutchinson and with his archives at Yale, a treasure trove of everything he had saved since his university years. There was too much material for one article. I worked in additional archives and recorded interviews, now at the Smithsonian Institution, with more than fifty Hutchinson students, scientific associates, and family members in several countries, much of which went into writing this book. It is both a chronological account of Evelyn Hutchinson's life and the people in it and a scientific account of the major role he had in the invention of modern ecology.

Acknowledgments

A great many people have given important help in the writing of this book. I want to thank those who read all or parts of the manuscript and gave me good advice and other much-needed services. These include especially Thomas Lovejoy, Eric Mills, Frederic Davis, Malcolm Willison, Derek Briggs, and W. T. Edmondson, as well as Clyde Goulden, Larry Slobodkin, Janet Browne, Gene Likens, Pamela Henson, P. Thomas Carroll, Christiane Groeben, and Joel Cain.

I interviewed more than fifty people in several countries for this book. Among them were many graduate and postdoctoral students of G. Evelyn Hutchinson, his Yale colleagues and associates, members of his family in England, as well as two of his fellow Cambridge University students and his one surviving Cambridge teacher, Joseph Needham. Almost all of these people provided recorded interviews, which are now in the Smithsonian Institution archives, and allowed me to use excerpts from them in this book. A number of people, including Yemaiel Aris, Donna Haraway, Phoebe Ellsworth, and Evelyn Hutchinson's sister, Dorothea Hutchinson, allowed me to read their letters from Hutchinson; a great many of Dorothea's letters are now in the Hutchinson Papers at Yale University.

The following people were especially helpful during my sabbatical year as a visiting scholar in the Section of the History of Medicine and Life Sciences at Yale: Professors F. L. Holmes, John Harley Warner, Charles Remington, and Willard Hartman, as well as Hutchinson's able assistant

Anna Aschenbach and G. Evelyn Hutchinson himself, in his last year at Yale. Archivist Judy Schiff and others at Manuscripts and Archives, Yale University Library, and at the Beinecke Library were also helpful; I could not have written this book without Hutchinson's very extensive archives, as well as his illuminating correspondence with writer Rebecca West, now housed at the Beinecke Library and elsewhere. Archivists at the Fitz-william Museum, Cambridge, England; the University of Tulsa; and the Smithsonian Institution were also very helpful.

Hutchinson's family offered hospitality while I was conducting inter-views in London and elsewhere in England. I particularly thank Andrew and Linda Hutchinson, Francis and Joyce Hutchinson, Rosemary and Peter Dodd, Yemaiel and Ben Aris, and Dorothea and Hannah Hutchin-son, but others in England and Wales helped me as well.

My own family and friends were important in many ways, from ar-ranging and transcribing interviews, providing lodging, finding and iden-tifying photos, exchanging ideas, and cheering me on. Some of these in-clude Helen Steiner, Toby Appel, Christina Spiesel, Naomi Rogers, John Warner, David Furth, Egbert Leigh, Ann La Berge, Sally Gibbs, Sherrie Lyons, Anne Cahn, Dian Deevey, and G. Dennis Cooke, as well as family members Mari Eggers, David and Jonathan Slack, Gina Feuerlicht, and Rachel Garbus. My husband Glen, critical reader and constant supporter, helped most of all.

Yale University Press deserves many thanks, particularly my editor Jean E. Thomson Black and in-house editor Margaret Otzel; they saw this book through many difficulties. I also thank Press assistants Matthew Laird, Joseph Calamia, and Jaya Chatterjee for their invaluable help.

I am indebted to the anonymous donor, who made a generous con-tribution toward the production of this book. I especially wish to thank Edward O. Wilson for writing the foreword.

CHAPTER ONE

A Man So Various

I f there is anything I thank the Gods for . . . it is a wide diversity of tastes. Barred from scientific work, I should be miserable, if there were not heaps of other topics that interest me, from Gadarene pigs to Gladstonian psychology. . . . The cosmos remains always beautiful and profoundly interesting in every corner—and if I had as many lives as a cat I should leave no corner unexplored." These words are a quotation from Thomas Huxley, writing about himself.[1] But G. Evelyn Hutchinson, born in 1903, just eight years after Huxley's death, could have written it about himself. He might have had to substitute pottos for pigs and parapsychology or evolutionary psychoanalysis for Gladstonian psychology, but this quotation, as well as the chapter title, equally well describe Hutchinson.[2] One can hardly imagine either Huxley or Hutchinson "barred from scientific work."

In his long life Hutchinson left very little unexplored, both within and outside of ecology. Spectator sports and popular music were perhaps unexplored, as someone suggested at his Yale memorial service, but he actually taught a course at Yale with jazz musician Willie Ruff. And he famously seemed to find whatever his students and colleagues brought to him—ideas, hypotheses, organisms, cultural objects—"profoundly interesting." Hutchinson was certainly a polymath, a rare scientist who bridged the two cultures, the scientific and the humanistic, so clearly portrayed in the books of C. P. Snow.

New Fields in Modern Ecology

Rare indeed among top scientists are those few who bring about paradigm change in their own fields. Most important in the history of ecology, it was Hutchinson who invented new fields within modern ecology and who strongly influenced others in these new fields. His roles in the development of ecosystem, radiation, population, and aquatic ecology, including not only limnology but also biological oceanography, have been noted by several historians of science.[3] The last field was one in which Hutchinson himself did not work, but one in which he had a strong influence, as Eric Mills has shown. He also strongly influenced biogeochemistry, making it an important part of ecology.

Ecosystem ecology became a new and vibrant field with work by Hutchinson and his postdoctoral student Raymond Lindeman. Lindeman's seminal paper, "The Trophic-Dynamic Aspect of Ecology," had what Robert McIntosh called "an extremely difficult birth."[4] Hutchinson both strongly influenced the development of this paper and acted as its midwife when more established limnologists turned it down as too theoretical.

Hutchinson pioneered the first use of radioisotopes in ecological field experiments, in the now famous Linsley Pond near New Haven. These experiments and publication spawned the new, and later expanded and well funded, field of radiation ecology.[5] The Linsley Pond experiments, a demonstration of phosphorus cycling, were also part of Hutchinson's and his early graduate students' venture into biogeochemistry. Hutchinson did not invent this field, which was not originally considered closely allied to ecology, but he was its pioneer in North America and later its authority in relation to lakes. His early graduate students used biogeochemistry in the study of lake ecology. Later it became essential to the study of marine ecosystems and of such terrestrial systems as peatlands.

Ecology went from an essentially descriptive field to a highly theoretical one from the time of Hutchinson's arrival on the scene in the late 1920s to the 1960s, in large part due to his work and that of his students. He was one of the first to wed ecology and mathematics, the "language" that was missing in ecological studies, as he wrote in a 1940 review of a book by Frederic Clements and Victor Shelford, the eminent ecologists of the period.[6]

The Many Decades of Hutchinson Graduate Students

Hutchinson spent almost all his career, over sixty years, at Yale University. He truly valued his role as teacher—especially as a teacher of graduate students. His students valued him as well. One of them, Larry Slobodkin, pointed out that although the students were very different from one another, they were uniform "in the conviction that Evelyn helped each of them. Anthropologists, entomologists, limnologists, geologists, primatologists all felt that he impelled them along their paths."[7] In addition to the varied areas of ecology in which his own graduate and postdoctoral students worked, Hutchinson served on the doctoral committees of Yale students in this whole range of fields cited above and influenced their research. He was once offered an excellent position at the U.S. Geological Survey, which would have provided him with all the research funds and facilities he could wish for and allow him to work on whatever interested him, but he turned it down because he would no longer have graduate students to teach.

Many of Hutchinson's students had their own graduate students, in some cases to the fourth generation. His intellectual family tree was tabulated and drawn in 1971;[8] by now there are several more generations, including many of today's top ecologists.

From Cambridge Childhood to Norfolk to Cambridge University

Hutchinson's early life was spent in Cambridge, England, where the intellectual community in the 1920s and 1930s has been compared to that of Athens in its early glory. Hutchinson, whose father was Arthur Hutchinson, a mineralogist at Cambridge University, grew up among famous scientists. Two of Darwin's grown sons came to dinner at his family's home. Eric Mills called him a "born ecologist," brought up as he was in the "intellectual and environmental riches of Cambridge."[9] It is indeed true that by the age of five or six he was not only collecting aquatic insects but studying their preferred environments in small aquaria he made at home. Thus started a lifelong interest in water bugs, on which he was considered an expert by his teenage years.

He went away to secondary school, Gresham's, a school in Norfolk. Gresham's was unusual in emphasizing science instead of Latin and

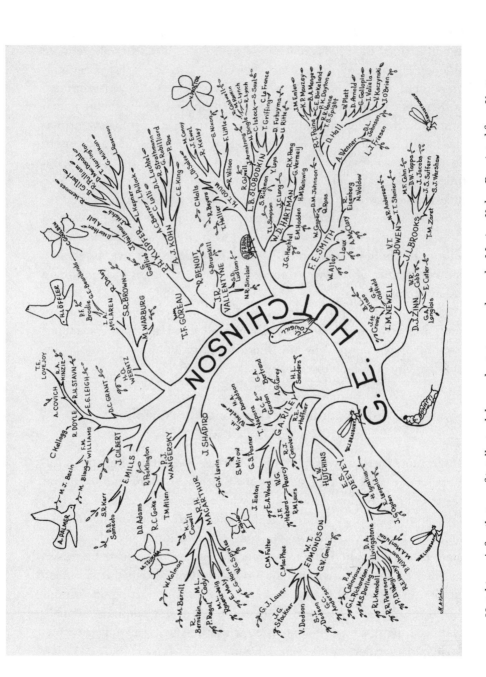

1. Hutchinson's family tree of intellectual descendants to 1971, drawn by Marion Kohn. Reprinted from Yvette Edmondson, 1971, page 163, fig. 9. Copyright 1971 by the American Society of Limnology and Oceanography, Inc.

Greek (which, however, he also learned). He was an active member of all the natural history clubs at Gresham's and published his first paper, about a swimming grasshopper, at age fifteen.

Later, at Cambridge University as a zoology student, he was a self-confessed "hunter and gatherer," searching out top people in such fields as biochemistry and genetics as well as in zoology. He received two firsts, the highest grades, on the tripos (final examinations), as well as a major zoology prize (only, Hutchinson insisted, because the top student was a woman and thus not then eligible for prizes). But he chose not to stay on at Cambridge for graduate work. He never earned a doctorate, of which he was rather proud in his older age, although he received many honorary degrees, including one from Cambridge University.

Naples and South Africa

In 1925, after his time at Cambridge, Hutchinson went to Italy, to the famous Stazione Zoologica in Naples with a Rockefeller Fellowship. There he studied octopi, which were caught and brought into the laboratory together with a great variety of sea creatures. He also fell in love with Italy and its culture, and later in life he returned there many times to study its lakes and enjoy its art and music.

From Italy it was on to South Africa, at age twenty-three, to explore the fauna of a new continent and to teach at a university under a professor even his parents had warned him against: Professor Fantham, who had physically thrown out the previous lecturer. The young Hutchinson did not fare much better; he was soon fired from his teaching post. But because he had a contract he was able to continue his research, together with his new wife, Grace Pickford, who had been his fellow zoology student at Cambridge. She had come to South Africa on a Newnham College research fellowship, and they married there. Together they worked on the shallow lakes or "pans" near Cape Town. Here Hutchinson found his first true field: limnology, the study of freshwater systems. Later he became an international expert on lakes, and beginning in 1957, the author of the four-volume *Treatise on Limnology*.[10]

The "Accidental" Instructor at Yale and the North India Expedition

When his appointment in South Africa ended, Hutchinson applied by cable for a graduate fellowship to work with the very well-known Yale embryologist and inventor of tissue culture Ross Granville Harrison. His cable arrived too late; the fellowships had already been given out. There was, however, an unexpected opening for a zoology instructor at Yale, and Hutchinson was offered this position by return cable.

He arrived in New Haven in 1928, together with his wife Grace. She became a doctoral student, but he went right into teaching several different undergraduate courses including embryology, the field to which he had originally applied for graduate study.

Like his predecessors, not only Huxley but earlier Humboldt, Darwin, and Wallace, Hutchinson went off by ship and various other means of transportation, even elephant, to an exotic, underexplored part of the world. Quite early in his career at Yale, in 1932, he was selected by Professor Harrison, then chair of the Zoology Department, to be the lead biologist on the Yale North India Expedition, including present-day Ladakh. On this trip Hutchinson collected everything: water bugs and other insects, spiders, fish, plants, even fossils. He skinned large animals for the Yale Peabody Museum.

His special study was the ecology of very high-elevation lakes, as well as the lives of the people who lived in Ladakh, a mountainous section of northern India. Ingenious at making equipment out of bits and pieces of hardware, he was able to scrounge; he took temperature, light, and other measurements and was soon sending experimental data back to Grace at Yale. His many evocative letters to her trace the expedition and his work.

This trip also marked the beginning of Hutchinson's literary career. His often lyrical book *The Clear Mirror* describes the colors, organisms, and ecology of the high-elevation lakes as well as the lives, religion, and art of the people of Ladakh.[11] Few scientists can write well, but Hutchinson could; his many subsequent essays and books were read by a wide audience. His *The Ecological Theater and the Evolutionary Play* captivated this author early in her ecological career.[12]

Hutchinson's Graduate Students

The first crop of Hutchinson's graduate students—Gordon Riley, Edward Deevey, and W. T. Edmondson, all of whom subsequently had distinguished scientific careers—appeared on the scene soon after he returned to Yale in 1933. Hutchinson was by then teaching graduate courses. In those days he was able to be out in the field with his students. As Riley wrote, field trips were not an idyllic pursuit: "A milestone was the first trip to Linsley Pond, which eventually became well known as a result of the work that we and succeeding generations did there. . . . It was a chilly March day in 1935. The ice was barely off the lake. We went out in a rubber boat that he used in his earlier expedition to North India. It leaked. Our bottoms were wet and cold. . . . I particularly remember one day early in the following winter when the ice was thin and rubbery. I frankly did not have the courage to venture out on it, but Evelyn did; he got our samples."

Riley also said that Hutchinson imbued students with a special feeling about doing science: "Science is an artistic achievement to be relished as such. The small logical deductions from hard facts are part of the structure, but unless we visualize a larger structure we can hardly hope to guide our search for further facts."[13] This larger structure, Riley thought, is in some ways analogous to the composition of a painting, but the scientist has stricter rules than a painter. In that sense science was, to Hutchinson, a work of art in the same way that a mathematician talks about "a beautiful theorem."

Hutchinson's three early students all worked on lakes, but their subjects of study were very different. Riley studied the copper cycle, one aspect of biogeochemistry; Deevey did pioneering work in paleolimnology and related subjects; and Edmondson investigated the ecology and population biology of rotifers, a phylum of tiny but complex water animals.

These early students did not come to Yale specifically to work with Hutchinson; later ones did. By the late forties he had attracted many more graduate students, who, together with former graduate students who became Yale professors, formed his research group, which continued on into the 1970s. It was not a classic research school, however; these were largely the province of German and other European universities, with one "Herr" Professor, one set of problems, and often similar research tools. Ross Harrison and his graduate students at Yale fit this research school pattern quite well. Harrison was "the Chief," and graduate students worked on problems of embryology using a set of research tools largely invented by Harrison and his students. Hutchinson's students interacted with him and often with each other, attending the same graduate courses and seminars, but they conducted a wide variety of research and used many different research methods.[14]

The next crop of students, including Donald Zinn, John Brooks, Vaughan Bowen, Willard Hartman, Fred Smith, Howard T. Odum, and Larry Slobodkin, all of whom earned Ph.D.s by 1951, studied a great variety of organisms, some in freshwater, some in marine ecosystems. Bowen and Odum did their doctoral research in biogeochemistry, although they worked in different areas later in their careers. Many of these men also became well-known ecologists and produced branches of the Hutchinson "tree"; several were elected to the National Academy of Sciences. Riley, Deevey, Brooks, and Hartman all spent some years on the Yale faculty.

The fifties and sixties were the golden age for both Hutchinson and his graduate students. He published an amazing number of papers and books during this period, several of them of great importance in terms of initiating new ecological research. Larry Slobodkin was one of the first to work in the field of population ecology, an important new field of Hutchinson's. Among his important graduate students of this era were Egbert Leigh, Peter Klopfer, Joseph Shapiro, Alan Kohn, Peter Wangersky, and Robert MacArthur. It was Slobodkin who sent Robert MacArthur, who had two degrees in mathematics, to Hutchinson for his doctorate. MacArthur subsequently became a leading scientist in population ecology.

Women were among Hutchinson's later graduate students and included Donna Haraway, Karen Glaus (later Porter), Maxine Watson, and Linda Carlson; among the men were Thomas Lovejoy, Alan Covich, and Patrick Finnerty. Hutchinson was sometimes called "Hutch" by his graduate students. He was never called by his first name, George; he was Evelyn from childhood. The response to his original letter to the Stazione Zoologica in Naples was addressed to "Miss Hutchinson." He later answered a letter to a correspondent who hadn't met him: "Please note that I am male. The E of Evelyn is long. As someone put it in introducing me, 'as in Evelyn Waugh.'"[15]

Hutchinson thanked several of his graduate students (in print) for help in developing the ideas in one of his most famous papers, the 1957 "Concluding Remarks," in which he introduced his multidimensional niche concept. These were students in his graduate seminar, which Hutchinson always made interactive.

Books and Papers of the Golden Age

The year 1957 also saw the publication of the first volume of *A Treatise on Limnology,* a seminal work still in use more than fifty years later. Three other volumes of the *Treatise* followed, making Hutchinson America's premier limnologist. Another very important Hutchinson paper that spawned a great deal of ecological research—"Homage to Santa Rosalia or Why Are There So Many Kinds of Animals?"—appeared in 1959. His 1961 "The Paradox of the Plankton" and another with the provocative name, "Copepodology for the Ornithologist," contained new provocative ideas. Almost all his ecological papers of this era contained data from Hutchinson's own ecological studies of freshwater organisms.

Not all of the papers of the golden age were about ecology. From their titles we can see Hutchinson the polymath: for example, "Religion and the Natural Sciences," "A Speculative Consideration of Certain Possible Forms of Sexual Selection in Man," "Anybody Seen an Abominable Snowman Lately?" "A Preliminary List of the Writings of Rebecca West," "The Naturalist as Art Critic," "Is an Electron Smaller than a Dream?"[16] These were published in a great variety of journals, but by this period he was well known for his frequent articles labeled "Marginalia" in the *American Scientist,* covering whatever topic he chose, all in some respect related to science. Through this column he became known to many nonbiologists.

Three Very Different Wives

Hutchinson's personal life was complex as well in that he had three wives. The first, his fellow Cambridge student Grace Pickford, became a well-known scientist in her own right. While he was on the 1932 Yale North India Expedition, he and Grace were writing to each other about their impending divorce. On the return boat from India to England he met Margaret Seal, an Englishwoman whom he married in 1933 after his divorce became final. Margaret was a musician not a scientist, but she and Evelyn shared a deep interest in art, literature, and music. She was his beloved wife for fifty years, but they had no children of their own. They did have a World War II half-Jewish foster child, Yemaiel Oved, who was evacuated from London for her safety at the beginning of the war at age two and stayed with Evelyn and Margaret until she was five. Yemaiel was Evelyn's great delight, both in those early years and again in the latter years of his life. Margaret died of Alzheimer's; Evelyn cared for her during the many years of her illness.

Hutchinson married a third time in his eighties to Anne Twitty Goldsby, a much younger biologist of Haitian descent. They traveled to Kyoto together, where Evelyn received the Kyoto award. He survived her as well as Margaret and Grace.

A Special Gift for Friendship

Hutchinson had a wide range of friends. His immense correspondence, preserved in his archives, indicates the breadth and variety of his friendships. There are letters from a great many former graduate students who were doing research all over the world. At Yale he had faculty friends in a wide range of fields, from art history to literature, psychology, and music. Former students and many others returned to Yale for Hutchinson's eightieth birthday party in 1983. Some had been fellow members of Saybrook College or the Elizabethan Club. Many women were among them, some of them his former students, but there were other friends as well, like Shakespeare scholar Marjorie Garber and psychologist Phoebe Ellsworth.

Famous biologists and anthropologists were also among Hutchinson's good friends. In England they were mostly fellow zoologists and ecologists; David Lack and Charles Elton especially influenced his ideas. Margaret Mead, anthropologist, and Gregory Bateson, biologist and anthropologist, were also his close friends: Bateson knew Hutchinson from their early years together at Cambridge, and Mead knew him from their time together on the staff of the American Museum of Natural History, she in anthropology, he as a consultant in biogeochemistry. In the 1940s all three of them were involved in the Macy cybernetic conferences in New York. He helped Mead a great deal with one of her later books.

Hutchinson's closest friend over a very long period of time was the English feminist and well-known writer Rebecca West (later Dame Rebecca West). After reading one of his "Marginalia" articles in 1947, she wrote asking to meet him. When she came to New York to research articles for the *New Yorker,* the *Herald Tribune,* and other magazines and newspapers, they often met. They exchanged hundreds of letters about their lives and work over a thirty-five-year period. When Dame Rebecca West died, Hutchinson became her literary executor.

Hutchinson was an inveterate letter writer. His wife Margaret and Rebecca West both died in 1983. He continued to write about his daily life in the following years, mainly to his sister Dorothea in England. In accordance with an old English tradition, he kept all his letters, professional and personal, all the memos from his many societies, essentially all his papers from the time he arrived at Yale in 1928 until he went back to England after the death of his third wife, Anne, in 1991. His is a very rich archival collection indeed, and most of it is in two libraries at Yale.

Hutchinson in the Eyes of Others

Thomas Lovejoy, a late graduate student who became a vice president of the World Wildlife Fund and subsequently has held other important conservation positions, was close to Hutchinson. They first met when Lovejoy was a freshman at Yale. As Lovejoy later recalled, "his deep cultured voice [which retained its English accent] asking penetrating questions during . . . Zoology Department seminars. . . . Uniformly, the first impression we all had of him was of incandescent brilliance. Stephen Jay Gould described him as 'the greatest American ecologist.'"[17]

Tom Lovejoy was surprised when a contemporary graduate student said that what made Hutchinson great was his students. Lovejoy wrote, "There was, in fact, extraordinary synergy between the great professor and ourselves which extended throughout our lives." Lovejoy quoted one Yale student who had taken only a single Hutchinson course in the 1950s yet said decades later, "My head is still full of him."

Hutchinson encouraged his graduate students to develop their individual talents and interests. The National Science Foundation was concerned in the 1960s that "graduate students were being co-opted by serving as research assistants for their professors," Lovejoy wrote. Hutchinson responded that he was not interested in making "smudged carbon copies" of himself. Hutchinson emphasized, as several others of his students have also related, the positive: one should always recognize the "good things in bad papers"—or in poor seminars. His students, Lovejoy felt, reflected "Evelyn's concern for women's and minorities' proper role in science." Hutchinson could often see far ahead. When he won the Franklin Medal in 1979 for both theoretical and applied contributions in science, he made the chilling remark that he hoped "the various things we are doing to the atmosphere will cancel each other out."

Many of his graduate student friends, as well as others whom he helped even after his retirement, felt that Hutchinson understood them as people, not just in their role as scientists. For example, Saran Twombly, future limnologist and professor, wrote a very long letter from Pellanza, Italy, where she had been working and was experiencing a degree of "ambiguity and turmoil" within the institute there. In her letter she wrote, "My dear Professor Hutchinson, . . . In just a few weeks [after she returned to Yale] we can sit and talk about everything, which will be much more satisfying for me than trying to write."[18] But she had also written earlier letters to him about her life and work in Italy, which she knew he would understand.

Evelyn Hutchinson made important contributions to scientific work in ecology and related fields, yet it is important to also have an understanding of Hutchinson's character. Two examples, one told by a limnologist who was not his student, the second from a letter Hutchinson wrote to one of his Yale graduate students, reveal Hutchinson as a person and teacher.

In the early 1960s a young master's student who was studying zooplankton in an Indiana lake went alone to a meeting of the international limnology society (SIL) in Wisconsin. He recounted: "On the first morning I went to the dining hall, picked up some food and sat by myself. An older man came to my table and asked if he could sit there. I agreed of course and he proceeded to ask me what I was doing in limnology. I told him of my project as best as I could, having had just one summer of lake research under my belt. He was interested, asked me questions and made me feel pretty special. The next thing I knew I was at the plenary session and he was introduced as the speaker. I was stunned and went over and over the things I had told him."

Four or five years later the student, who was by then a postdoc with Eugene Odum, attended the American Association for the Advancement of Science (AAAS) meeting in Chicago. There he received an invitation, he knows not how, "to join a smallish group in a very nice Chicago hotel to listen to a lecture by Hutchinson. There was a string quartet and wine and snacks before the lecture. I remember the evening as one of the most civilized I have ever attended. The man had a type of class that was all his own." This student, now a well-known limnologist, added, "I still use Hutchinson's masterful books, the *Treatise on Limnology*. Volume 2 is on my desk as I write this."[19]

In summer 1969 Hutchinson wrote from an inn in Gloucester, England, where he was vacationing with his wife Margaret to Karen Glaus (Porter), one of his graduate students. After thanking her for her "most interesting letter," Hutchinson went on: "I am now writing in the New Inn Court c. 1450, founded to accommodate pilgrims to the tomb of Edward II who was murdered, probably as a result of his homosexuality, but became thereby a martyr. The inn has about 12 bars now and lots of noisy teenagers." Hutchinson seemed to be enjoying this scene. He continued with his serious thoughts and advice:

> I have thought a lot about your project. First a piece of general observation. Practically nobody has ever done any good work by setting up a detailed project ahead of time and then carrying out all the projected details in order. One can start out and get a lot of data; when it has become familiar[,] the new and exciting implications appear while drying a cup or getting into a car, in a quite unpredictable way. This I fancy only happens if one really knows a lot of factual material got with our own hands and allowed to dance around in our unconscious. . . . If you really get started and are as good as I think you are . . . the original ideas will come. Until you get them you don't have any idea what they will be.

Hutchinson then proceeded to give her several pages of detailed advice, starting with, "The little evidence available for the Daphniidae [water fleas] suggest no great likelihood of morphology regulating food habits, but apart from Carolyn Burn's work, nobody had done anything very good." He added that the group Glaus was working on was likely to be "much more diversified . . . and range all the way from almost pure filter feeders . . . to predators. I think there is much to be done there. . . . I like much of what you are thinking about here."[20]

Hutchinson had a sign on his office door that read: "Do not discourage students. You are almost certain to succeed." Far from discouraging his students, Hutchinson encouraged them to follow their own paths to discovery.

Hutchinson's Impact as Environmentalist and Ecologist

Hutchinson's environmental work is less known than his innovative re-search in ecology and limnology. Some of it was widely recognized by prestigious awards, but other aspects, including his important part in pre-serving the island of Aldabra and his contribution to the banning of Agent Orange, were carried out behind the scenes.

Hutchinson received many honorary degrees and even more awards from many countries for both his ecological and environmental work. It is also of interest that he turned down one award: the President's Medal of Science during the Nixon administration.[21] He was especially proud of his Franklin Medal and wondered what it was he had invented that had put him in the company of previous medalists Albert Einstein, Niels Bohr, Max Planck, Enrico Fermi and Thomas Edison, as well as of Benjamin Franklin. But Hutchinson had invented much of modern ecology, as we shall see.

By considering Evelyn Hutchinson's life and work and attempting to enter the intellect of this complex, diverse, and innovative man, we will gain a clearer understanding of his role in the development of modern ecology and his views on challenging ecological problems. We will also gain insight into educating today's science students in innovative ways that succeeded for Hutchinson and the Hutchinsonians.

CHAPTER TWO

"The Circumstances of My Upbringing"

W hy is it that childhood experiences, painful or enriching, affect one child in a particular way and a brother or sister quite differently? It is difficult to separate out the individual child from the external influences, including the attitudes and expectations of parents. Moreover, biography is influenced by hindsight. The end of the story is known; we can only try to sort out the influences that seem to be important in their early aspects.

G. Evelyn Hutchinson was born in Cambridge, England, in 1903. He lived in and near Cambridge until 1925, when he completed his university education. Cambridge in that era was full of scientific and literary ferment and innovation. Hutchinson's early years were quite extraordinary, and in some manner they surely influenced his future life and work.

Both of Hutchinson's parents were unusual people, distinct from others of their time and class. His father, Arthur Hutchinson, in particular opened many doors for Evelyn and provided him with companionship and instruction from an early age. His mother was a serious, strong-minded woman who fought for women's causes, including suffrage and fair treatment in the police courts. The Hutchinson family, academic and decidedly upper middle class although not rich, gave Evelyn, their older son, a privileged childhood.

"Hutchinson was a born ecologist," wrote one historian of science.[1] It does appear that he started in his chosen field at an unusually early age.

2. Evelyn at eleven months old with his mother, Evaline D. Hutchinson.
Courtesy of Francis Hutchinson.

Many small children are collectors; Evelyn, however, was a remarkable one. By the age of five he was collecting sticklebacks, water spiders, and water mites and making aquaria. Their denizens came from creeks and ponds near his Cambridge home. At five Evelyn realized that different kinds of animals lived in different kinds of water.[2]

From his early school days Hutchinson always had a peer group that pursued natural history. For the Cambridge Junior Natural History Club he recruited, along with several schoolmates, his younger brother and sister, neither of whom, however, continued in this kind of activity. He also seems to have become, in these early years, an "experimentalist," for example, trying to find ways to vary the environments of caterpillars to succeed in raising them in order to determine what sort of moth or butterfly might emerge. Unfortunately for the caterpillars, this endeavor taught him that experiments often fail. Cambridge, both during his childhood and later during his university years, was full of natural wonders, scientific discovery, and extraordinary people. His parents were two of them.

Arthur Hutchinson

Arthur Hutchinson, born in 1866, was not only Evelyn's role model but his mentor and advisor as well, the person whose approval Evelyn needed even if he did not always follow his advice. Evelyn's paternal grandmother, who lived next door during Evelyn's childhood, had been widowed early. She moved to Clifton so that her son Arthur could attend a science-oriented prep school and then moved to Cambridge, when he reached university age. Arthur was a scholar at Christ's College, Cambridge, and received two first-class degrees in natural science and in chemistry. He then traveled to Würzburg in Bavaria, where he completed a doctorate degree under Emil Fischer, and studied at the University of Munich. When he returned to England he was elected to a fellowship at Pembroke College, Cambridge. He started to work in mineralogy, and he is best known for his research in this field. He was credited with such exceptional powers of lucid exposition and such a keen sense of humor that he made even mathematical crystallography interesting.[3] Walter Campbell Smith of the British Natural History Museum called Arthur Hutchinson the best living teacher of mineralogy.[4]

Arthur Hutchinson also had great success in the design and construction of research apparatus, including Hutchinson's goniometer, named for him. One of these was set up in the Hutchinson house, where he used it for experiments. Evelyn later remembered watching his father producing thallium, sodium, or lithium flames to determine refractive indices of the crystals he was examining.

During World War I Arthur Hutchinson did important work on gas masks. He received many honors, including election as a fellow of the Royal Society (FRS) and presidency of the Mineralogical Society (1921–24). For much of his career, from 1896 to 1923, however, he was only a "demonstrator" of mineralogy, the lowest rung on the academic ladder. It was not until 1926, after Evelyn had graduated from Cambridge, that his father became professor of mineralogy and not until 1928 that he was elected master of Pembroke College. He was then in his sixties. Dorothea Hutchinson, his daughter and Evelyn's younger sister, commented that only after her father became both a professor and master of Pembroke were there finally plentiful family funds.[5] Evelyn's letters reveal much about his father. He wrote to Dorothea that he had discussed the careers of several German nineteenth-century chemists, including Emil Fischer, with Joseph Fruton, biochemist and historian of science at Yale: "Joe Fruton concluded that Fischer, under whom Father took his Ph.D. was very egocentric, compared even with his German contemporaries. Although lots of foreigners worked in his lab, they all had singularly undistinguished careers on returning home—except one, a mineralogist . . . called Hutchinson. Fruton was quite amazed when I told him that he was my father. Father had an immense admiration for Fischer as a scientist. . . . I think Father being able to learn so much from Fischer speaks very much in Father's favor. . . . The kind of attitude we knew, on which I have tried to model myself, must have been fully developed when he was a very young man."[6]

In Evelyn Hutchinson's last year of life he reminisced about his early years with father. They walked together in the hills of Cumberland where the family had inherited land. His father's knowledge included wildflowers, geology, and meteorology; he taught Evelyn a good deal about these subjects. He also had a wide knowledge of British archeology, which they frequently explored together. Evelyn said that when he explored new sites unaccompanied, he always tried to think about what he was looking at in his father's terms. Evelyn went on to cite his father's extreme helpfulness to him through his entire life, specifically his deep chemical insight. This was particularly important in Evelyn's work on chemical variation of lakes. "He undoubtedly lay behind me in a rather modest way in South Africa and Central Asia . . . and materially advanced my whole attempt to make a satisfactory classification of nutrients in terrestrial waters."[7] Dorothea also wrote about her father's influence.[8] Arthur Hutchinson died in late 1937, shortly after returning from visiting his son at Yale, where Arthur lectured on the history of science at Cambridge University. Evelyn's filial sentiments are apparent in a letter he wrote to the editor of *Nature* in response to his father's obituary published there: "I need hardly stress my Father's great interest in the history of science, as his last published work in your pages bears witness to. . . . [The] establishment of a permanent

collection of historic instruments at Cambridge would be a fitting memorial to him."[9]

In answer to a question about Evelyn's great respect for his father,
Willard Hartman, Yale zoologist, answered: "Respect is too cold a word. . . .
I just wouldn't have used it in that relationship. He was very proud of his
father and very close to him."[10]

Evaline Demezy Shipley Hutchinson

Evelyn's mother also played a significant role in Evelyn's life, albeit not as
a scientific mentor. Evaline Demezy Shipley married Arthur Hutchinson
in 1901. The Shipley ancestors had had a saddler's shop in London. One of
Evelyn's uncles, his mother's brother, was a well-known Cambridge zoologist, Sir Arthur Shipley, F.R.S. The Demezy name comes from an Italian
ancestor (originally de Mezzi) who came to England in the 1700s. Family
tradition maintained that he had left Italy after becoming a Protestant,
but it is now thought that he fled to England after killing a man in Italy.[11]

Evaline Hutchinson was a strong personality who exerted both positive and negative influences on all three of her children. Evelyn recalled
his mother as an ardent but strictly nonmilitant feminist. According to
her daughter, Dorothea, she was a keen suffragist but not a suffragette;
that is, she wanted women to get the vote by legal and peaceful means,
not by smashing shop windows or chaining themselves as the suffragettes did. Evelyn recounted his mother's appearance in Cambridge police
court. She was much concerned about the treatment of women there.
While she did much good, he recounted in *Kindly Fruits*, she possibly also
did some harm by her rather dominating and at times intolerant personality.

Dorothea characterized her mother as a strong personality who had
had a difficult time in her own late teen years. While her older sister,
Emily, married at an early age, Evaline was sent to an excellent boarding school. "Mother was an intellectual; her great ambition was to go to
Oxford University. But my grandmother died when Mother was barely
eighteen, and this meant that she [Evaline] had to give up all thoughts of
university and go home to keep house for her father and two younger sib-
lings."

Evaline Shipley found tending house utterly boring. The family had

3. Hutchinson's mother with his younger sister, Dorothea.
Courtesy of Francis Hutchinson.

a cook and maids; there was little for her to do. She later told Dorothea
how frustrated she had felt, but Dorothea was not sure if this experience
increased her mother's disapproval of her sister Emily, "known to us privately as 'wicked sister Emily.' If she hadn't been, as Mother said to me
years later, 'so highly sexed,' things would not have been so bad, but when
her three children were in their teens and her husband was a flourishing
barrister, Emily ran away with an artist and had an illegitimate child by
him. In a respectable, professional class family this was appallingly dis-

graceful. Mother was by nature something of a Puritan, and Emily's behavior directed this especially toward matters of sex."

But all the same Dorothea described Evaline as a warm, loving person
to her children—Evelyn, brother Leslie, and herself: "a splendid mother
to us when we were young, enjoying our games, encouraging all kinds of
out-of-door activities, and giving me the chance of sharing everything the
boys did." In Dorothea's view, Evaline Hutchinson was a fine and admirable mother "until we reached our teens, but she then had a rather inhibiting effect on me, and I had little help in developing from a tomboy to
an adolescent female."[12] She had an inhibiting effect on Evelyn also, well
into his twenties. Evaline Hutchinson later wrote a feminist book titled
Creative Sex. Her daughter, Dorothea, studied at Cambridge University,
became a psychiatric social worker, but never married.

English feminists of all types abounded in the years before and during World War I. The suffragettes were in a different world from that of
Evaline Hutchinson. These women were truly radical, often supporting
more than one unpopular cause. Alice Wheeldon was a suffragette in the
Women's Social and Political Union, whose members opposed conscription in World War I in addition to their views on women's suffrage.[13] Alice
Wheeldon's pacifist views were not unusual; Bertrand Russell was also a
peace advocate against conscription. Sylvia Pankhurst, a radical feminist
in today's terms, wrote several books about World War I and the suffragette movement. She was scornful of wartime policies of the Tory government under Lloyd George. Alice Wheeldon, not one of the elite, ran a
secondhand clothes business. She was accused, almost surely wrongly, of
conspiring to kill Lloyd George. The English spy who made the accusation
against her, her daughter, and her son never appeared at their 1917 trial.
Nevertheless, all three went to prison.

Evaline Hutchinson never went to prison for her beliefs, nor was she
a militant feminist like Sylvia Pankhurst, but neither was she a typical
upper-middle-class homemaker. During World War I, all of her children
were young; Evelyn, the oldest, was eleven. Many women were doing war
work, but conservative forces in England were worried about this trend.
One government spokesman even proposed that motherhood be treated
as a productive trade. A demographic crisis was seen to be generated
by the war and pronatalist policies were proposed. Doctors wrote that
war work, especially in industry, could damage women's health. Women
risked making themselves unfit for their "predestined role," to replenish

the human species. Yet there was also a negative view of women as the instigators of immorality, the bearers of illegitimate children, in wartime Britain, and the government tried to regulate women's bodies in terms of sexuality and venereal disease.[14]

Evaline Hutchinson supported women's right to do war work but harbored her own worries about venereal disease and prostitution. As Evelyn wrote about syphilis and his mother: "When salvarsan was developed some prospect of treatment seemed possible. . . . A considerable moral problem now arose. Two societies were introduced to combat the disease, one using only moral pressure, the other salvarsan. My mother was an ardent supporter of the moral society, though for most of my youth I was protected from all this sort of thing."[15]

Evaline Hutchinson's 1936 book, *Creative Sex*, written under the name E. D. Hutchinson, contained an introduction by a clergyman, Canon Charles E. Raven, D.D.[16] The origin of the book, according to the author in her foreword, was a luncheon in Cambridge four years earlier, for a group of women "well known beyond their town for social and public work" with a guest speaker from London whose topic was "Sex Life under Modern Conditions." The question was raised as to why some of the deviations from conventional morality advocated by many modern, that is, post–World War I, thinkers and writers should not be tolerated.

The book is a curious one, at least from a contemporary viewpoint. It is clearly a book about married sex, but even so it must have surprised, perhaps shocked, some of Evaline Hutchinson's Cambridge friends. She must have been aware of this likelihood since she used only her initials, not her first name, and had a clergyman write the introduction. The book includes New Testament quotations and references, but other references are to Havelock Ellis, Sigmund Freud, A. N. Whitehead, Bertrand Russell, Joseph Needham (*The History of Embryology*), the *Census of India*, and Rathbone's *Child Marriage: The Indian Minotaur*. Such a list indicates Hutchinson's wide reading and also her concern about maternal and infant mortality. She wrote extensively about sex as a creative force in marriage, but also in the arts and elsewhere, and gave advice to young engaged and married couples. She argued against premarital sex and trial marriages—but also against long engagements. She condemned prostitution ("vice"), legal or otherwise, and masturbation, although here she warned that its evils had been exaggerated. She was, however, in favor of contraception. Married women, she felt, should have a wider sphere than motherhood, and she cited evidence in both England and India of the toll of repeated childbearing. She argued for easing the divorce laws, particularly within the Anglican Church. At the time it was harder to have an Anglican marriage dissolved than a Catholic one.

Many women wrote marriage manuals in the twenties and thirties. So did men. "The wife must be *taught,* not only how to behave in coitus, but above all, how and what to feel in this unique act." So wrote Theodor Van de Velde in 1928.[17] These early marriage manuals spread the ideas of Havelock Ellis and other sexologists, whose writings were not readily available. The sexologists thought that their new scientific information should be translated into practical advice, but that this advice should be given by doctors. They did not consider women like Marie Stopes to have the proper credentials. Stopes wrote five books on sex and marriage between 1918 and 1935, the first being *Married Love: A New Contribution to the Solution of Sex Difficulties.*[18] This book sold more than a million copies and went through twenty-eight editions by 1955. It was translated into twelve languages. Stopes, born in 1880, earned degrees in botany and geology at University College, London, and was the first woman to receive a doctorate in botany in Germany. She was also the first woman appointed to the science faculty at Manchester University, but a doctor of science, not an M.D. She wrote in the preface of her first book that she had had a failed marriage and had paid a terrible price for her sexual ignorance. She wished to spare other women the pain of her experiences by writing this book. In preparation, she went to the British Museum and read every book she could find on sex. She was particularly influenced by Havelock Ellis's theories and by Margaret Sanger's views on female sexual pleasure and on contraception. Sanger, America's premier birth control advocate, had come to England in 1915 to escape prosecution for illegally disseminating contraceptive literature in the United States. Stopes herself later organized the first birth control clinic in Britain. *Married Love* purportedly made Victorian husbands gasp. In it Stopes argued, for example, that if a husband insisted on his marital rights against the wife's wishes, it constituted rape.[19]

The religious overtones of Evaline Hutchinson's marriage manual were not unusual. Many referred to sex in marriage as a sacrament; other books included prefaces by clergymen. Attitudes of both churchmen and of doctors were often negative toward contraception, which, however, her book advocated. Sexual disharmony was thought by Evaline Hutchinson

and other writers of these manuals to be a major cause of unhappy marriages. Margaret Jackson has suggested that the real problem that the
writers of marriage manuals were grappling with was feminism. Men perceived a threat to their power in the heterosexual relationship of marriage
in an era when wives were, at least relatively, more liberated than in previous times.[20]

Some of the marriage manuals were written by doctors; some of these
were explicitly addressed to husbands, who were supposed to guide their
wives' sexual education. Van de Velde's *Ideal Marriage* was one of these.
He was a Dutch gynecologist, whose book was translated into English by
Stella Browne and was in print even in the 1970s. Women doctors wrote
marriage manuals as well, most notably Helena Wright, also a gynecologist. She was, like Sanger and Stopes, a birth control pioneer. She wrote
The Sex Factor in Marriage in 1930, not long before Evaline Hutchinson's
book appeared. Hutchinson made no pretension to medical training; her
book is less theoretical, offering practical advice and psychological coun-
seling, both in terms of marital sex and of parenting.

Evaline Hutchinson herself seems to have failed in regard to both
Dorothea and Evelyn, in terms of advice she gave in her book. She wrote
of the effort and pain involved for the mother when the child is "ready and
anxious to break away from her side and lead its own independent life."
The reward for such a mother, she wrote, is the enhanced love of the children she has set free.

But did she set them free? Dorothea worked hard at her profession
and did eventually get away to London. Evelyn accepted an instructorship
in South Africa in spite of warnings from his parents about the professor
there. Did he do this, at least in part, to get away from his mother? From
the evidence I have gathered he married in South Africa, rather than at
home in England. Earlier his mother had been negative about one of his
prospective marriage partners.

Evaline and Arthur's third child, Leslie, went away to boarding school
at the age of nine. He had had rheumatic fever and the damp air of Cambridge was supposed to be harmful. He later joined Evelyn at Gresham's
School and then at Cambridge University. Perhaps he was not as influenced
by his mother. Unlike Evelyn, he married a fine though less well-educated
woman, a marriage of which Evaline may perhaps have disapproved. It was
a happy marriage and produced five children, thus fulfilling both aspects of
"creative sex" as advocated by E. D. Hutchinson in her book.[21]

Evaline Hutchinson fought for her causes all through Evelyn's childhood. He later recollected: "One's mother's friends . . . seemed essentially nonvisual, dressing in brown, gray and dark green, uncertain procryptic colors that faded into the landscape. They themselves were large boned and vigorous, with a strictly constitutional passion for female suffrage. When they rode through the town on their bicycles it was difficult not to absorb the impression that men were responsible for all the evil in the world. . . . They felt very secure, traveling to London for meetings, knowing few people outside the university."[22]

Childhood in Cambridge

Evelyn was born in 1903 on the top floor of the family home at 3 Belvoir Terrace, Trumpington Road, Cambridge, and lived there until 1912. It was a tall but narrow brick house built in 1825. In Evelyn's early years his was not a wealthy family. Nevertheless, there were nurseries and maids. He remembered his childhood vividly: "The room in the front was the study. . . . Father had the west side, mother the east. I learnt to read with her sitting under the northeast window, where I had a jam jar aquarium. I am surprised that this was allowable; it may be a fantasy. . . . I remember being picked up before a dinner party to see a dish of red fruits. . . . In my memory they are now persimmons, which I think cannot have been the case because they were new to us in Venice in the 1920's. . . . Anyhow it is probably my oldest . . . memory."

Young Evelyn showed the usual traits of a devilish boy. Three Darwin sons lived in the neighborhood. One night, Evelyn related, Sir George Darwin and Lady Darwin, "in a magnificent cerise dress," came to dinner at the Hutchinsons. Two small boys watched the guests from the top of the stairs. According to Hutchinson's Kyoto Prize speech in 1986, Evelyn and Leslie "crept down the stairs quietly, locked the dining-room door, turned out the electric light at the main and threw away the key in the garden." The outcome of this experiment is unknown. Evelyn added in his speech only that he was most happy that no anti-evolutionary group had honored him for this anti-Darwinian act.[23]

Similarly, Dorothea remembered that Evelyn and his first cousin, G. R. S. "Bill" Stewart, a little younger than he and a close boyhood friend, gave performances in this house on Sunday afternoons. They invited Bill's two younger sisters and the two live-in maids. Dorothea was very small and was supposed to hand out tickets. Evelyn was about eight; he did conjuring tricks.[24]

After reading lessons by his mother and arithmetic taught by his father, Evelyn went, still very young, with Christina Innes to be taught by her aunt, Miss Innes, in a two-child school in Christina's home. They learned English poems by heart and stories of Greek and Roman mythology and were taken in 1909 to an academic procession on the centennial of Darwin's birth. Even more memorable to all the characters involved was another of Evelyn's early experiments. The Hutchinsons had a private key to the University Botanic Gardens. One Sunday Evelyn pushed Christina into a pond in the garden "to see if she would float, an early interest in the hydromechanics of organisms of which I am not proud."[25]

Christina, later Christina Innes Morley, was one of Evelyn Hutchinson's earliest friends. Her father was senior bursar of Trinity College. Evelyn was about a year younger. In her nineties Christina wrote to this author that Evelyn

> was like a younger brother to me, and I a rather bossy older sister to him. . . . He did indeed share lessons with me at our house, where we were taught by my father's sister, an early "kindergarten" teacher.
> Evelyn and I also met very often on Sunday mornings in the Botanical Garden . . . open on Sundays only to senior members of the University. Our chief interests there were flowers and birds and small creatures of all kinds. The well-remembered episode of my being pushed into the pond by Evelyn is not wholly correct. . . . Actually it was not into the pond that I was pushed, but into a very small stream . . . on the bank of which, as I can clearly remember, I was crouching down hoping to see some tadpoles, when Evelyn came from behind and pushed me in. . . . This could not possibly have been to see if I could float . . . but it was a golden opportunity to get the better of a very bossy little friend. . . . I had to be hurried home by my nurse, drenched from head to foot, and (because it was a Sunday morning) I was wearing my best coat and bonnet.[26]

In Dorothea's version of the story, Christina was wearing a green coat and bonnet, and the nurse took her home dripping and screaming.[27]

So much for hydromechanics. Memories written long afterward in autobiographical memoirs are often faulty. Perhaps some are fantasies, as Evelyn admitted above (though not in print). Even when memory serves well, the incidents are surely highly selected. One knows with hindsight which were "important," in that they affected the psyche of the future naturalist, for example, E. O. Wilson in his fascinating autobiographical *Naturalist* or, in this case, the future aquatic ecologist Hutchinson. Nevertheless, even if the story is not entirely accurate, it is clear that at an early age, four-and-a-half according to Dorothea, Evelyn had a companion in searching for tadpoles.

A few years later, when Evelyn was at Saint Faith's, one of the three private schools for boys in Cambridge, he had a whole group of such companions. The family had moved to a large house they called Aysthorpe, on Newton Road. At his old house his hunting grounds were Coe Fen and a ditch containing common newts, whereas at the new house there were meadows with kestrels, skylarks, even a nightingale, and a variety of moths, as well as wetlands with aquatic creatures. Saint Faith's was a preparatory school for English public schools, in American terminology a private secondary school. Saint Faith's taught the elite of Cambridge families, but according to Hutchinson it was not as elite as King's Choir School. The boys who went to that school developed a superior attitude and "always seemed to carry umbrellas." However, boys with good voices, picked from all over England, could acquire a free education at King's Choir School.

Large quantities of classics, English poetry, history, and geography and also mathematics were taught at St. Faith's. On the advice of his father, Evelyn did further mathematics instead of Greek. Cambridge University at that time was in the throes of a major controversy about the necessity of Greek for university entrance. Cambridge finally dropped this Greek requirement after World War I. It seems likely that Evelyn showed some talent for biology even at Saint Faith's, as he won a prize, a book by Jean Henri Fabre on the lives of insects, that he mentioned in a letter nearly seventy years later.

Hutchinson must have remembered Saint Faith's with fondness because in 1984, when he was about to visit Cambridge, he wrote to the headmaster, M. P. MacInnes, and offered to visit and speak to the boys (still all boys). MacInnes responded that he hoped Hutchinson would speak about St. Faith's in his day, a subject of interest to the boys. The speaker who had been the great highlight of the previous term was the great-great-aunt of one of the present boys. She was one of the few still-living survivors of the *Titanic*. It is not known whether Hutchinson gave a talk to match hers.

Although Saint Faith's claimed to teach everything that a young gentleman should know, there were other aspects to Evelyn's early education. He was a leader of a Cambridge junior natural history society that included his closest boyhood friend, Jim Pearce, as well as David Pollock and younger brother Leslie. Were girls allowed? Yes, there was Audrey Lloyd Jones, whose father was a paleontologist, and Dorothea herself, who went along with Leslie and collected shells.

Evelyn had earlier collected butterflies, using a killing jar made for him by his father. He now decided that butterflies were too mundane. Flies were too difficult, but true bugs or Hemiptera were appealing and good books on them were available. He and Jim Pearce, who collected beetles, especially water beetles, went to ponds together. The water bugs that Evelyn collected became a lifelong fascination and an important research subject for him.

Another fascination for Hutchinson that dates to this period was geology. The Cretaceous rocks around Cambridge were exposed in road cuttings and chalk pits. Fossils could be found there, and a little further north, corals and sea urchins fossilized in Jurassic lime pits. Evelyn often traveled with his father. Arthur Hutchinson and two family friends helped Evelyn with his fossil hunting and identification. One of them was an eminent paleontologist, Cowper Reed, curator of the Sedgwick Museum of the Cambridge University geology department. All of this occurred while Hutchinson was still at Saint Faith's.

Among the scientists who helped him as a boy was Dorothy Elizabeth Thursby-Pelham. She worked on mammalian embryology, particularly of the hyrax, a very small African relative of elephants. She took him to her laboratory and showed him her research techniques. Early exposure to capable women scientists, both in this case and as a student at Cambridge University, was surely important in his later positive attitudes toward women in science. Another very early friend was Anna Bidder, later his classmate in zoology at Cambridge and a lifelong fellow scientist and friend.[28]

World War I brought great changes, even to children. Dorothea remembers that the war dominated their lives. She was seven, Leslie nine, and Evelyn eleven when the war started. According to Dorothea's recollections: "We lived in a lovely house with a big garden just off the main road to London going south. . . . Leslie and I would be in the garden and we would hear drums and bugles sounding. . . . We leapt up and went around to the corner of the main road. . . . There would be miles and miles of men marching and marching and marching and going somewhere. . . .

4. Hutchinson as a Boy Scout during World War I while
at Saint Faith's School. Courtesy of Francis Hutchinson.

I felt and Leslie did as children quite oppressed by this sense of impending disaster and doom."[29]

During World War I, Evelyn became a Boy Scout orderly at the First Eastern General Hospital. More important for Evelyn's later life, the family took in a young woman named Gwendoline Rendle, known to them as Gwendle, who was eighteen and became a sort of older sister to Evelyn. During World War II, Evelyn and his wife Margaret took in Gwendle's very small daughter, Yemaiel, who was evacuated from London to Connecticut for several years of that war.

5. Hutchinson's parents, Arthur and Evaline Hutchinson, at Pembroke College,
Cambridge University with Ghandi and his helper, Mira Bey (left).
Manuscripts and Archives, Yale University Library.

World War I did indeed bring doom and disaster to the Hutchinson
family's friends in Cambridge. Christina's two brothers and William Bate-
son's son John, an older friend of Evelyn's at Saint Faith's, were killed.
A surviving Bateson son, Gregory (later an anthropologist and Margaret
Mead's third husband), became a close friend of Evelyn's at Cambridge
University.

Cambridge between 1894 and 1914 was by some considered the intellectual capital of the world. It produced scientists like J. J. Thomson, E. R.
Rutherford, and A. S. Eddington in physics and astronomy; F. G. Hopkins
in biochemistry; A. N. Whitehead and B. A. W. Russell in mathematics;
J. G. Frazer and Jane E. Harrison in classical literature. Evelyn later wrote
that the intellectual atmosphere of Cambridge was all-pervasive. This
atmosphere included his family. His father, by profession a mineralogist,
carried a pocket geological atlas and Bentham and Hooker's *British Flora*
wherever he went in Britain. Evelyn grew up enjoying the pleasure of

learning, of fitting things together, both in terms of nature and of human artifacts.

Away from Home at Gresham's School

In 1917 fourteen-year-old Evelyn left Cambridge for Gresham's School, an English public school founded in the sixteenth century. Located in Holt in Norfolk, four miles from the sea, it afforded much new territory for natural history exploration, which became an important part of Evelyn's four years there. The boys, and there were only boys, were more humanely treated there than at many other English public schools. But it was war-time and the food was awful. "For pudding we have nothing but rhubarb and it is so sour it makes your teeth curl." There were also the "cook's farm eggs." Chemistry was one of Evelyn's subjects and he took a sample of the scrambled "eggs" to the chemistry lab. They gave no reaction for sulfhy-dryl; they were not eggs of any kind.[30]

The school was unusual and was probably chosen by Arthur Hutchin-son because it was less classical than most other public schools, omitting Greek and including more mathematics and science and also emphasizing modern languages and history. It was a good place for Evelyn, although perhaps less so for his brother Leslie. Leslie, who had woken up one Christmas morning with scarlet fever, developed rheumatic fever and, as noted above, was sent away at age nine to a boarding school where the cli-mate was better than that of Cambridge. When he was fourteen he joined Evelyn at Gresham's, and they had two years together there, but by that time if not earlier their interests had diverged; Leslie's were not in science.

Boys' boarding schools were constantly concerned about homosexu-ality. At Gresham's both the headmaster and housemaster were bachelors in Evelyn's, and later in Leslie's, house. "Evelyn obviously thought they had homosexual tendencies," Dorothea related. "I remember going with my mother to see the school. . . . Leslie took us up and showed us his dormitory . . . and said there was a rule that no boy was to touch another boy in the dormitory. He said it was to stop the fighting and wrestling. . . . As I'm quite sure Evelyn would say, it was to stop any homosexual acts. He said his brother was quite unaware of this. I think Evelyn was a bit too aware." Homosexuality and its evolutionary implications were an aca-demic interest of Evelyn's in his later years. Dorothea said he repeatedly referred to this subject when she visited New Haven.

The concern about homosexuality at English public schools was relatively recent in the histories of these schools. Sexual experimentation, which had gone on without much negative comment in these all-male schools until the Victorian period, was suddenly looked on with horror by headmasters. Some of these, like schoolmaster F. W. Farrar, wrote widely read Victorian novels on the subject. In one of his books an older boy shows his ruin in his face. He has "lost the true joys of youth, and knows . . . the shame of the unclean."

Many of the dire warnings to the boys were so vague that they were often not understood, especially by the young prep school boys. When the public school boys arrived at Winchester, at least one housemaster explicitly warned of self-abuse, mutual self-abuse, and "buggery." From collected memoirs of former public school boys it is clear that all these did indeed occur, the first two very commonly. The evil consequences of masturbation were expounded at many schools. At Rugby one housemaster told the students, "If you touch it, it will fall off," and at another public school, boys were warned they would lose their ability at sports, and at yet another school, that anyone engaging in such acts would "find an early grave, killed by his own foul passions."

The Victorian strategy against such vice was to toughen up the boys and keep them always busy. Cold baths and unheated dormitories were the usual Spartan measures. Endless games and no free time were the rule in most schools. This regime differed from that of previous centuries, when studies were under strict supervision but there was also considerable unregulated free time during which students could explore the countryside.

It was a society without girls or women, apart from the house matrons and the occasional wife of a housemaster, although a process of "de-bachelorization" eventually took place. Physical beauty, especially of younger boys, was much admired and crushes were rampant. They were called "pretty boys." Hutchinson remembered one such blond boy playing a female role in a school play; this was specifically forbidden in some public schools.

Many currently well-known Britons, and some who wrote anonymously, reminisced in print about their experiences of homosexuality in particular public schools. Only a few people have denied that there was any. What about Gresham, Evelyn's boarding school? The school emphasized intellectual achievement; in the words of one headmaster, "It was not the kind of school where, if a boy is not good at arithmetic, he is allowed to keep rabbits instead." Boys were allowed some free time without too many restrictions, at least on Sundays, to judge by Hutchinson's wanderings. Poet W. H. Auden, a Gresham graduate, said that each new boy was interviewed by the headmaster and had to swear not to smoke and not to say or do anything indecent. They were supposed to report themselves and others if these rules were broken—others only after first asking the boy to report himself. All the boys had their trouser pockets sewn up, reported another former student. One graduate, currently a filmmaker, who was at Gresham's not long after Hutchinson, reported that when he was a house captain he found a boy in tears because he had masturbated. He went to the housemaster and told him that he thought it was wrong to put such fear into the boys.

Only in the past thirty-five years has there been dismantling of the means used to attempt to control homosexuality at English public schools. The public school society is changing; nearly all the schools have been opened to coeducation. Cold showers, sewn-up pockets, and toilets without doors were not medieval phenomena but late nineteenth-century ones that continued through Hutchinson's time at public school.[31]

The Curriculum at Gresham's School

What about arithmetic at Gresham's School? It was indeed serious. Mathematics went as far as differential equations and linear algebra. One mathematics master who was deep into Einstein invited Evelyn for weekly talks about relativity. The chemistry teaching was excellent, according to Evelyn, the physics rather less so. One of his chemistry masters named Hammick, who did his own research, was able to excite a receptive group of schoolboys, including Evelyn, about van der Waal's forces.[32]

Largely as a result of this schooling, mathematics and the physical sciences were an integral part of Evelyn Hutchinson's early and continuing knowledge of science. They served him well in his later research from limnology to biogeochemistry and mathematical ecology. In fact, he did not take the Cambridge University entrance scholarship examination in biology at all, even though he was planning to study zoology. He took the exam in mathematics, physics, and chemistry.

Biology was not taught until his fourth year at Gresham's. New laboratories for both biology and physical geography were opened that year. Evelyn's zoologist uncle, by then Sir Arthur Shipley, officiated. Evelyn's father, a strong supporter of the school, then and later, was also present.

As noted earlier, access to professional scientists was part of Hutchinson's privileged childhood and teenage years.

Evelyn did not need the new laboratories to continue his earlier entomology studies, however. While he was at Gresham's he began a longtime correspondence with E. A. Butler, a retired schoolmaster and leading authority on British bugs, in the scientific sense (Hemiptera-Homoptera).[33] Even earlier, at the unusual age of fifteen, Hutchinson published his first scientific paper, about his observation of a swimming grasshopper.[34] Butler was his mentor in entomology, helping Evelyn identify water bugs he had collected around Holt. Butler was essentially teaching him by mail.

Hutchinson was particularly interested, as a St. Faith's schoolboy, as a Gresham's School student, and throughout his life, in two groups of aquatic insects, the Corixidae, which are now generally called water boatmen, and the Notonectidae or backswimmers. The Corixids are shiny oval-shaped bugs with legs that are specialized for different functions. These water boatmen store air under their shiny wing covers, and as they are thus less dense than water, they must be attached to some underwater object in order to stay submerged. The long, thin middle legs anchor the bug to submerged plants. The front leg of the male is modified to hold the female in mating. The hind legs have evolved into oar-like structures, thus the term "boatman." They are common and easily observed even in small ponds. Members of this group "sing," some using a structure called a strigil on the male's sixth abdominal segment. This had been first observed in 1845. Hutchinson observed singing Corixids during his Cambridge years, and one of his students later studied their genetics. These water boatmen fascinated him: "These most attractive insects pose a great many problems in their asymmetry, and their different modes of song, as well as in their ecology and distribution, and . . . they have been a continual source of interest and delight since I first met with them on Sheep's Green [as a child in Cambridge]."[35]

The other major group of water bugs is the Notonectidae or backswimmers. Those of the genus *Notonecta* had early been called "boatflies" and later "water boatmen" in England, but they are now called by their original North American name, backswimmers. They do indeed swim on their backs, prey on mosquito and other surface larvae, and also have oar-like back legs. The taxonomy of the genus *Notonecta* had been studied as early as 1909, and Hutchinson first became interested in the question of speciation in relation to these water bugs.

Hutchinson and E. A. Butler carried on extensive correspondence

6. Evelyn Hutchinson while a student at Gresham's School.
Courtesy of Francis Hutchinson.

during Hutchinson's Gresham's School period. By September of 1920, Evelyn had discovered a species new to Britain, *Nabis boops*. Butler sent him his hearty congratulations. By 1921 they were discussing color forms and breeding experiments: Butler wrote to Evelyn that many questions can be answered only in that way and encouraged him to continue his experiments, asking to what extent the colors of the parents were produced in the offspring. "I sincerely hope," he wrote (August 2, 1921), "that you will continue your work on this insect and find out all that is to be found about it. There is quite enough for a lifetime!" Evelyn sent him his own observations, and Butler incorporated them into his major work, *A Biology of the British Hemiptera-Heteroptera*, which gave the available life history, behavior, food habits, and distribution for each species.[36]

In his public school years Hutchinson also corresponded with F. H. Day in Carlisle. Evelyn wanted to come to Carlisle to see Day's insect collections. This could be done only on Saturdays after two; at other times Day was engaged in business. He invited Evelyn to come and sent directions to a collecting place, Cumwhitten Moss, "not very easy of access, an hour's walk from a RR station, 10 miles from Carlisle." Day bicycled there to meet him. Day also encouraged Evelyn in his collecting near the Hutchinson family place in Culgaith.[37]

Hutchinson moved from student to colleague and eventually to mentor of others in entomology by the time he left Gresham's School for Cambridge University. Winifred E. Brenchley, D.Sc., at Rothamsted Experimental Station, Harpendon, conveyed her interest in the list of true bugs (Hemiptera) from the Holt district that Evelyn had published in the Gresham's School *Natural History Society Annual Report.* She was working on the insects of Blakeney Point on the North Sea, which Evelyn had visited. She asked if he had a list of species he had collected there that she could use. "It would be a great acquisition, as my own visits are necessarily few and far between, and doubtless large numbers of species of every order are still awaiting capture."[38] Hutchinson had become the local expert while still in public school.

The Gresham's School *Natural History Society Annual Reports* for 1919, 1920, and 1921 witness Evelyn's participation in all of the various sections of this society.[39] Both the society and its rules were highly formalized. There were thirty-eight rules occupying three pages of small print in 1919. Among the objects of the society were to encourage the study of natural history by meetings, including papers and discussions, arranging exhibitions of research, and publishing an annual report when finances permitted. Another prime objective was to carry out a natural history survey of the Holt district. Evelyn was active in all of these endeavors. Those who wished to join the society as associates applied to a section (zoology, geology, etc.) and had to be approved by the section and elected by the membership. There were honorary members, corresponding members, members, and associates, making it sound like the Royal Society, except that there were no foreign members. Only twelve members were allowed at any time, new members being elected to fill vacancies when members graduated or resigned. In 1919, only the second year of the society, it held two exhibitions of its work. The botanical, entomological, and zoological sections had already published lists from their survey activities. The Natural History Society also gave out yearly prizes using the interest from a 100-pound war loan. They were not to be awarded except for "very distinct merit."

Although membership in the Natural History Society was not as prestigious as membership on the cricket team at Gresham's, it was an important part of the school. Evelyn did not excel at cricket. What were his accomplishments in natural history? In 1919, G. E. Hutchinson (no first names used) was a member of the Architectural and Archaeological

Section, the Botanical Section, the Chemical and Physical Section, the Entomological Section, and the Zoological Section, besides serving as secretary of the Geological Section. The Botanical Section had made two field expeditions and published a list (Latin names only) of all the plants found in the Holt region for 1918–19. The list contains nearly three hundred species and would be useful to a visiting botanist today. The Botanical Section also carried out physiological experiments.

The Entomology Section made three expeditions and also published a list, but only of Lepidoptera (butterflies and moths) and of Hemiptera (bugs); Hutchinson was probably instrumental in compiling the latter. Many aquatic insects were collected, including water boatmen, backswimmers, and also six species of water striders. Rare species were found, especially one that had previously been collected only in Scotland. Several fine collections of minerals and fossils had been presented to the Geological Section of which Hutchinson was the newly elected secretary. The Zoological Section had done a study of rook flights and behavior on the school grounds and elsewhere. Their report included many other birds and noted that several expeditions were made to Blakeney Point to see coastal birds, but "nothing worthy of record can be learnt of these in one or two afternoons." More important, all nests found in 1918 and 1919 in the extensive school woods were mapped; 132 nests of thirty species were found. Whoever wrote up this study, and Hutchinson was one of only four actual members of this section, raised interesting quantitative questions about population biology in relation to the nest data.

The secretary's report for 1920 recorded the three most noteworthy achievements of the past year. Number 1 was the "capture of a very rare insect" by G. E. Hutchinson. This was the true bug *Nabis boops,* new to Britain. Numbers 2 and 3 were the invention of a self-tuner for a wireless installation and the independent discovery of Nova Cygni III.

Hutchinson was also the co–first prizewinner of a natural history (Holland Martin) prize for his paper on variation in the spittle bug. The secretary's report for this section reads, "Much work has been done by a member of this section on the variation of . . . Philaenus spumarius L., [which] has been shown to depend to a large extent on locality and perhaps on habitat. This work is being continued and it is hoped will prove of great value."

The 1921 report lists the names of the members of each section. Hutchinson is the only one who was a member of all seven sections. Among others listed, the best known is W. H. Auden, who belonged

7. Hutchinson collecting insects at Cherryhinton Chalk Rt. in Cambridge, 1920.
Manuscripts and Archives, Yale University Library.

only to the Architectural and Archaeology Section. Evelyn once went to Auden's study to read some of Auden's very early poems.

In 1921 the rarest insect specimen was caught by someone else. But the secretary's report also featured a paper read by G. E. Hutchinson on "Variation and the Origin of Species." Amusingly, since Hutchinson was later a strong adherent of natural selection, he concluded that mutation could account only for superficial changes in species and that the true cause for structural variation had to come from the inheritance of acquired characteristics, orthogenesis, or polyploidy. All of these subjects were much under debate in 1921 as well as during Hutchinson's Cambridge University career.

That year, 1921, Hutchinson was chairman of the Entomological Section at Gresham's. Masters, indicated by "Mr." in front of their names, were usually the chairmen of sections. Hutchinson exhibited extensive collections of the variation in *Philaenus spumarius* and published "Hemiptera, 1917–1921" (Hutchinson's years at Gresham's) in the 1921 *Annual Report*. In addition to his lengthy list of species, his ecological introduction read, "Perhaps no order of insects shows forms confined to and specialized . . . for life in so many environments as the true bugs or Hemiptera," after which he described the many habitats in the immediate vicinity of Holt and the particular insects found in them. He had already noted that closely related species occurred in very different habitats, including aquatic and semi-aquatic habitats, and asked why this should be, a "question our present knowledge does not permit us to answer."

By this period, as the question exemplifies, he was certainly thinking ecologically. He himself noted, but many years later, that he had begun to realize while at Gresham's that organisms had different chemical environments. "I remember one Good Friday, probably 1919, when the day was free after chapel, using much willpower to practice titrations in the chemistry laboratory over and over . . . before I allowed myself a visit to my favorite ponds and woods, feeling that I should have to be an adequate chemist if I were to become a good field zoologist."[40]

Culgaith Explorations

While Hutchinson was at Gresham's, his father inherited the house belonging to his Aunt Jane. It was in Culgaith in the north of England, where Evelyn's grandmother had grown up. The family traveled there on

8. Hutchinson shortly before entering Cambridge University.
Courtesy of Francis Hutchinson.

holidays, and Evelyn explored the small lakes or tarns and the zoogeography of the north country, with its different species of aquatic insects, further expanding his familiarity with water bugs. These tarns were on Dufton Fell in the Pennines. T. T. Macan, in a 1986 article, recalled seeing these tarns for the first time—with Evelyn Hutchinson fifty years earlier. Hutchinson himself recorded his schoolboy excitement in finding an unusual water bug, *Artocorisa carina,* in these tiny lakes, which were teeming with them. The species is also found in Iceland, the Faroes, and the Alps, but not in the vicinity of Gresham's School.

Hutchinson's expertise with water bugs was highly developed during these early years, and it enabled him to use this group in much of his later path-breaking work in ecological theory. Years later he advised his Yale graduate students to develop expertise with some group of organisms. His own interests and publications encompassed many aspects of entomology besides taxonomy, particularly life cycles and reproduction (later population biology), behavior, genetics, ecology, and biogeography. "My mind full of problems of distribution and variation, in a great state of excitement, I left [Gresham's] school to start a real scientific career at Cambridge."[41]

Hutchinson compared the extraordinary richness of his pre-university environment, both intellectual and artistic, with that of people whose work he admired. One of these was Maria Sibylla Merian, an entomologist and artist who worked in Surinam in the early eighteenth century. Merian's family were painters, not scientists. Both Hutchinson and Merian, however, were exposed quite early in life to both insects and art. She, like he, traveled to an exotic new continent, though at age forty-two rather than at twenty-three. Hutchinson described Maria Sibylla Merian as a "very late flowering of the Middle Ages." Later Hutchinson became an expert on medieval animal art. Unlike Hutchinson, eighteenth-century Merian could still believe that small frogs could turn into large tadpoles, though she raised and closely observed insects.[42]

Much later in life Hutchinson wrote: "I realize that the circumstances of my upbringing led me to acquire an enormous amount of information about natural history, geology, archeology, both British and exotic, and art history when not more than 17, before I entered the University."[43] The circumstances of Evelyn Hutchinson's upbringing were indeed unusual.

CHAPTER THREE

Becoming a Zoologist:
The Hunter and Gatherer

utchinson never did graduate work at Cambridge; in fact, he never earned a doctorate. His first two post-Cambridge ventures as a zoologist, at Naples and in South Africa, could both be deemed failures. At Naples his research efforts were not successful. At Witwatersrand University in South Africa, he was fired as a teacher. One could not have predicted his extraordinary future career from the several years that followed his Cambridge graduation in 1925. Yet his Cambridge undergraduate years were fundamental to that future success. His experiences at Cambridge were also the source of his views on education, particularly graduate education, and on the abilities of women scientists.

Cambridge University in Hutchinson's Time

How did all of Hutchinson's extraordinary future career in ecology come about? Surely Cambridge University itself was the prime actor. Never before or since were so many brilliant men and a few brilliant women gathered in one place as in Cambridge in the interwar period. In his book *Cambridge between Two Wars*, T. E. B. Howarth described the renaissance in science at Cambridge after World War I, in which a third of the Cambridge men who fought were killed or wounded.[1]

Kathleen Raine, the poet, who studied natural sciences at Cambridge

at the same time as Hutchinson, described "the civilization of the Cambridge I knew . . . with all that subtle beauty and knowledge, the mental spaciousness and freedom, the breadth of humanity, the never-to
be-repeated quality of life. . . . We children of the *entre-deux-guerres* enjoyed our world without any sense of guilt or doom." She described the
standards of excellence in both the arts and the sciences, high standards
set for all who studied or practiced them, both students and faculty.[2]

Hutchinson, one of those students himself, described Cambridge University in his day as the best in the world, and not only in science: "Witness the discovery of electrons, protons and neutrons, sex-linked inheritance, glutathione, cytochromes, the writing of Frazer's *Golden Bough*,
Jane Harrison's *Greek Religion and Themis*, Marshall and his successor,
Keynes, on economics, Adrian on the nervous system."[3]

Witness the range of Hutchinson's interests in this enumeration,
interests that probably predate his university years, 1921–25. Cambridge
University in 1921 was composed of eighteen colleges, the oldest founded
in 1286. The colleges actually make up the university but are separate enti-
ties, contributing their own money and personnel to teaching. At Oxford
and Cambridge much of the teaching was and is still done by the tutorial
system in the colleges.

The majority of Cambridge students "read for honors"; less serious
students did a lot of punting on the river and took "pass" degrees. To receive an honors degree in the 1920s one had to pass at a particular level,
three subjects in the part 1 tripos (final exam). In natural sciences the subjects included zoology, botany, physiology, and anatomy. The other natural science subjects were physics, chemistry, geology, and mineralogy. The
inclusion of mineralogy had much to do with the earlier teaching and research of Hutchinson's father, Arthur Hutchinson, in this field.

There were also part 2 tripos, taken in the third or in some cases a
fourth year. A student received a first, second, or third class degree depending solely on the results of the tripos. Throughout their university
careers, as was true at Cambridge and Oxford even in recent times, attendance at lectures was optional and there were no laboratory exams. This
system gave students much more personal responsibility than the American university system. Hutchinson reported himself a "hunter and gatherer" among Cambridge's rich course and laboratory offerings, a pursuit
that Cambridge University made possible.

The awarding of first, second, or third class degrees was not based on relative exam results in Hutchinson's time or is it now. One year, if there were several brilliant students, they all might receive firsts; another year no one would (as was true the year I spent at Oxford as a demonstrator). One usually needed to receive a first or an upper second to be accepted as a graduate (often called research) student. Earlier at Cambridge, there had been a more competitive system. Students had been listed in order of success in the tripos. The top student in the mathematical tripos was named senior wrangler. The second best was second wrangler. Lord Kelvin, then William Thomson, was second wrangler in his year; the senior wrangler that year apparently was best only at taking exams. One year a woman scored higher than the man who became senior wrangler, but as a woman she could not be recognized as such.

The only special honors for zoology and botany in Hutchinson's time were the Frank Smart prizes for the top Cambridge student on the part 2 tripos in each of these subjects. Women, however, were still not legally Cambridge students and therefore were not eligible for the Frank Smart prizes. Hutchinson won the zoology prize but said, as mentioned earlier, that a woman student, Sydnie Manton, not he, should have had it.

Until quite recently, Cambridge colleges were all single-sex. The two original women's colleges, Newnham and Girton, go back to the nineteenth century. Beginning in 1881, women were allowed to take the Cambridge exams (tripos), even though they were not eligible for Cambridge University degrees. In fact, women students were not granted Cambridge University degrees until 1948. The opposition to granting degrees to women students came both from some though not all of the dons and from a great many of the male students. Emmanuel College, to which Hutchinson belonged during his university years, was the only one whose resident students voted, in 1897, in favor of women's degrees. Professor of archeology Sir William Ridgeway strongly disagreed, warning in that same year, "Our pilots have given ear to the sirens of Girton and Newnham and unless we take heed will wreck this great University."

According to Howarth, the *Cambridge Review* also railed against young fellows in colleges who married, although the statute that fellowships had to be resigned upon marriage had been rescinded in 1882! The *Review* referred these fellows to an obscure seventh-century B.C. poet, Semonides of Amorgos, who wrote:

For this is the greatest ill that Zeus hath made, women.
Even though they may seem to advantage us,
a wife is more than all else a mischief to him that possesseth her;
for who dwelleth with a woman, he never passeth a whole day glad.

When the subject came up again in 1920, this time with the support of many important Cambridge dons, including the biochemist F. Gowland Hopkins and the economist John Maynard Keynes, the *Cambridge Review* denied its support: "Not that we wish . . . to minimize the good work often done by women students, but 'so long as the sun and moon endureth' Cambridge should remain a society for men."[4]

Oxford admitted women to full membership in the university that year, but Cambridge did not. Much of the resistance to granting degrees to women had to do with the fear that women, once official members of the university, would try to invade the men's colleges. (The greatest fear of these men has indeed come to pass; there are now a number of coed colleges at both Cambridge and Oxford.)

Although Girton College was described by Virginia Wolf in *A Room of One's Own* as quite primitive as compared to the better endowed men's colleges, photographs of Dorothea Hutchinson's room and those of her friends in Newnham in the 1920s make them look quite elegant by today's standards. Moreover, Kathleen Raine's account of living in Girton in the mid-twenties sounds similar, the "very realization of Tennyson's *The Princess*." Raine was a natural sciences student at Girton, where she had two rooms of her own, the summer that Virginia Woolf delivered her paper. A Girton staff member of the twenties reported that there were forty or fifty on the staff, including twenty-six housemaids. At tripos time cups of tea were brought early in the morning to each student taking her exam.[5]

Hutchinson reported that when he arrived in 1921, women students were segregated in university zoology classrooms, sitting in the front rows. Big and popular undergraduate lectures like Lord Adrian's on the central nervous system had a ratio of ten men to one woman student. Women were still excluded from some scientific organizations at Cambridge. Nevertheless, by 1921, the science students from Newnham and Girton were working in the same science laboratories as the men. Kathleen Raine wrote that every woman who was admitted to Girton or Newnham had to have reached what was, for men's colleges, scholarship standard, and that they all considered themselves the mental equals of the best of the men. Raine also provided evidence that the labs were themselves egalitarian: "I was on surer ground with my fellow-scientists, for in the labs I was accorded my due, no more, no less."[6]

Hutchinson's women classmates included Grace Pickford, Anna Bidder, Penelope Jenkin, Sydnie Manton, Cecilia Payne, Anne Hastings, and Joyce Barrington, all of whom later made significant contributions to science. Most of them remained Hutchinson's good friends and correspondents; he admired their abilities in science.

Emmanuel College, Cambridge

Emmanuel College at Cambridge was Hutchinson's home from 1921 to 1925. It was founded by Puritans in 1584; Christopher Wren later built a chapel for it. It is known to Americans because John Harvard was a student there, though apparently not a very distinguished one. Hutchinson was a scholarship student, and these students lived in college throughout their Cambridge years; other students lived in lodgings for one or more years. All students ate dinner, in their gowns, "in hall," the college dining room, but not breakfast or lunch. The wealthy got these meals sent up from the college kitchen, but others, like Hutchinson, did their own cooking in their rooms. "Most of us cooked eggs and made toast and tea on a gas stove or on the fire."[7]

Science was not the forte of Emmanuel College, although J. D. Bernal, a pioneer in X-ray crystallography, was an Emmanuel student beginning in 1927. During Hutchinson's time a mathematician named Bennett was the college's only fellow of the Royal Society. Many Cambridge dons had recently returned from World War I. One of these at Emmanuel was Charles E. Raven, who later wrote the foreword to Hutchinson's mother's book (see chapter 2). He had been a chaplain on the Western Front and became a leading Cambridge pacifist.[8]

Hutchinson characterized other Emmanuel fellows as "adequate scholars and devoted mentors of youth." He claimed never to have been taught by science fellows at Emmanuel; the scientists who lectured and taught him in the zoology laboratories had a much greater impact on him, as did some of his fellow science students.

The students of Emmanuel College were an interesting lot from Hutchinson's description. One played women's roles in Cambridge theater productions. Another was an "exquisite with immense floppy silk

9. Emmanuel College, Cambridge University.
Photograph by M. O. Hill, with permission.

bow ties" but was mistreated by other students. In addition there were
musicians, one of whom, W. A. H. Rushton, had also been at Gresham's
School. He later became an authority on the physiology of vision. At college Rushton played the viola and was an important source of Hutchinson's lifelong enjoyment of the music of Bach and other Baroque com-
posers.

Then as now much science teaching was done as department lectures
and laboratories. Hutchinson's recollections of the Zoology Department,
and of its professor, J. Stanley Gardiner, who was an early explorer of coral
reefs in the Indian Ocean, were not enthusiastic. He thought the department was suffering from the aftermath of World War I and that it was not
until James Gray became head of department in 1937 that Cambridge became a flourishing center for experimental zoology.[9]

Anna Bidder, who studied zoology while Hutchinson was there, disagreed. She pointed out that Hutchinson left Cambridge immediately
after receiving his degree in 1925. She thought that J. Stanley Gardiner
was "an absolute wizard in getting good people to work in his department"

and that he built a superb zoology department.[10] Joseph Needham independently made a similar comment about Hutchinson's negative impressions of the department under Gardiner.

The Cambridge Zoology Department in the early twenties included, in addition to Professor Gardiner, L. A. Borradaile, F. A. Potts, and J. T. Saunders as senior lecturers. Hutchinson essentially damned all these men with faint praise, but other contemporary students disagreed with this assessment. Bidder said, "Evelyn talked a good deal of nonsense about the department of zoology in his book [*The Kindly Fruits of the Earth*]." Limnologist Penelope Jenkin was considerably influenced by J. T. Saunders. She took her first course in limnology, the study of lakes, with him. It was then called hydrobiology and may well have been the first such course in Britain. Nevertheless, Hutchinson's rather negative views of the Zoology Department might be supported in part by the complete omission of that department from Howarth's chapter about the twenties in his book about Cambridge University between the wars.

Other Cambridge dons did influence the young Hutchinson. Hans Gadow was remembered as a marvelous nineteenth-century relic who taught vertebrate zoology. F. Balfour Brown, teacher of entomology, was a student of water beetles. Brown did not make the entomology course exciting, but he did send Hutchinson off to Scotland on a collecting trip in the Hebrides, an unusual opportunity for a Cambridge undergraduate. This trip, in 1922, created its own excitement for Hutchinson, which he conveyed to me in a conversation sixty-eight years later.[11] He made extensive collections and found many species entirely new to him.

The Cambridge Zoological Museum was also important to Hutchinson during his university period and even earlier. He is reputed to have known all the specimens by heart by the age of fourteen. While he was at Cambridge University the insect collections became professionally important to him. The curator, Hugh Scott, though much older than Hutchinson, became his close friend and coworker on the Hemiptera or true bug collections.

But it was E. A. Butler, rather than Scott, Brown, or any other Cambridge don, who had the most important influence on Hutchinson's research in entomology, research that continued throughout his life. Butler, a retired schoolmaster, continued to correspond with Hutchinson throughout his Cambridge years. These letters, unlike the earlier ones written while Evelyn was in public school, were discussions between colleagues. Questions were raised about the true bugs they both studied, questions relating to color phases and sexual dimorphism, for example, and about ecology. When Hutchinson wrote that he was going to the Hebrides, Butler advised him to look carefully for water bugs in *Sphagnum* (peat moss), which should be carefully shaken. He also suggested habitats for many other insects, reinforcing Hutchinson's developing ecological viewpoint. By September of 1922, at age nineteen, Hutchinson had been made a fellow of the British Entomological Society and was able to write "FES" after his name.

A sample of Butler's letters in reply to Hutchinson's from 1923 to 1925 covers many topics, including population biology and life history questions.[12] In August 1923 Butler wrote, "I trust you had a good time at the Tripos [part 1] and got a good degree," then went on to discuss four different color variations in females of the genus of the water bug *Notostira*. The question was whether they represented seasonal changes or species differentiation. An October letter of the same year related to species of another genus of water bugs, *Saldos,* each of which had a distinctive habitat. These included marshes, mountain torrents, and moorlands. In other, more modern scientific terms, there was intrageneric niche differentiation. A 1924 letter described Butler's breeding experiments; he often encouraged Hutchinson to use experimental methods. Hutchinson's own expertise with water bugs became highly developed during these early years. It enabled him to use this group in much of his path-breaking work in ecological theory. In later years Hutchinson advised his own students to develop expertise with respect to some particular group of organisms.

A Most Exciting Cambridge Teacher, George P. Bidder

Once Hutchinson got to part 2 of honors zoology, after receiving first class honors in part 1, there were several scientists he found exciting and whose research he praised. H. Munro Fox had worked on lunar periodicity in sea urchins. He and James Gray both did experimental work on invertebrates. Gray taught an experimental cytology course. It is now difficult to imagine a cytology course at a time when most cytological structures could not yet be visualized. In what was an exciting event for cytology, however, S. T. P. Strangeways showed films he had just made of mitotic figures in living cells.

But the zoologist about whom Hutchinson wrote and spoke most highly was George P. Bidder. Bidder, an independently wealthy freelance zoologist, taught at Cambridge University but was not a regular faculty member. He did research in his own laboratory and gave a famous course on sponges. It was this course that most influenced Hutchinson's thinking in zoology. George Bidder's daughter, Anna, said that her father and Evelyn Hutchinson, despite the difference in their ages, were intellectual kindred spirits; their brains seemed to work in similar, perhaps odd, ways. "Their minds asked the same kinds of questions. That is why he responded so readily to father's questions."

In fact, the whole Bidder family influenced Hutchinson. The Hutchinson and Bidder families knew each other from Evelyn Hutchinson's earliest days. Anna's mother had carried out her own independent research project, which had to do with food processing in the amoeba. She was then Marion Greenwood; she had completed her Cambridge University studies in 1882. Later she was the director of the Balfour Laboratory for the women science students of Newnham and Girton. She married George Bidder in 1899. Their daughter Anna Bidder was Evelyn's first friend. They lived next door and had had tea in each other's nurseries as small children in the early years of the century. In the 1920s they were zoology students at Cambridge together, although Anna was a year behind Evelyn. At their last meeting, in Cambridge in 1991 shortly before Hutchinson died, Anna reported that they had a long and insightful talk about how each did zoological research.[13]

George Bidder contracted tuberculosis early in his marriage. Therefore, he and his wife had to leave England for the winter, to Aswan for the dry air of Egypt and later for Naples, where Bidder bought Parker's Hotel, where Hutchinson later stayed. The story goes that when staying at the hotel, Bidder didn't get out of bed one morning and was told, " 'You have to get up now.' He answered, 'Well, I'll buy the place.' So he did."[14] When Bidder recovered from tuberculosis he and his wife moved back to Cambridge shortly before Anna was born.

Hutchinson described George Bidder as being "largely nocturnal with beard, moustaches and cloak." The course that Hutchinson took with Bidder mainly concerned calcareous sponges. In this course Bidder introduced ecological questions relating largely to the marine environment, although there are some freshwater sponges. The ecological significance of size was the idea that particularly intrigued Hutchinson. He recalled that the course was "a series of footnotes on scientific civilization," which sounds much like Hutchinson's much later population biology course and textbook. To be a good zoologist, Bidder advised, "One had to know everything except possibly irregular Greek verbs," advice Hutchinson seems to have internalized.

Hutchinson called Bidder a greatly gifted writer of English prose and quoted him on the life of a sponge in its "eternal abyss, with its time-like stream, there is no hurry, there is no return."[15] Hutchinson emulated him in this too, in his "Marginalia" column in *American Scientist*, in his essays elsewhere, and in his books.

Biochemistry at Cambridge:
F. Gowland Hopkins and Joseph Needham

It is not surprising that with his strong chemistry background from Gresham's, Hutchinson became interested in the relatively new field of biochemistry. This subject was originally part of physiology, a separate Cambridge department from zoology. Biochemistry itself became a separate department at Cambridge, and a very famous one, under Professor F. Gowland Hopkins. Hopkins, together with C. Funk, was a co-discoverer of vitamins. He had done this pioneering work between 1906 and 1912. Later, just before Hutchinson listened to his inspiring lectures in introductory biochemistry, Hopkins had discovered glutathione, important in cellular respiration.[16] Hutchinson was not impressed by the rote biochemical exercises set up by Sydney Cole, who was in charge of the laboratory. But he was much impressed by Joseph Needham, one of the demonstrators and Hutchinson's biochemistry teacher.

Needham was hardly older than Hutchinson but in addition to biochemistry, he was already engaged in his fascinating work in chemical embryology. Needham's wife, Dorothy Needham, was also a biochemist at Cambridge; they both became fellows of the Royal Society. Needham later became a leading embryologist and historian of embryology, and even later the most eminent historian of Chinese science and medicine. Discussing Cambridge in 1921, the year Hutchinson arrived, Needham told me in an interview that he was working under F. Gowland Hopkins on the biochemistry of development in mammals. Hopkins was "the great man on the subject [biochemistry], president of the Royal Society and the one who wangled me into biochemistry, because I came up as a medical student." Needham also discussed many of the people Hutchinson studied with, including Punnett and J. B. S. Haldane, Needham's predecessor as reader in biochemistry.[17] Needham said that he had a very clear recollection of Hutchinson as a student. He often went to the Biological Tea Club, of which Hutchinson was a founding member. Needham was a close friend of Yale's great embryologist, Ross Granville Harrison, who was later to be important in Hutchinson's life. Needham visited Ross Harrison at Yale when Hutchinson was a newly arrived Yale zoology instructor. Joseph Needham and Evelyn Hutchinson kept in touch for many years.[18]

Of Hutchinson's other biochemistry teachers at Cambridge J. B. S. Haldane, the reader (the university rank just below professor), gave a course on enzymes and worked on salt metabolism. He later became famous in a seemingly unrelated field, the mathematical theory of natural selection, closely related to Hutchinson's later interests in population ecology. Haldane had become director of the John Innes Horticulture Institute where he worked in genetics in 1926, the year after Hutchinson left Cambridge. During this period, another biochemist, David Keilin, who worked on parasitic insects, discovered cytochromes. It is now difficult to imagine biochemistry before cytochrome, but this was clearly an exciting time and place for that field. Another biochemist was Huia Onslow, named Huia for an extinct New Zealand bird. He was part of an earlier English tradition; he worked as an "amateur" at home. He did early work in biochemical genetics both on melanin in rabbits and on the inheritance of color in the currant moth, *Abraxis grossulariata*. Hutchinson later worked on variation in this moth. Muriel Wheldale, Onslow's wife, was also a biochemical geneticist (all this twenty or thirty years before the term was in common use). She worked on anthocyanins, the pigments that provide the orange and reds of our brilliant fall colors. Hutchinson obviously found the biochemists, or at least later remembered them as, more exciting than the zoologists, other than George Bidder.

Biochemistry at Cambridge was lauded by both Hutchinson and by T. E. B. Howarth, who wrote extensively about many aspects of science at Cambridge University. Howarth related that William Whewell, master of Trinity, proclaimed in 1847 that to preserve undergraduates' reverence for their dons, a century should be allowed to elapse before any new discovery in science was admitted to the curriculum. Howarth commented, "Such a prescription would have been no less inappropriate for the study of biochemistry and physiology in Cambridge in the twenties than it would have been for nuclear physics."

Not only is F. G. Hopkins credited with laying the foundations of "our whole modern knowledge of biochemistry of the human body, with all the clinical applications involved," in addition he was a universally revered figure, "loved no less than he was admired." Hopkins had a Horatio Alger life history, if that term can be applied to a scientist born in 1861. Hopkins became an insurance clerk at sixteen, then a forensic medicine laboratory assistant for five years. After that he acquired a London external degree and medical qualifications. Hopkins arrived at Cambridge in 1898 at the age of thirty-seven to supervise physiology at Emmanuel College. By 1914 he was Cambridge's first professor of biochemistry, and by 1925 had his own department and a new laboratory. Cell chemistry moved from the era of protoplasm with mysterious special properties to the modern era of the cell as a chemical machine. By Hutchinson's time at Cambridge, Hopkins, in addition to being the co-discoverer of vitamins ("accessory food factors"), had worked on the chemistry of muscular contraction and, as noted above, isolated glutathione. He also was the leader of a research group, including Marjorie Stevenson and Joseph and Dorothy Needham. Another member was F. J. W. Roughton, a fellow of Trinity from 1923, who worked on hemoglobin and lectured in both biochemistry and physiology.

Physiology and Hutchinson's Other Fields at Cambridge

Physiology as a field had been founded at Cambridge by Michael Foster, prelector in physiology at Trinity College. It was Foster who established the first university biological laboratories in Britain and essentially introduced experimental biology at Cambridge. Foster was also a prime mover, together with embryologist and animal morphologist Francis M. (Frank) Balfour, who died young in a climbing accident, in the establishment of the Balfour Biological Laboratory for Women. In this laboratory Cambridge women studied science from 1884 until 1914. By the 1920s the professor was Joseph Barcroft, who was well known for living for a week in an airtight glass room at oxygen tensions equal to those at 15,000 to 18,000 feet. E. D. Adrian, the popular lecturer on the nervous system noted by Hutchinson, was experimenting with electrophysiology. Howarth attributes the success of physiology (which included biochemistry during Hutchinson's student years) to the way it "boldly attacked general biological problems and only indirectly those of human physiology," a largely nonmedical approach that Hutchinson shared.[19]

Hutchinson acquired strong interests in other fields at Cambridge as well. James George Frazer, author of the *Golden Bough,* was at Trinity College, Cambridge, from 1879 to 1941, long before and after Hutchinson was a student there. A. C. Haddon, Cambridge's leading field anthropologist in the early years of that subject, retired in 1925, the year Hutchinson left Cambridge. Hutchinson developed a lifelong interest in anthropology.

Geography, or at least biogeography, was a subject that had fascinated Hutchinson since his first serious insect studies, even before coming to Cambridge. A tripos leading to an honors degree in geography was not instituted at Cambridge until 1921. Two Antarctic explorers were involved in the department in the 1920s, Frank Debenham, an Australian who survived Scott's ill-fated expedition, and R. E. Priestly. These two men did a geological survey in the vicinity of an old hut of Shackleton's on Mount Erebus. While in the hut during a blizzard, Debenham got the idea for the Scott Polar Institute. It was Hutchinson's uncle, Arthur Shipley, master of Christ's and by now vice chancellor, who helped them with funding to start the institute in 1921. Shipley, a keen zoologist and patron of science, was described at that time as so fat that his butler had great difficulty getting him in and out of the bath. This does not sound like the man who was so influential in Hutchinson's early life, but clearly Sir Arthur Shipley, by this time knighted (as were J. G. Frazer and F. G. Hopkins), was still keenly interested in science.

Genetics was taught by R. C. Punnett, who studied the genetics of sweet peas. He is the Punnett of "Punnett squares," on which students of elementary genetics were brought up for decades thereafter. He had worked closely with William Bateson. Bateson's two older sons, who had introduced Hutchinson to beetle collecting while he was still at St. Faith's, had both met early deaths, one in World War I, the second by suicide. But the youngest, Gregory Bateson, later a well-known anthropologist, was Hutchinson's fellow zoology student and friend at Cambridge. Genes were still referred to as "dominant and recessive factors," and the chromosome theory was still under debate; Bateson himself was skeptical.

What about ecology? Animal ecology was limited to the hydrobiology course taught by Saunders. Hutchinson wrote that the course included both limnology and oceanography and was excellent but was too short. Sir Arthur Tansley, an extremely influential plant ecologist, was at Cambridge then, but during this period he essentially left this field to study psychoanalysis. Hutchinson wrote: "The plant ecology course, which had been epoch-making under Tansley, had declined as it became more and more weakened by Tansley going into psychoanalysis. He wrote a good book on the latter, one of the first in English, and then went [back] to botany some years later at Oxford."[20]

During his part 1 years Hutchinson did study plants. He had kind words for H. Gilbert Carter, whose "marvelous course on flowering plants" combined great knowledge both of plants and of the classics. Hutchinson already knew much about plants from the Natural History Society surveys at Gresham's, and he delighted in the field trips led by Carter and in Carter's labeling of plants in many languages at the Botanical Garden.

In the last analysis it may have been Needham, the young lab instructor in biochemistry, who had the most lasting effect of all his Cambridge teachers on Hutchinson's ideas. In 1983, long after his own retirement from Yale, Hutchinson wrote a letter supporting the nomination by the Yale biology faculty of Joseph Needham for an honorary degree. He pointed out that Needham's work on the chemistry of embryology and particularly his concept of biochemical allometry had greatly influenced not only developmental biology but also ecology, including Hutchinson's own concerns. He wrote, "[Needham's] suggestions as to the barriers that a marine organism has to face in entering fresh water are almost certainly correct and have a curious bearing on the effects of human pollution on rivers and streams.[21] Needham was at the time of this letter (1983) the only person still living who had taught Hutchinson as a student. Needham was still actively working in his own institute in Cambridge when interviewed by this author in July 1991, after Hutchinson's death.

The "Hunter and Gatherer" outside of Class: Student Clubs

In addition to what he learned in Cambridge lectures and laboratories, Hutchinson, the self-proclaimed "hunter and gatherer," gathered a lot of knowledge outside of class. It was an exciting time to be a biology student. In addition to biochemistry, whole other fields were in the making. Hutchinson's fellow student and first wife, Grace Pickford, later became an eminent comparative endocrinologist. The surprising discovery that thyroid gland tissue (actually thyroxin as was later revealed) could cause premature metamorphosis of tadpoles into frogs was made in two different laboratories in 1912 and 1913. Hutchinson claims to have tried this experiment himself on tadpoles in his first year at Cambridge, using dried-up thyroid prescribed for his sister, Dorothea. He did not record his results.

Drosophila genetics was also underway. Thomas Hunt Morgan and Calvin Bridges were doing work on sex determination in *Drosophila* and in 1919 had published on gynandromorphs, flies that were male on one side and female on the other as the result of an early loss of an X chromosome. This subject was much discussed by Hutchinson and his classmates in a variety of important student clubs.

The Cambridge zoology students were a talented group. Robin Hill, the man who ten years later carried out the "Hill reaction," the light reaction of photosynthesis, in a test tube, was a classmate of Hutchinson's and a fellow member of Emmanuel College. Gregory Bateson, who later changed fields, was a fellow zoology student and remained a close friend. As already indicated, there were outstanding women science students as well, at Newnham and Girton Colleges. Grace Pickford, Hutchinson's constant companion on numerous Wicken Fen collecting expeditions, was also his co-president of the Biological Tea Club.

During the 1920s, and long before, Cambridge was the home of many exclusive clubs. "The twenties were the age of exclusivity par excellence," wrote Howarth. He cited especially the Heretics, founded in 1909 and "conspicuous in the twenties for total freedom of thought." The Apostles were "more exclusive than the Heretics . . . probably a great deal less lively and certainly less heterosexual." Among the Apostles in Hutchinson's time were the mathematician-philosopher, originally a student of Bertrand Russell, Ludwig Wittgenstein and Frank Ramsey, a childhood friend and neighbor of Hutchinson's, a mathematical genius who died at age twenty-six. Ramsey was also a member of the Heretics.

Haldane was also a Heretic, whose "capacity to outrage contemporary canons of respectability was surpassed only by his brilliance as a biochemist." He never got a science degree but became a reader in biochemistry in 1923 thanks to F. Gowland Hopkins. For the next ten years friends attempted to have him elected to a fellowship in spite of his outrageous behavior. Hutchinson's first term lecturer on the nervous system, Lord Adrian, tried especially hard and nearly succeeded until Haldane, hurrying to dinner at High Table from his laboratory, brought with him "a gallon jar of urine, forming part of his current experiment, which he had placed on the table amid the college silver." Haldane subsequently got into much greater difficulty because of his adulterous affair with Charlotte Burgess, whom he later married. He had warned both his professor, Hopkins, and the vice chancellor of his intention to commit adultery, seeing it as the only route (under 1924 law) to Charlotte's divorce. Nevertheless he was brought up by an august Cambridge University body, the Sex Viri ("sex" referring not to the subject matter of the case but to its six members, the "viri"). Haldane was deprived of his readership but appealed the decision to another august body, this one including Sir William Bragg. Haldane won the appeal and kept his readership until 1933, when it passed to Joseph Needham.[22] Yet Haldane was clearly a brilliant scientist. In the year of his much-publicized adultery, he published a paper that Hutchinson cited as an even earlier contribution to the competitive exclusion theory in ecology than the better-known but later published work of Volterra.

Evelyn Hutchinson was not elected to these more renowned and exclusive clubs, but science clubs abounded at Cambridge and he belonged to several. They may have taken up more of his time and thought than the science laboratories. There was the very selective Cambridge University Science Club, to which students were elected in their third year, and two science clubs at Emmanuel College, one for scholars like Hutchinson and Robin Hill and another for all others. Hutchinson also belonged to the Cambridge Natural History Society, which included both professionals and amateurs and did not require a university connection, a truly town-and-gown organization. It was through one of its members, a taxidermist, that Hutchinson became interested in the inheritance of coat color in stoats and other animals, a continuing interest. Hutchinson became senior secretary of this organization and was present when Paul Kammerer gave his famous demonstration of what he claimed was the inheritance of acquired characteristics in the midwife toad, *Alytes obstetricans*. Hutchinson contributed what he recalled of Kammerer's demonstration to Arthur Koestler's book on the subject.[23]

Science seems to have been broadly defined in many of these clubs; the talks given by club members varied widely. Robin Hill gave one to the Emmanuel Science Club on why art involves an element of illusion, and Evelyn himself delivered one to the Cambridge Natural History Society having to do with cultural diffusion, a subject about which he had done a lot of independent reading after hearing a lecture by Grafton Eliot

Smith. Smith's lecture was presented at a British Association meeting in Liverpool during Hutchinson's first year at Cambridge. Was it unusual for a first-year student to attend a scientific meeting in a distant city? In any case, "the study of man" was added to that of water bugs, stoats, and other animals. Hutchinson's lecture to the Natural History Society was about the mandrake root, which Rendel Harris, one of Smith's supporters, thought was the origin of the cult of Aphrodite (Venus). Hutchinson became interested in these plants, of the genus *Mandragora*. They contain narcotics and are reputed to promote fertility.[24]

Perhaps the most important club for Hutchinson and his friends was the Biological Tea Club. Joseph Omer-Cooper was the original convener. Joyce Barrington, Omer-Cooper's future wife and yet another beetle expert; Evelyn Hutchinson; Grace Pickford; Jim Pearce, the water beetle enthusiast of St. Faith's; and Gregory Bateson were among the founding members in 1922. Omer-Cooper, an older student who worked on crustaceans, became one of Hutchinson's most faithful lifelong correspondents.

Evelyn Hutchinson was recorder for the Tea Club, which was extant as late as 1979. Many now well-known zoologists attended the Tea Club meetings and are duly recorded in the minutes. Everyone gave talks on a great variety of subjects. Evelyn's classmate Penelope Jenkin (known to her Newnham College friends as "Clops," from Cyclops because she was five foot eleven) said the club was elite, and she was not invited to join. Her advisor, however, told her it was just as well she was not invited because they "talked a lot of hot air" and that she would be better off getting on with her work.[25] That she surely did; she became a well-known limnologist.

Much tea was consumed by Biological Tea Club members; both Hutchinson and Pickford poured tea. Men and women students seemed to have had equal roles in this club, despite the university discrimination against women students in awarding degrees and prizes. The topics of the members' "discourses" survive. Joyce Barrington talked about sex determination in *Drosophila*. Bateson gave a talk on contemporary arguments for and against the inheritance of acquired characters. He found the evidence in favor unconvincing but thought other theories of evolution lacking as well. New ideas were also discussed. In another talk Bateson invoked geographical isolation in the evolution of snails; Hutchinson spoke about the species question, including the role of reproductive isolation.

The students' talks did not necessarily address their own research but were often reviews of the recent work of others. Delcourt's 1909 French publication on water bugs of Europe was used by Hutchinson to discuss "what constitutes a species," using the genus of backswimmers, *Notonecta*, as an example. Delcourt cited regional "amixia" where two species occurred together and interbred; it would now be called introgression or introgressive hybridization, a subject later studied in depth in plants by Edgar Anderson. From Hutchinson's notes of the 1922 meeting, the discussion sounds quite modern apart from the terminology.

The Tea Club talks and some of George Bidder's ideas were the nearest that Cambridge biology seemed to come to biological theory, with the exception of D'Arcy Thompson's work *On Growth and Form,* which had been published in 1917. D'Arcy Thompson had been a friend of Arthur Hutchinson, and Evelyn Hutchinson had apparently read his work before coming to Cambridge. Only later did he see how Thompson's book might be related to evolutionary ecology.

Many other subjects were examined by the Biological Tea Club. Vitalism, the doctrine that living things must possess some special vital force, that their functions could not be explained by physics and chemistry alone, was still widely discussed in the 1920s. In addition, "psychical" research, now known as parapsychology, had been in vogue in Cambridge earlier. This was a subject Hutchinson later took quite seriously and wrote about, although his own experiments came out negatively. As with the midwife toad experiments, experiments on psychic phenomena are subject to fraud, as Hutchinson later discovered about some that had originally impressed him.

Freudian psychoanalysis was a new subject of discussion at Cambridge, and Hutchinson wrote a paper about it later; it continued to interest him throughout his life. Hutchinson was also influenced by the 1920s writings of W. H. R. Rivers, which advocated a sort of physiological approach to the unconscious. Multiple personalities interested Hutchinson too, as did the possible connection between Freudian theory and evolutionary processes. He specifically mentioned reading in his Cambridge days Freud's *Beyond the Pleasure Principle* and thinking about it in terms of instinctive behaviors of animals. Hutchinson related this to the behavior of hermaphroditic snails, not exactly what Freud had in mind.[26]

In addition to this type of speculation, shared with other members of his various clubs, Hutchinson continued his empirical water bug research. He was most intrigued by the "singing" corixids or water boatmen and how the males made their sounds, generally related to mating. Where did Hutchinson do his studies? As he had done when he was a small child, he kept aquaria right in his (Cambridge) room, but now with a microscope mounted so that he could watch the movements and "song" of a particular species of *Corixa*. He sent his observations to E. A. Butler, who published them in his encyclopedic book, under the species *Corixa panzeri*. The "songs," however they were accomplished anatomically, proved to be important in mating in these insects and thus, from Hutchinson's point of view, for population ecology.

A Student Trip to the Channel Islands

Not all of Cambridge life was so serious. In 1925, before the tripos, Hutchinson and Pickford invited Penelope Jenkin to go on a trip with them to the Channel Islands off the north coast of France. The fourth member of the expedition was Robin Hill. Penelope Jenkin recounted the story: "Grace was a noticing woman and did take great care of her friends. She suddenly asked me if I'd like to go to the Channel Islands just before the exam. I sort of blinked at her and said, 'What, why, how?' 'Well, Evelyn and I are going and we thought perhaps it would be nice if you would come, too.' A pseudo-chaperone, I suppose . . . and another fellow student of Evelyn's from Emmanuel College, Robin Hill. . . . So that funny party of four went for a fortnight to the Channel Islands. Went to Alderney . . . went on to Sark. There was no motor traffic allowed on the island at all by the Dame of Sark, the lady in charge. The job, which was Evelyn's idea, was to collect rats."[27]

Jenkin was not sure of the reason for this, but she thought it had to do with color variation in the black rat on the different islands as a result of isolation. Hutchinson had been interested in coat color in various animals and had long been interested in the black rat in relation to the history of bubonic plague. Jenkin continued: "What we found was that there were some rats on Alderney which had white fronts and the Sark ones were ordinary. Our vocation was to set traps in the evening, go visit in the morning and take the skins off if we caught any. Otherwise we were completely free to enjoy the sunshine and the sea."

Most of the rats they caught were skinned by Evelyn; taxidermy was one of Hutchinson's early skills and proved useful in 1932 on the Yale North India Expedition. However, Evelyn managed "to concoct a cage in which he kept one of them alive. . . . We took it back on the boat to Guernsey; we had one night there before we got the boat back to Southampton. . . . We were in digs of some sort and the wretched rat got out. . . . It got behind a large cupboard and somehow without a moment's thought, Evelyn picked up a poker from the fireplace and stuck it in behind the cupboard. The rat jumped on it, being a climbing rat. Evelyn drew it out, pushed the rat in the cage and shut the door."

Hutchinson was often reported in later years to have been terribly clumsy in the laboratory, always breaking and bumping into equipment. Jenkin said she did not remember this in the laboratory at Cambridge, and in this incident manual dexterity seems to have won the night.

There is still more to the story. When they got to the customs in Southampton, there was this rat in a cage with straw to keep it quiet and comfortable. The customs official asked what it was. "We said, 'A rat.' He suspected we were illegally bringing in alcohol in a bottle and disguising it as a rat. 'I want to see that.' We all shrieked at him, 'It will bite you!' He desisted when he saw enough of it to save his own skin. We got it in and took it along to the physiology lab as soon as we got back to Cambridge. The next morning it produced a large family." This story is not apocryphal. There is an extant notice of a rat named Maude being duly deposited in the Cambridge museum after their return.

Clearly, life among student zoologists was enjoyable, even just before the tripos, which Hutchinson did not seem worried about. When Jenkin was asked what made Hutchinson stand out during the Cambridge years, she replied, "He started with an encyclopedic memory. Absolutely amazing! As a student he read, and he had read since he was that high. Enormously widely read. Some subject cropped up and he'd say, 'Oh yes, so-and-so in 1857 in the Journal of something showed that . . .' He remembered it all and could produce the complete references for the wretched things."

Penelope Jenkin herself must have done well on her tripos since she stayed on at Cambridge University to do research with J. T. Saunders in limnology. She was the only woman to work in Saunders' group. She later did important research at Plymouth, on the channel coast of England. It was Penelope Jenkin who passed on knowledge of the Thienemann school of limnology to Hutchinson; she was his early mentor and lifelong correspondent.

One noteworthy person and event from Hutchinson's Cambridge

years was not mentioned in his autobiography of his early years, *The Kindly Fruits of the Earth.* That person was Grace Pickford; the event was their engagement. Grace was a founding member, with Hutchinson, of the Biological Tea Club and his constant companion on field trips and excursions, including the one to the Channel Islands. Two of Grace's Cambridge friends whom I interviewed told me that they had certainly expected Grace and Evelyn to marry but could not remember a formal engagement. Hutchinson did write, however, to E. A. Butler, his mentor in entomology, about his engagement. Butler wrote back to him before Hutchinson left Cambridge, "You are fortunate in having a fiancée who is also biologically inclined. I congratulate you."[28]

CHAPTER FOUR

From Naples to South Africa:
A Failed Career?

Naples, Italy

Hutchinson at age twenty-two was about to start on his first real venture outside of England. He had been awarded a Rockefeller Higher Education Fellowship to do research at the Stazione Zoologica. "You ought to have a glorious time at Naples, with splendid opportunities. Sardinia should also yield well." So wrote entomologist E. A. Butler to Hutchinson before he left Cambridge. Butler was writing about insects, of course.[1] Hutchinson was bound for Naples, however, not to look for water bugs but to investigate the branchial gland of the octopus.

The Stazione Zoologica in Naples was a mecca for zoologists who wanted to do research on marine animals. The station had been founded by Anton Dohrn, a modestly wealthy German zoologist, in 1872. Dohrn was a pupil of Ernst Haeckel and a docent at the University of Jena. As an enthusiastic young man he had had a mission: to solve important problems in biology that the theory of evolution by natural selection had brought to light through the study of seashore animals. On his own initiative, Dohrn, who corresponded with Darwin, decided to establish and equip a major marine laboratory on the shores of the Mediterranean.[2]

In 1867 Dohrn traveled to Sicily and started a laboratory for his own research at Messina, on the north coast of the island. He went back to Germany to raise funds; subsequently the laboratory was moved to Naples. In 1870 he bought land on the waterfront of the Bay of Naples. It was to belong to him and to his heirs for ninety years and then would revert to

10. Naples Zoological Station in the 1930s. Courtesy of Historical Archives,
Stazione Zoologica Anton Dohrn.

the city of Naples, but the site was always to be used as a biological station. There were delays due to the Franco-Prussian War, but the building
was finished and the station opened in 1874. Dohrn contributed his entire
fortune and raised additional funds in Germany and elsewhere. A thousand pounds came from England through the efforts of Frank Balfour and
Thomas Huxley. Comparative physiology, which would become an area of
interest to Hutchinson, was included in the original plans, but equipment
for studies in this area was not funded until 1903.

Both the German and Italian governments contributed funds for the
enlargement of the station and for equipment. By 1909 Reinhard Dohrn,
Anton's son, was director, and there was already a large, mostly German
staff, including Hugo Eisig, who had come in 1871. Paul Meyer, who had
been there since 1877, was in charge of zoology. Meyer had already guided
and counseled a whole generation of zoologists. The station's staff also included a librarian and an expert photographer, Emil Schobel, in addition
to three artists, four preparateurs, and mechanics, pump men, masons,
plumbers, a carpenter, and a "keeper of the aquarium." Nine fishermen
were permanently employed to provide "living booty" from the sea for

study and research. Charles Kofoid, a visiting naturalist, wrote in 1909: "Upon the morning of arrival, the expected naturalist finds upon his table a collection of preserved material . . . or a dish of fresh sea water with some brilliantly colored squirming inhabitant of the subtropical Gulf of Naples, brought that morning by a barefooted fisherman. . . . Every morning at 10 o'clock the ten or more fishermen of the station, with buckets or baskets full of glass jars poised gracefully on their heads, march into the court and receiving room of the station with their prizes from the sea— scarlet starfishes, orange feather stars, red and black sea-cucumbers, sea-urchins, squirming bits of red coral or a wriggling creeping octopus." This operation was directed by Dr. Salvatore Lo Bianco, a native of Naples who had been hired years before as a laboratory boy at age thirteen. He was taught by Anton Dohrn and others and became a well-known expert on the fauna of the Bay of Naples.

The Stazione Zoologica was a private institution, the property of its director, but it was funded from a variety of sources, including rental of research "tables." In addition, tourists to Naples paid admission to visit its aquaria. The aquaria were quite remarkable, although in 1909 small in comparison with the well-known New York Aquarium. They contained two hundred different genera of marine animals, all local Naples Bay fauna. Invertebrates were featured: an "exhibit of pelagic coelenterates and mollusks, including the delicate siphonophores and Venus' girdle, the cephalopod tank with its ever restive squids . . . the electric ray, the gorgeous tank of tube-dwelling worms and brilliant *Cerianthus.*" One specimen of the latter had survived in the aquarium for twenty-six years. There were also movable aquaria in the physiological laboratory for use by investigators.[3]

The research table system enabled both recent graduates and established scientists to work there. Research privileges were granted to each person for the equivalent of $500 per year. Most of these tables, and thus the investigators, were supported by state funding, universities, or scientific organizations. The station early became an international center, with researchers from "practically all the civilized lands." In 1910 Germany controlled twenty-two tables; Italy eight; Russia four; Austria, Belgium, and the Netherlands each two; Switzerland and Romania one apiece. There were three English tables, sponsored by Cambridge and Oxford Universities and the British Association for the Advancement of Science. There were also five American tables including those sponsored, though not all at the same time, by the Smithsonian, the Carnegie Institution, Williams

College, the University of Pennsylvania, and by Alexander Agassiz, the Harvard zoologist.

In addition there was an American women's table. At the time of the twenty-fifth anniversary of the founding of the Stazione in 1897, Dr. Ida Hyde and other women who had worked at Naples had started a drive for such a table. They formed the Naples Table Association for Promoting Laboratory Research by Women. From 1898 they maintained a table paid for by annual subscriptions of $50 each. Twenty American women had worked there by 1910. They were warmly welcomed. Anton Dohrn wrote to Dr. Hyde in 1897: "Let me openly and sincerely confess that it has taken long years to persuade or rather convince me that the modern movement in favor of women's emancipation is a sound one. . . . There is one part of it for which I have not hesitated to feel and confess a strong sympathy, that is, the throwing open to women the pursuits of science and the higher intellectual development. . . . Nevertheless I have always felt it my duty to act upon my convictions, and have always received ladies in the zoological station with the same readiness as men, and from the first, ladies have worked in it, and at the present one is here at work."[4] Women did indeed come; Ida Hyde had come in 1896, the same year as Hutchinson's future department chairman at Yale, Ross Harrison. Two women, Mary Wilcox and Florence Peebles, came in 1898, and three in 1899. Two well-known American women scientists, Nettie Stevens and Cordelia Clapp, came in 1903. By Hutchinson's time women from many countries came to the Stazione and had a large claim on scarce marine resources, such as octopi.

George Bidder, Hutchinson's favorite teacher at Cambridge, had been an investigator at Naples in the early 1900s, working on hydrodynamics in sponges. Bidder had bought a Naples hotel, Parker's, and held it until Mussolini outlawed foreign proprietors; he then sold it to an Italian. Bidder's daughter Anna, Hutchinson's fellow Cambridge zoology student, also traveled to Naples to do research, but at a later date.

Both the Cambridge Table and the British Association Table were available in 1925 when Hutchinson was awarded the Rockefeller Fellowship. In April 1925 Hutchinson wrote to the Stazione on the recommendation of Cambridge Professor J. Stanley Gardiner: "I am anxious to do a piece of research at Naples next winter. I intend to work on the physiology of the branchial gland and the white body of cephalopods [octopus, squid and relatives]."

Hutchinson wanted to know what physiological apparatus and tanks were available and what he should bring with him. For his work he would need a rotating drum and a mercury air pump, and he asked what the laboratory fees would be. A prompt reply came addressed to "Miss G. Evelyn Hutchinson," welcoming "her to spend next winter in our laboratory for physiological research work." They would provide "her" with the rotary drum and air pump and anything else needed. If occupying either the Cambridge Table or the British Association Table, there would be no lab fees that "she" needed to pay.[5]

Outside of England, Hutchinson often had problems with his middle name, Evelyn, which he preferred and had always been called. Early in his Yale career, when he submitted a paper on the Burgess Shale for a Smithsonian publication as "G. Evelyn Hutchinson," he was asked if he would not prefer to use George E. or at least G. E. Hutchinson because otherwise, in the United States, everyone would assume he was female.[6] Nevertheless, he proceeded to publish as "G. Evelyn" then and thereafter.

However, his next extant letter to the Naples Station was signed "G. E. Hutchinson." He wrote that he would arrive September 30 and would hold the Cambridge Table. "I want to work on the branchial gland and other organs of doubtful function," he wrote, and he hoped that *Sepiola* or other small cephalopods would be available all winter. The reply, addressed to "Mr. Hutchinson," assured him that they would be.[7]

In the last of these letters to Dohrn (September 25, 1925), Hutchinson said that he would stay a few days at a hotel and find a room afterward. According to Hutchinson's own story, he arrived nearly penniless because of high import duties on his microscope and soaking wet because his raincoat had been stolen. But thanks to a letter from George Bidder, former owner of Parker's Hotel, he was able to spend a week or so at this elegant hotel before moving to lodgings.

Hutchinson had undertaken his research in hopes that he could establish evidence of endocrine function in the lives of higher invertebrates. Cephalopods, such as octopi and squids, were much more capable in terms of brainpower than other mollusks, such as clams or snails, and seemed likely candidates. After the excitement of the tadpole metamorphosis into a frog using thyroid tissue, Hutchinson and other Cambridge students had tried using endostile tissue, thought to be homologous to the thyroid, to induce metamorphosis in the tadpoles of ascidians. These small animals are considered higher invertebrates than the cephalopods. Nothing happened; this experiment was a failure.

11. Naples Zoological Station Library in
Hutchinson's Day. Courtesy of Historical Archives,
Stazione Zoologica Anton Dohrn.

Nevertheless, the octopus's branchial gland seemed worth investigating. The function of this structure, with its own blood supply and
no duct, was unknown. Hutchinson thought it might be an endocrine
organ. He found the same cytological structure and staining properties in
a variety of cephalopods. His blue and red stained sections were spectacular. In order to go further in experimental endocrinology, a field in which
Hutchinson's first wife, Grace Pickford, later made great advances, one
needs to remove the organ whose function one is studying. At this point
there was a shortage of octopi. One had to arrange with the fishermen to
bring in the needed catch. Hutchinson lamented that the octopi, the only
cephalopod he found that would survive both the operation and the lab

conditions, went to a beautiful Dutch woman, a visiting zoologist in the laboratory. It was also reported that the Dutch woman and others requiring octopi did not actually operate on them so that the fisherman could have them back for dinner!

Whatever the reason, Hutchinson's research did not succeed. Important hormone control in invertebrates was later discovered but primarily in insects and in crustaceans. Lancelot Hogben, who only a year or two later became very important in Hutchinson's life in South Africa, had already published work on endocrine function in lower vertebrates.[8] His work may have influenced Hutchinson and, later, Grace Pickford. Hutchinson did manage to write one paper, a 1928 note in *Nature,* from his cephalopod work at Naples, but he admitted to failure in his physiological research.

Many other foreigners, including Americans, were working at Naples while Hutchinson was waiting for octopi. Paul Welch, the limnologist who was later to become something of an antagonist to Hutchinson in limnology, was there, working on a limnology textbook that was not published until 1935. Zoologists Newton Harvey and Ethel Browne Harvey were both there. Hutchinson described a trip he took with Newton Harvey to Messina on the north coast of Sicily to see deep-sea animals. He was excited about a squid that when disturbed produced not an ink cloud but a ball of light. He reported little to his correspondents on the work that other zoologists were doing, however.

Hutchinson's Love for Italy

The Naples stay was not a failure in all respects, for it greatly affected his later life. He fell in love with Italy, particularly with Italian art and folklore. These experiences were enhanced by Hutchinson's interest in his Italian ancestry, through the de Mezzi (later Demezy) family. He was to return to Italy later in his life and even to do ecological research there.

After his stay at Parker's Hotel, Hutchinson had been living in a pensione called Alexandria House. It was run by Winifred Allen, a woman of Evelyn's mother's generation and a member of the English colony in Naples. Her family had lived in Naples for three generations. Winifred Allen, who was extraordinarily knowledgeable about Naples, became Hutchinson's tour guide about everything from Baroque art to Roman

Catholic festivals to werewolves. A letter she wrote to Hutchinson in 1927, by which time he was in South Africa, began: "Many werewolves still exist in Naples. Had I known [of your interest] last year we would have gone to seek for one."[9] Hutchinson was interested in everything, and whether believing in werewolves or not, he would certainly have consented to seek them out. Allen apparently did not believe herself as there is a remark at the end of her letter about its being a form of epilepsy. During his time in Italy Hutchinson became an expert in a subject far from octopus endocrinology or any of his subsequent research, the liquefaction of the blood of saints. Religious and sexual imagery, popular cults, pagan aspects of religious festivals, which he often attended with Allen, all fascinated him.

Winifred Allen's letters to Hutchinson in South Africa in 1926 discussed the other people who had boarded at Alexandria House with him. The two of them shared interests in people's lives as well as in customs, paintings, churches, plants (many of which are named), and animals, almost everything Italian.[10] Several of Hutchinson's extra-scientific interests, especially in art, including church art, relics and folklore, developed during his stay of less than a year in Naples. Hutchinson also learned Italian that year; he was later able to deliver an entire lecture in Italian at an international meeting.

Sicily provided great artistic experiences during Hutchinson's Naples year. He was invited to visit Cefalu on the Sicilian coast near Palermo by Matthew Prichard, an English expert on Byzantine architecture.[11] Much later in Hutchinson's life, when he became a close friend of novelist Rebecca West, he learned more about Prichard. He had been interned with Rebecca West's husband, Henry Marwell Andrews, in the German concentration camp at Ruhleben during World War I. There Prichard had lectured to the inmates on Byzantine art. Hutchinson returned to Sicily later in life. That visit was the source of one of his most famous papers.

Shall we believe Hutchinson's statement that he was unable to do more on his branchial gland studies because of the scarcity of octopi? Was his failure the result of a poor choice of organisms? Or was he irresistibly drawn to those nonzoological Neopolitan studies of which he later wrote? What did the Rockefeller Foundation think of his report? He wrote little about the science going on at the Naples station. It appears that his interests were elsewhere that year. No letters from Naples to his Cambridge friends or teachers have been found. He must have written to his fiancée Grace Pickford, but she later destroyed almost all of her correspondence.

The South African Experience

In early 1926 Hutchinson saw an advertisement in *Nature* for a lecturer at the University of Witwatersrand in Johannesburg. He decided to apply. Although his parents warned him about H. B. Fantham, the professor of zoology there, who was known to be very difficult, Hutchinson persisted and was appointed. The need to go back to England to prepare for this major move caused him to leave Naples sooner than he had intended.

The years that Hutchinson spent in South Africa were two of the most important but also two of the most difficult of his life. As at Naples, he failed in what he had set out to do. He had just turned twenty-three when he accepted the South African position, but apart from the year in Italy he had spent his entire life in Britain, and most of it in Cambridge. He undoubtedly wanted to get away—away from Cambridge, away from his parents. However enriching his life with them had been, it was also rather stifling. His mother was surely overbearing, and she had disapproved of one of his earlier female friends. His father, Arthur Hutchinson, was probably too important a figure in the university for the younger Hutchinson to find his own independent place there, although with two first-class degrees there is no question that Evelyn could have gone on as a research (graduate) student there, studying for a doctorate.

Hutchinson, explaining his lack of an earned doctorate many years later, wrote: "The Ph.D. had not been firmly established as a British institution, at least at Cambridge. It was possible to take one but it was solely a trap to catch Americans before they spent their money in Germany."[12] This is not the whole story behind Hutchinson's lack of a doctoral degree. A Cambridge graduate automatically received an M.A. from Cambridge University with the passage of three years and the payment of a fee. It was also possible to continue doing research in England or elsewhere and later send your published papers or books to Cambridge University for consideration for a doctoral degree, an earned rather than an honorary doctorate. For reasons that will be discussed later, Hutchinson never did this.

Why a lectureship in South Africa? In the first place there is no evidence that he was offered other positions. Many Oxford and Cambridge graduates, even today and including those who complete a doctorate, accept teaching positions in secondary schools in England, many of which have excellent science programs. Some of Hutchinson's contemporaries followed this route, but he had higher ambitions and undoubtedly felt that more was expected of him by his father.

What did South Africa mean to Hutchinson? It represented a whole new biological world to explore. What young person who had spent all his life steeped in natural history, and Hutchinson started unusually young, would not want to take the opportunity to visit another continent? Some of this continent is tropical; all of it is very different from England. South Africa itself is, moreover, a very varied and beautiful country, containing an unusual array of different ecosystems, each with its unique flora and fauna.

Hutchinson wrote that he put the opportunity to experience a wholly new fauna ahead of the horror stories his parents, his uncle, and others in zoological circles told about Professor Fantham. In terms of South Africa's animals—and the lowly Onychophora, or velvet worms, may have impressed him more than the lions and leopards—Hutchinson was amply rewarded. There was probably another important reason for choosing South Africa: Grace Pickford was there. During her first year in South Africa Grace was doing research on earthworms and their aquatic relatives; she made extensive collections of these and other aquatic invertebrates in the Cape region.

While still at Cambridge, Hutchinson had referred to Pickford as his fiancée in a letter to E. A. Butler. His sister, Dorothea, and his old friend Anna Bidder, good friends of Grace, had assumed they would marry but didn't seem to know about the engagement. It was probably not official, although the two sets of parents had visited each other before either Grace or Evelyn had departed for South Africa.[13]

There is considerable folklore about the Hutchinson-Pickford marriage. The story most often heard from Hutchinson graduate students and others was that Grace had married Evelyn so that she could go to North India (Ladakh) on the Yale North India Expedition. This story has no foundation at all. In the first place, she never went on this expedition. Moreover, it occurred more than six years after they married. Others have said that she married him after he received his instructorship in order to go to South Africa. This story is also false. Grace Pickford was awarded her own Mary Ewart Scholarship in 1925 for travel in South Africa. Although a printed account after Pickford's death says that they married and then went to South Africa together, it is almost certain that they did not marry in England. Grace Pickford went to South Africa first. Dorothea remembers reading a letter Grace sent to Evelyn from South Africa while he was still in England. They were married in Cape Town; the Hutchinson family received a cable announcing their marriage.

Hutchinson had to begin his zoology teaching duties at the University of Witwatersrand in July. He went from Naples to Cape Town via England, where he arrived with an infected foot and in the middle of the Great General Strike of 1926. When these problems were resolved and the necessary items packed or sent, he set off by boat for Africa, stopping at the island of Madeira. On his arrival in South Africa, he first stayed near Cape Town, where he saw the wonderful and tremendously diverse cape flora and the impressive botanical gardens at Kirstenbosch and delighted in the proteas there. These spectacular flowers of the paleobotanically ancient family Proteaceae are pollinated by sunbirds and by small mammals in South Africa. Their relatives are pollinated by small marsupials in Australia. Since Pickford took Penelope Jenkin to see both the cape flora and the Kirstenbosch Gardens when Jenkin later visited South Africa, she may likely have been Hutchinson's tour guide when he first arrived.

The animal that first captivated Hutchinson was a velvet worm (a *Peripatopsis* species) along the Table Mountain trail outside of Cape Town. This small animal with many peg-like legs is a member of the phylum Onychophora, the first of which was not discovered until 1825. It was thought to be a slug with legs. This group may have evolved when all the continents were joined into the supercontinent Pangaea at the end of the Paleozoic era, more than 200 million years ago. There are two families, the second looking more like an arthropod than a slug with legs. Both occur in South Africa as well as in South America and Australia; one occurs in India. Hutchinson found several species and later wrote a paper on the South African Onychophorans. Sydnie Manton, his classmate at Cambridge, later did more significant work on them.

Africa has the most fantastic and diverse birds of any continent and these, too, fascinated Hutchinson. He did considerable traveling in South Africa during his two-year stay, largely to aquatic areas. He made bird observations at lakes near Johannesburg and elsewhere and watched the behavior of the crested coot, described in a letter he sent home to his family. He also watched the "dance" of the hamerkop, a Dr. Seuss–type bird that appears, as in its name, to have a hammer on its head. One dances while standing on the other bird's back. Their behavior was described by Hutchinson in rather Freudian terms; psychoanalysis was a continuing interest of his and the subject of another of his early papers.

The teaching of zoology at Witwatersrand was very strange by any standard. There were three women lecturers, but they did not give their own lectures; they read lectures written by Professor Fantham. Hutchinson described this process in a 1986 letter: "They were actually in the medieval sense of the word, readers. They were supplied with lectures written out by the professor and read aloud by the lecturers. Unlike the practice in the 14th century, this had nothing to do with the scarcity of manuscript books but solely with the head of the Department controlling all the teachers. The lecturers did research, directed by the professor."[14] This was Fantham's world; lecturers in other departments, such as botany and geology, did independent work. One of the women zoologists, Johanna F. M. Schuurman, apparently staged a small revolution and did her own work, aided by Hutchinson, on the plankton of a small lake. It is not known why Fantham hired a male lecturer and allowed him to lecture—for a while. He came to listen to some of Hutchinson's lectures. Whether he did not like what he heard or he needed to be in full control, probably the latter, Fantham fired Hutchinson from his teaching duties in his second year. At least he was not forcibly ejected from the laboratory as his predecessor was said to have been. Hutchinson could not legally be let go during his two-year contract. Therefore, although he felt humiliated, he was able to continue his research, and he now had much more time for it. He was trying to work on the taxonomy and evolution of South African water bugs. He planned a work including chromosome analysis, geographical distribution, and ecology. Ethel Browne Harvey, who had been at Naples at the same time as Hutchinson, had already done some chromosome work on American water bugs. Hutchinson started on the collecting and taxonomy but never got to the cytology.

When Hutchinson began his research in 1926, only fourteen species of water bug from three families were known for South Africa. Before he left two years later there were nearly forty known species of these three families: Notonectidae, Corixidae (both described above), and a very small family Pleidae, which was increased from one to three species through his study. Temperature tolerance and dispersal ability both seemed important to their ranges. By this period Hutchinson was an established and published water bug expert. Entomologists in Sweden, Hungary, and the United States as well as in England and South Africa wrote to him. Specimens other than insects—for example, polychaete worms and water mites—were sent by both Evelyn and Grace to experts at the British Mu-

seum in London and elsewhere. People sent Hutchinson water bug specimens from all over Africa to be identified. This went on after he arrived at
Yale. In 1930 the British Museum sent him 723 specimens of water bugs
to identify. These had been found during a mosquito breeding study in
Uganda.[15]

Hutchinson's major paper on the water bugs of South Africa was not
published until 1929. In it he revised the systematics of the families Notonectidae and Corixidae. South Africa was here defined as "that part of
Africa south of the Zambesi and Kenene Rivers," then and now comprising several countries, but now with different names since independence
from the colonial powers. Some of his specimens came from museums in
South Africa and Britain. Specimens referred to in the paper included,
in addition to his own collections, a large collection made by Grace Pickford in the East Cape Province in 1926 and another from the Orange Free
State, which she made together with a Miss D. F. Bleek in 1928. Zoogeography, including probable dispersal mechanisms, and ecology are discussed.

Hutchinson noted that since European colonization, natural aquatic
habitats had been supplemented by reservoirs and other man-made wetlands. The latter actually constituted over 36 percent of all aquatic localities studied, showing "the importance of such constructions in opening
up the country to water bugs as well as civilized man."

Hutchinson also discussed what would now be called niche separation in terms of "the partition of species between moving and standing
water," with specific examples. Species diversity was also of interest: one
pool contained ten species including five species of one genus, *Anisops*.
The distribution of nineteen species of *Anisops* for various geographical
locations, including Kampala, Uganda (data of G. L. R. Hancock), and
what was then Southern Rhodesia (Zimbabwe), showed that the vicinity
of Kampala contained the most species and both primitive and specialized forms. Hutchinson postulated that Central Africa might be the center of origin of this genus of water bug. He had also studied colonization
of new localities, such as a large pan, which had been dry for several years
but filled up with water in December and January of 1927–28. By April it
contained large populations of water bugs of six different species, a very
rapid colonization.[16] All of these subjects, which included community
structure, species diversity, and dispersal, as well as niche diversification,
later became important topics of ecological innovation by Hutchinson.

1. Hutchinson Lake, section between Sta.9 & Sta.10, Vallentyne 1971
2. Protziella hutchinsoni Lundblad 1934
3. Pamachlus hutchinsoni Hora 1936
4. Eocyzicus hutchinsoni Bond 1934
5. Daina hutchinson Brehm 1947
6. Davicula hutchinsoni Patrick 1970
7. Lophocharis hutchinsoni Edmondson 1935
8. Cyclops hutchinson Kiefer 1936
9. Bryocamptus hutchinsoni Kiefer 1929
10. Kashmirothyas hutchinsoni Lundblad 1934
11. Gomphonema elivacioides var. hutchinsoniana Patrick 1971
12. Keratella quadrata var. pyriformis f. hutchinsoni Ahlstrom 1943
13. Hutchinsoniella Sanders 1955
14. Paraschinorhyncus hutchinsoni Datta 1936
15. Isobactrus hutchinson Newell 1947
16. Halmopota hutchinson Cresson 1934
17. Seng-ge Hutchinson
18. Machlanus hutchinson Silvestri 1936
19. Apantaniana hutchinsoni Mosely 1936
20. Athela (Dimetiota) hutchinsoni Cameron 1934
21. Ilydonius hutchinsoni (Denis) 1936
22. Bambidion hutchinsoni Andrewes 1934

M.J.Kohn

12. Drawing of all the organisms named after Hutchinson, by Marion Kohn. Reprinted from Yvette Edmondson, 1971, page 477, fig. 10. Copyright 1971 by the American Society of Limnology and Oceanography, Inc.

Adding to his excitement with the fauna of South Africa, Hutchinson met R. C. A. Dart, an anatomy professor at the University of Witwatersrand. Dart let Hutchinson hold the original skull of *Australopithecus,* an early member of the family of man. Dart got the skull from R. B. Young who got it from a mine manager. Hutchinson had known about Dart and his fossil skull before going to South Africa.[17]

The First Limnology Research

Lancelot Hogben now entered Evelyn Hutchinson's life. Hogben had been a lecturer at Imperial College, London, and had worked on both the physiology of color response and on comparative endocrine physiology in lower vertebrates. He had published books on these subjects, the first while Hutchinson was at Cambridge. Hogben did important research at the Plymouth Laboratory and was one of the founders of the Society of Experiential Zoology and of the *British Journal of Experimental Zoology*—also while Hutchinson was at Cambridge University. Hogben was, by 1927, professor of zoology at the University of Cape Town. Hutchinson must have known about Hogben's work, and Hogben had first learned about Hutchinson from someone at Gresham's School. After Hutchinson had been fired from the teaching portion of his two-year position, Hogben invited him down for the summer (the Northern Hemisphere winter) of 1927–28 to do research. Grace Pickford came as well.[18]

Hogben had some interest in water and had done some pH studies of ocean waters south of the Cape of Good Hope. He suggested that Hutchinson and Pickford study the chemistry and biology of the shallow lakes, called vleis, near Cape Town. These included somewhat saline, eutrophic coastal lakes that contained a diverse flora and fauna and contrasting reservoirs on Table Mountain. The latter were much less fertile. Later, when back in Johannesburg, Hutchinson, together with Pickford (in South Africa, but not afterward, called "Mrs. Hutchinson") and Johanna Schuurman, studied shallow lakes or pans in that region. They received help and lab space from several people, especially a former student of Evelyn's father, J. A. Wilkinson, who was then professor of inorganic chemistry at Witwatersrand. Hutchinson was not as far from Cambridge as he had envisioned. Another student of his father's, named Rogers, director of the South African Geological Survey, had studied the geology of these pans. These depressions had been formed earlier in geologic history when the climate was more arid and were probably the result of wind erosion. Some of the larger pans contained water all year. Hutchinson, Pickford, and Schuurman studied some of these in the Ermelo district of the Transvaal in February and May of 1928. In only one, Lake Chrissie, did they have a boat available for collecting aquatic specimens. They also visited some smaller pans near Johannesburg.

It was in these studies that Hutchinson first experienced the excitement of limnological research. The pans were extremely varied in both chemistry and in flora and fauna. Very little biological work had been done on them previously. An English amateur invertebrate zoologist, P. A. Methuen, had been there in 1908 and had found some interesting crustaceans, including a large water flea, *Daphnia gibba,* and two copepods of different sizes, a large one of the genus *Lovenula* and a small one, a *Paradiaptomus.* These lakes have no fish. In the absence of such predators, these usually submicroscopic crustaceans can attain larger sizes. Hutchinson later studied both other arid lakes and those at high latitude and found similar situations.

Hutchinson, Pickford, and Schuurman studied pans throughout the southern half of the Transvaal as well as elsewhere in the country, both temporary and permanent pans. The temporary ones dry up near the end of the winter season and fill up again after the summer rains begin. The temporary pans can be divided into grass pans and mud pans; the grass pans have a pH below 8.0 and support abundant plankton typified in the phytoplankton by the green alga *Volvox,* together with a particular rotifer. These pans have a rich terrestrial vegetation when dry and a rich aquatic flowering plant vegetation when filled with water. The mud pans have a pH over 8.2 (highly alkaline) and support a much more restricted flora and fauna with few phytoplankton or rotifers; crustaceans are the main fauna.

The permanent pans were studied in the Lake Chrissie region, but even among these there were many differences in water chemistry resulting in differences in flora and fauna. One type contained two copepod species and the large *Daphnia gibba* but few phytoplankton or aquatic vascular plants. These pans were less than a mile in diameter, ten to twenty feet deep, slightly salt, with an alkaline pH of about 9, and much humic material.

Later, in their papers based on this research Hutchinson and his coworkers tried to relate these varied pans or lakes to types established by Naumann or to somewhat different types suggested by Thienemann (oligotrophic, alkalitrophic, dystrophic).[19] At the time of their research, Hutchinson did not know about Thienemann's book, published in Germany in 1925.[20] The group did have a methods book, H. W. Harvey's *The Biological Chemistry and Physics of Sea Water,* just published, as well as *Standard Methods for the Examination of Water and Sewage,* published earlier in New York, to help them with techniques.[21] Harvey's book apparently worked well for fresh water as well as for seawater.

Hutchinson, Pickford, and Schuurman noted that alkaline dystrophic lakes (containing large amounts of humic matter) had been reported earlier, notably in Maine. The reservoirs Hutchinson and Pickford had studied earlier on Table Mountain are also mentioned in their first paper (1929). A fuller account of the plankton was published later, when the species had been worked out by a number of specialists; that work was already underway and was published in the second paper by Hutchinson, Pickford and Schuurman (1932).

I go into detail about these lake studies here because of their importance to Evelyn Hutchinson's life in science. Working on the South African pans, Hutchinson had found the kind of research that he truly enjoyed. Limnology combined his early interests in natural history, particularly of aquatic invertebrates, and his excellent training in chemistry. When he struggled with those titrations at Gresham's in order to become a good field zoologist he was already thinking ecologically, though he would not have used that term.

Limnology, the scientific study of fresh waters, from the Greek word *limne,* pool or lake, was likely at that time to be called freshwater biology, or as in Saunders's Cambridge course, subsumed under hydrobiology, which included salt as well as fresh waters.[22] Although Hutchinson was later much influenced by Thienemann's 1925 book in interpreting their South African data for publication, beginning with his first work with Grace Pickford, the physical factors, or hydrography, were part of his research.

In limnology Hutchinson had at last found what he really wanted to do. He had found a way to combine his talents, interests, and training successfully. His mathematical training and interest in biological theory were to give him a new vantage point in this field, although one that was not looked on favorably by older American limnologists. This new vantage point greatly influenced the younger generation of limnologists, especially Hutchinson's future graduate students.

But who influenced Hutchinson? What about Hogben? Although he started Hutchinson and Pickford off, there is no evidence that he was involved with the Transvaal pan work. Hutchinson wrote that although Hogben was extremely able intellectually, and had done excellent research in England, "When I knew him he seemed to have lost all sense of scientific direction."[23]

What about Hutchinson's co-researchers, Grace Pickford and Johanna Schuurman? Schuurman wrote many letters to "Mr. and Mrs. Hutchinson" after they left South Africa, asking for advice on research techniques. This was a new field for her, and she seems to have been more like a research assistant or early graduate student. Grace Pickford, however, was Evelyn's full coworker. Did some of Hutchinson's early ideas about limnology come from Pickford, as some have suggested? It is hard to assess this claim. Joint research, especially in the field, involves a great deal of interactive working out of both ideas and methods. There is no doubt that they did this together while studying a great variety of lakes, types that neither had ever seen before. They had both been top Cambridge students and had very similar training before and during their Cambridge years, and they had done joint fieldwork at Wicken Fen and elsewhere. No letters from this period seem to have survived. Thus we can only imagine the development of their ideas and techniques. Pickford in later years was always unusually adept at inventing and making equipment, Hutchinson less so. He did mention "homemade equipment" used in the pan study. What we do know is that in 1932, when Hutchinson was the biologist on the Yale North India Expedition, doing limnology at high altitudes, Pickford was his most important research consultant and critic. His letters to her from that expedition have survived (see chapter 6).

Limnologist Penelope Jenkin

The primary influence from an experienced limnologist during Hutchinson's early years was surely that of Penelope Jenkin. She had continued in this field after taking Saunders's course at Cambridge together with Hutchinson. She had this to say about Saunders: he was an odd character, but he was on the Cambridge zoology department staff after going through World War I. He got interested in the development of freshwater ecology in Europe through the writings of Thienemann and Ruttner.[24]

In 1924–25 Saunders gave the lectures in a course entitled "Hydrobiology," the first time such a course had been given in the British Isles. Although called "Hydrobiology," according to Jenkin, the course "was mainly related to fresh water, as the European work had been. There was a party up in Sweden that was busy with it, too." Jenkin stayed on at Cambridge for a year after graduation, with Saunders as her research supervisor.

When the Lake Windermere station in England was established, Jenkin went there in its first year, 1926–27: "Saunders started the freshwater lab on Windermere; he sent me out." She worked on the protozoan *Spirostomum* and related its physiology to the pH (acidity) of the water. Saunders was "very thrilled with the pH effects; they were much more variable in fresh water than they are in sea water." This freshwater pH work had been influenced by the work of W. R. G. Atkins at the Plymouth Laboratory, started much earlier in 1886. Jenkin also did a survey of Lake Windermere. The laboratory was small then and at the top end of the lake; "It's grown into an enormous place, and they've moved down to the middle of the lake." Jenkin then spent a year teaching at the University of Birmingham and was also appointed to work with Atkins at the Plymouth Laboratory. Here she did some of the first studies on the effect of light penetration on the growth of diatoms in the English Channel "running about in a trawler by the Eddystone Light." She later did some of the first light and dark bottle experiments in the channel relating the results to light measurements (see chapter 8).

Penelope Jenkin spent most of her career teaching at the University of Bristol, though she continued to work at Plymouth. Much later she worked on the filter-feeding and food of flamingoes. In 1928, however, she went on the Percy Sladen Expedition to work on the Rift lakes in Kenya. Hutchinson later published on the water bugs she collected on that expedition. But before she left for Kenya, Hutchinson saw her in England when he was on his way to Yale in 1928. He wrote, in a 1986 letter:

> After I left Cambridge a loose but officially more or less recognized group of limnologists grew up around J. T. Saunders. Clops [Penelope Jenkin] was the main woman involved and what she learned of the Thienemann school, influencing English limnology while I was in Africa, she passed on to me. It was of enormous importance to me. I felt part of a definite school of a quite informal kind. . . . Practically, I had become a limnologist in South Africa; in a sense the first one in that country. On the way to America I had many hours' conversation with Clops in Cambridge and

learned all about the new German literature . . . so in some ways I am really one of her students. J. T. Saunders was still teaching at that time, . . . and T. T. Macan was doing admirable work in Windermere. I was intellectually close to all of these people and would regard myself as a student of all of them, particularly Clops [Jenkin] even though we were working considerable distances away.[25]

During his 1928 visit to Cambridge, Penelope Jenkin showed Hutchinson August Thienemann's *Die Binnengewasser Mitteleuropas*. Reading this, he realized how the Lake Chrissie pans that he had studied fit into the scheme of lake classification that was being developed in Europe. At that time he also read Charles Elton's newly published *Animal Ecology*, another book that greatly influenced him. He wrote: "All the ways of looking at nature that I had acquired in a random disorganized way could be focused together on lakes as microcosms. I had, in fact, become a limnologist."[26]

The Unlikely Path from Witwatersrand to Yale

Limnologist or no, it was not at all clear what Hutchinson would do when his contract with the University of Witwatersrand ended. Hutchinson had been dismissed from his duties at the university by Fantham in November 1927. A professor of organic chemistry told Hutchinson that he thought it was illegal to dismiss him before his two years were up. The Lecturers' Association, which looked into violations involving what were called nonprofessional faculty, took up the issue. The principal of the university agreed to an inquiry, which marked Hutchinson's last appearance at the University of Witwatersrand, on April 25, 1928. There was a large committee, including the Anglican bishop of Johannesburg; Professor R. B. Young, who was the intermediary through which the *Australopithecus* skull came from Taungs to Dart; and the university registrar. Hutchinson did not keep his notes of what he said at the inquiry, but his sister kept a letter he sent home about it. He had to make his presentation in Fantham's presence, some of whose testimony seems quite ludicrous. Fantham claimed that Evelyn would not listen to the ends of his conversations.

One can picture the two of them left to walk on the lawn together while the committee deliberated. Hutchinson essentially won—he was granted his request for pension premiums and passage money—but Fantham could not legally be removed nor could Hutchinson continue to teach there. Fantham left several years later and spent the rest of his career at McGill University, where in spite of a favorable obituary quoted by Hutchinson, uncomplimentary things were said and written about him in private.[27]

Lancelot Hogben, although he did not wish to hire Hutchinson for his Zoology Department, kindly propelled him toward Yale University. He could hardly have guessed that it would be for sixty years! He knew of a Sterling Fellowship available at Yale for 1928–29 and wrote Hutchinson an excellent letter of recommendation. Hutchinson applied by cable as soon as he received the application form. By then, however, the deadline had passed. But a termite man named Harold Kirby had unexpectedly left Yale for the University of California, providing an instructorship opening, which Hutchinson was offered.

That story is obviously too brief. Why should Yale have hired, sight unseen, a young man with no advanced degree, who had been fired from his only teaching job, and who had published nothing but some small papers on water bugs up to that time? That question can perhaps be answered because all the relevant documents, even a cable from South Africa to Yale, are still extant. In this case it was Ross Granville Harrison, not Hutchinson, who preserved them.

On February 2, 1928, Lancelot Hogben wrote a letter to Ross Harrison, the chairman of the Zoology Department at Yale. Harrison was a renowned experimental embryologist who had made many important discoveries and had invented the technique of tissue culture. He was well known to European scientists. Hogben also knew him personally but noted in his letter that he had not seen him since taking the chair of zoology at Cape Town. In this letter he wrote that he wanted to send his "young friend," age twenty-five, to the United States "to see Experimental Zoology in action." He wrote that Hutchinson had taken a double first at Cambridge in zoology and was also knowledgeable about chemistry and physics. "If you could offer him the prospect of an instructorship with a small salary say $1200 . . . he could keep himself for nine months or 10."[28] According to Hutchinson's eventual application he had conducted experiments in Hogben's laboratory on the activity of crustaceans and polychaetes in relation to temperature.

Ironically, Harrison, who was on sabbatical in Europe that year, was at the Naples station when this letter arrived. It was therefore answered by another Yale zoologist, Lorande L. Woodruff, who worked on protozoans, specifically the question of immortality of cell lines in *Paramecium*. Together with Rhoda Erdmann, Woodruff had made an important discovery in 1914, the process of nuclear reorganization known as endomixis. Woodruff, like Harrison, was well known in Europe in the twenties, although his work on sex in *Paramecium* was superseded by the research of others. Woodruff replied to Hogben on February 29 that Hutchinson would have been a strong candidate for a Sterling Fellowship but that these would be awarded within a month and that he did not think it possible for Hutchinson to send the necessary data by that time. However he was sending application blanks and a Yale catalogue in case a vacancy might occur later. He asked Hogben to send "immediately full particulars in regard to Mr. Hutchinson as well as several supporting letters and a file of his reprints."[29]

Hutchinson actually applied for the fellowship by cable. He had received the application form, and on March 31, 1928, he answered all the questions according to the numbers on the form and sent the cable. He also wrote a letter to Woodruff on April 4 and sent the required photo, taken that day. He said in this letter that Professor Hogben had advised him to apply by cable even though the appointments for fellowships had already been made.[30]

Cambridge University letters of recommendation from Professor J. Stanley Gardiner and from Dr. L. A. Borradaile were on their way. These two letters are quite remarkable. Borradaile wrote: "He was at once brilliant and hard-working. He has an intense interest in the science and a keen desire to further it by research. He has had an education both wide and thorough in every branch of it, and I am convinced that he will one day take a high place amongst zoologists. He is of high character and agreeable personality."

Gardiner wrote a cover letter to Woodruff as well as his official recommendation. The cover letter read: "G. Evelyn Hutchinson . . . is *absolutely* first class and entirely devoted to his subject. His fault used to be that he had too much personality, making him a little difficult; it was not conceit, in any way, but a difficulty in understanding the position of other people; I understand that this has been entirely knocked out of him in South Africa" (emphasis in original). Gardiner added that had he a lectureship vacant, he would be happy to hire Hutchinson for his department.

Gardiner's actual recommendation is worth quoting as it provides a contemporary view, unlike what Hutchinson or others later wrote or told the author about this period:

1. The above student was a most thoughtful and quiet person, but he was the leader of his class and looked up to by every member of it. He was devoted to his subject but he worked on it on his own plan and on his own ideas, being mentally peculiarly independent for his age. He had read very widely not merely in zoology but in all science. He had none of the frills which make for popularity but yet he led his class and he led also the large natural history society of the University and organized its lectures, conversazione and work. I have seldom met a man of his age with more character and personality.

2. He started by being an Entomologist rather concerned with the questions of adaptation to environment and with physiology of insects. Of publications here there is a notable one on the water Hemiptera of Wicken Fen. He then went to the Stazione Zoologica, Naples, and tackled a semi-experimental and chemical problem of great difficulty; he got as much as anyone could expect. In S Africa he has shown great promise in research and I prophesy a high place for him someday.

3. I rate him as absolutely first class and should put him as a student among the first half dozen best men I have had in the last 18 years.[31]

The third letter was an official one from Hogben himself, who reported that he had followed Hutchinson's career "during his studies as an undergraduate." He wrote that Hutchinson had "unique qualifications for fundamental research in biology because he has an exceptionally good all round scientific training." Hogben also stressed Hutchinson's exceptional knowledge of field zoology on two continents, and his "great personal charm." He praised him as a "young man of all round culture," extremely well read, and suggested that his presence at Yale "would be an asset to the University as a whole," a prescient view. Finally, he was "convinced that he has before him a distinguished career as an investigator."[32]

Hutchinson filled out the application form. On this he gave some personal information: he was married, but neither his wife nor anyone else was dependent on him for financial support. He had an annual income, presumably from his father, of 50 pounds per year. He also said that he read "French easily, Italian slightly and German with difficulty."

He described his past, present, and future research. He put the Naples research in a better light than it probably deserved, "Naples 1925–26. Worked on the histology and physiology of the branchial gland of Cephalopoda, which is apparently an endocrine organ [which he certainly did not show it to be] — short account should appear in *Nature* in a few weeks."

He also noted his monograph on the Corixidae and Notonectidae of South Africa and his current "investigation of the ecology of freshwater 'pans.'" He later added "joint" before "investigation" to acknowledge his coworkers, Pickford and Schuurman. His experiments on crustaceans and polychaetes (segmented worms) were included as well.

As to the future, to the question, "Is it your purpose to enter teaching or research as a life work?" he answered, "research" and indicated the "special fields" of his interest in the following order: ductless gland problems especially in invertebrates, freshwater ecology, and animal behavior.

He then had to write a statement about the research he wished to undertake at Yale, including its importance, methods, and needed facilities. This clearly had to be in experimental zoology—the field for which Hogben had recommended him to Ross Harrison. Hutchinson wrote that he was "anxious to investigate the temperature characteristics of the long latent period of delayed or 'trace' conditioned reflexes." He thought this would elucidate the physicochemical nature of the functioning of the nervous system. He planned to do this work on cold-blooded vertebrates, probably tortoises, but "interesting results might also be obtained from invertebrates." The apparatus required was chiefly a thermostat; electrical stimulators he could make himself from ordinary laboratory electrical equipment.[33]

The written application form and Borradaile's letter are both stamped as received on May 8, 1928. That day or shortly thereafter, Woodruff sent Hutchinson a cable telling him that an instructorship was unexpectedly available and it was being offered to him. This cable is not extant, but Hutchinson's answer is. On May 11, Hutchinson sent a second cable to Woodruff. This one reads "DELIGHTED ACCEPT." It was followed by a May 13 letter to Woodruff, thanking him for letting him know so rapidly by cable, that Hutchinson was delighted to accept the instructorship, and that it would be a "great privilege to be associated with a department from which so much fundamental work has proceeded." His Witwatersrand contract would terminate on July 31; he would arrive at Yale via England in September. He wrote that his wife was working on a monograph of South African earthworms and asked whether she could work at the university museum.[34]

Woodruff's May 17 letter (following the cable) informed Hutchinson

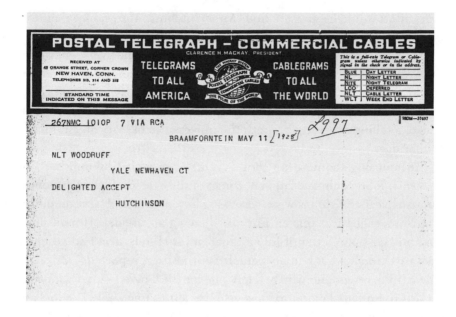

13. Hutchinson's cable from South Africa accepting the instructorship at Yale.
Manuscripts and Archives, Yale University Library.

that his fellowship application had indeed been received after the awards
had been made, but the instructorship in zoology was offered to him at a
salary of $2,100. The position was for one academic year from September
1928 to June 1929. Hutchinson would have two lecture sessions of General of about sixty men each in the fall and three in the spring, the rest of
the total of twenty hours was laboratory instruction; one teaching assistant would be provided for each twenty students. Woodruff also sent a
course description and the names of the texts: *Foundations of Biology* by
Woodruff himself and Baitsell's *Manual of Biological Forms*, presumably
for the laboratory. Woodruff added that Hutchinson would find every encouragement in developing his special field of research. Nothing was put
into writing about his wife's use of a laboratory for her research, or even
acknowledging that she was coming as well.

He asked Hutchinson to cable back "satisfactory," but as of June 18
had not heard from him. On this date Woodruff sent Hutchinson's formal
appointment notice from the Yale Corporation and advised him to come
a week early to find accommodations, since "the best apartments in town
are all taken before the day Yale opens." In this letter he did write that

every effort would be made to find "Mrs. Hutchinson" suitable laboratory facilities for her research.[35]

Hutchinson sent the requested cable saying "satisfactory" on June 20. He wrote that he hoped to find the required textbooks in England and would arrive at Yale about September 20. "I shall be only too glad to be in a thoroughly modern department again; we are years out of date here," he added, a telling comment on both the universities of Witwatersrand and Cambridge.[36]

An amusing footnote on these events is provided by another Yale zoologist, John Nicholas, later chairman of the Yale Zoology Department for many years. Nicholas wrote very long letters to Ross Harrison during Harrison's sabbatical year in Europe, relaying all the department problems and gossip. A letter of July 16, 1928, after Hutchinson had accepted the instructorship but had not yet left South Africa, reported: "Kirby [the Yale termite researcher who had left Yale for the University of California] returns to Kofoid's laboratory next year. He left us two weeks ago and as a temporary fill-in Hutchinson, an Oxford [sic] product, has been selected from the field. I shall be interested to see what happens."[37]

The "temporary fill-in" stayed at Yale from 1928 to 1991.

CHAPTER FIVE

First Years at Yale

A British Cantabrigian both by birth and by training, he has chosen to spend the greater part of his professional life in such alien provinces as Italy, Africa and Yale." Hutchinson arrived at Yale in 1928. He stayed until 1991 in this "alien province"—virtually the rest of his professional life.[1]

When Evelyn Hutchinson and his wife Grace Pickford arrived at Yale University in September of 1928, the Zoology Department was a separate entity, housed in the Osborne Zoological Laboratory (OZL). The zoology chairman, Ross Harrison, was back from his year in Europe. Other zoology professors were L. L. Woodruff, the protozoologist with whom Hutchinson had had much correspondence during Harrison's absence; John S. Nicholas, an experimental embryologist; Wesley Roswell Coe and Alexander Petrunkevitch, both invertebrate zoologists; and G. A. Baitsell, who had written the laboratory manual Hutchinson was to use. Baitsell had been a student of Woodruff's but also worked with tissue culture, as Harrison did. Apart from Harrison and Petrunkevitch, Hutchinson was not enthusiastic about any of these zoologists.

In addition, there were zoologists at the Peabody Museum, first opened in 1876, which contained the wonderful vertebrate paleontology collections of O. C. Marsh. The museum had just relocated to its new building, not far from the Osborne Laboratory, when Hutchinson arrived. It was mostly the domain of paleontologists, including the director, Richard Swann Lull, and invertebrate paleontologist, Charles Schuchert. The curator for zoology was S. C. Ball, an excellent field zoologist, according to Hutchinson, who, however, had little else to say in his favor. Ball appears later in relation to the Yale North India expedition (see chapter 6).

Yale at this time was an all-male bastion, much more so than Cambridge University, in spite of the difficulties women students there had in becoming full members of the university. Cambridge University women, at least in zoology, were very much present and doing well by the 1920s. In contrast, Hutchinson's zoology classes at Yale literally contained "sixty men." Many Cambridge students were both rich and upper class, but at least after World War I there were also many who were not. Hutchinson's family, at the time he started at Cambridge, was not rich; he boiled his own eggs for breakfast, unlike the gentlemen scholars.

Hutchinson felt that many of the Yale faculty were very conservative. He described Professor Woodruff as "conventionally conservative in his political and religious loyalties" and noted that he had "never ventured across the Atlantic." Baitsell was "a Republican Masonic Baptist and an ardent prohibitionist," who had "strong sympathy for the sons of wealthy alumni."[2]

Both Woodruff and Baitsell were involved in the beginning biology course, in which Hutchinson taught on his arrival. The organization and teaching of this course in 1928 does not sound too different from most such courses at American universities many decades later. There were lectures, some by the professors, as well as laboratories and laboratory examinations. "Lecturers," like Hutchinson, or graduate students were in charge of most of the latter two. The whole system was quite different from that at Cambridge (or Oxford University), then and now.

Hutchinson did not approve of a system in which there were a required textbook, assigned chapters, laboratory exercises, and frequent examinations. It was impossible to be a "hunter and gatherer" of what was best, as he had been at Cambridge, impossible for an undergraduate student to be "self-educated" in this sense, although Hutchinson did admit that this might be an unproductive freedom for many students.

The other difference between at least Oxford or Cambridge and nearly all American universities was in the preparation of the students before entering the university. Students from any good grammar school in Britain, let alone public schools like Gresham's, had much more training in the sciences than those at most American high schools, even the elite private schools from which many Yale men came at that time. Hutchinson himself said, as did his graduate students later, that intellectual achievement was much less important than social position among Yale undergraduates. He compared the Yale junior faculty and graduate teaching assistants to "highly intelligent Greek slaves serving Roman patricians." Such "patricians" certainly attended Cambridge in the twenties (and Oxford in the author's experience in the late sixties), but they did not generally compete for honors degrees. Hutchinson visited Cornell University in 1929 and found it less elitist than Yale, with "good workers, good collection, and good libraries, rather than endless mock gothic palazzi." He was probably correct in his general assessment. Cornell was known to some as the "cow college," not entirely deserving of its Ivy League designation. Hutchinson saw the "greater insistence on essentials" there as very positive in comparison with Yale.[3]

Alexander Petrunkevitch

Hutchinson felt greater affinity with Alexander Petrunkevitch (known affectionately to faculty and graduate students as "Pete" and to other students as "Dr. Pete") than with most other Yale zoologists. He himself wrote Petrunkevitch's National Academy biographical memoir.[4] Petrunkevitch was born in Russia and came to the United States in 1903. In Russia he had attended Moscow University with the geochemist V. I. Vernadsky, whose work was very important in Hutchinson's later biogeochemical studies. Petrunkevitch was forced to leave Russia in 1899 for protesting the treatment of students. He went to Germany, where he completed a doctoral dissertation on bees with the famous August Weismann. When Petrunkevitch came to the United States, spiders became the subject of his research. He was appointed a Yale faculty member in 1910. Fossil spiders, especially those preserved in amber, were his special interest. He, like Hutchinson with his water bugs, was interested in phylogenetic and zoogeographical problems, not entirely popular in a department known for experimental zoology.

Grace Pickford did her doctorate at Yale with Petrunkevitch, continuing her study of the South African earthworms and their relatives. She later worked with Petrunkevitch on physiological problems in spiders, which required her to make ingenious apparatus to measure important aspects of spider blood. She and Petrunkevitch became close lifelong friends.

Hutchinson wrote later about Petrunkevitch: "I had for many years as a colleague and close personal friend the late Professor Alexander Petrunkevitch who, as a young man in Weismann's laboratory in the beautiful German city of Freiburg, put the finishing touches on the roles of the queen, males and workers in the development of the honeybee."[5]

Petrunkevitch often strongly disagreed with other faculty members, for example, later, over Hutchinson's tenure and promotion. In a July 1928 letter to Harrison, John Nicholas wrote: "Petrunkevitsch [sic] has tamed down quite a bit although he has been a great trial during the year. I do not see how Woodruff could have handled the package of dynamite with more care and tact than he did." At that time Petrunkevitch had withdrawn from the regular faculty meetings over some issue. Nicholas continued, "I must say that I feel greatly relieved not to have Pete's presence in the regular meetings." These events happened before Hutchinson had left South Africa, but all indications are that Petrunkevitch, the "package of dynamite," had not "tamed down" in the interim.[6] During this period and later, "Pete's Tea" was held every Monday afternoon in his laboratory, attended by many graduate students, including those working on experimental embryology. Both Hutchinson and Pickford were regular members.

Ross Granville Harrison

Ross Harrison was the person who most influenced Hutchinson at this stage of his life and work. As Hutchinson acknowledged: "Harrison was the greatest living scientist I have known continuously . . . over a long period of time. In age, I fitted into the span of his own family, who became friends. . . . He probably represented as well as or better than any other immediate predecessors or contemporaries, the spirit of American biology in the first third of this century."[7]

Most of the graduate students in the Zoology Department during that period were studying with Harrison. In Hutchinson's first year at Yale, three doctorates were awarded, all of them under Harrison. Between 1928 and 1938 about twenty-five zoology graduate students were in the department, half of them working with Harrison and a fifth with Woodruff. The other faculty usually had only one student at any given time.

14. Professor Ross G. Harrison in his laboratory at Yale.
Manuscripts and Archives, Yale University Library.

Born in Germantown, Pennsylvania, in 1870, Harrison completed his undergraduate study at Johns Hopkins University in 1889. He was a graduate student of William K. Brooks at Johns Hopkins, as were Thomas Hunt Morgan and E. B. Wilson. Harrison received his doctorate in 1894. His early research was on fin development in fishes. In addition he received an M.D. degree in 1899 from the University of Bonn, following further postdoctoral studies in Germany. He then returned to Hopkins as associate professor of anatomy. There he was influenced by F. P. Mall's teaching methods. Mall considered the student his own best teacher. The professor should give a student only enough assistance in research to help with problems the student had already met and partly analyzed. This was the method Harrison used with his own graduate students. It became the one that Hutchinson used as well.

Harrison came to Yale in 1907 and developed a true "research school" in experimental embryology there. Harrison had transplanted amphibian limbs and grown nerve fibers in tissue culture during the first decade of the twentieth century. He was still publishing innovative papers in the 1940s and gave the Yale Silliman Lectures in the 1960s; his was a long and fruitful career.[8]

T. W. Twitty, who arrived at Yale as a graduate student in 1925 to work with Harrison and subsequently did important work in salamander development, population biology, and ecology, wrote of Harrison and his laboratory in the late 1920s:

> How varied and unpredictable the circumstances that shape the course of a scientific career may be. . . . Possibly my greatest piece of good luck came when I wandered into the orbit of Ross G. Harrison at Yale.
>
> Its central figure and father image, Harrison himself, occupied a position of respect and affection that had already become legendary. To the generations of students who passed through the OZL during his reign, there could be only one "Chief" at Yale or elsewhere. His position in science was such that he was naturally held in considerable awe. (The story is told of the graduate student who was so afraid to address him that she could only watch in horror while bite by bite his spoon came closer to a fly imbedded in his banquet dessert.) But no one ever strove less to intimidate or impress, and if he seemed aloof it was merely that he was shy. I can remember the long and painful silences that sometimes developed during the bag luncheons over which he presided daily and the Chief's obvious gratitude to any of us brash enough to end them by some desperate conversational gambit.[9]

Twitty reported that in this very exciting period in embryology, "The parts of the embryo were soon parceled out, like mining claims in a new gold field, among Harrison's students and followers, and eye, ear, limb, nervous system . . . each became a surgical specialty yielding a rich bonanza of information." Harrison did not teach his graduate students techniques, however; they were left to resolve all the difficulties themselves. "I was not the first to learn that the Chief, kindly as he was, did not exactly smother his students with care and supervision." Hutchinson, when he later had his own graduate students, followed this same style of supervision.

What motivated Harrison's scientific work? Twitty mentions "consuming curiosity, thirst for truth, desire to aid mankind, or just the plain fun of research and discovery." Institutional motivators would today be added by contemporary social historians of science, but Twitty considered aesthetics as the likely "single factor that made Harrison's work great." Twitty wrote, "He was constitutionally incapable of leaving a project until all its pieces had been fitted into a unitary whole whose composition met his artistic requirements. . . . More than most scientists, I believe, he both required and created an element of beauty in the form and perfection of his experimental analyses."[10]

These sentiments, too, are close to those later seen in Hutchinson's approach to science. Did he consciously or unconsciously imbibe Harrison's outlook, or did they understand each other—despite their different research interests—because their approach to science was already so similar? They were both innovators in the realm of ideas. One of Harrison's original ideas of this period was to transplant organs between embryos of related salamander species, such as spotted and tiger salamanders, which grew more slowly or more rapidly. This idea (and laboratory technique) yielded "unexpected results."

Twitty completed his doctorate in 1929 and stayed on at Yale as an instructor in zoology for two additional years. During this time he must have interacted with Hutchinson, also an instructor. Subsequently Twitty went to Germany, to work with Otto Mangold, a student of Hans Spemann, who was at the Kaiser Wilhelm Institute in Berlin. Harrison thought his graduate students, most of whom were from "the hinterlands," according to Twitty, should acquire more high culture. What better city for cultural enrichment than Berlin? Twitty remembered that, "Dr. Harrison always seemed to delight in ribbing me gently about my Hoosier background, and especially in reminding me of Indiana's notoriety at that time as a leading stronghold of the Ku Klux Klan."[11]

Hutchinson, too, did not care about his students' backgrounds but expected his graduate students to become intellectuals and to absorb a certain amount of culture at Yale, introducing some to the music of Bach, to Latin sources, and to other wonders that could be found without even crossing the Atlantic, although some did that as well.

The difference between the atmosphere of German and American scientific institutions at that time is evident in an incident observed by Twitty: "It was evident that Harrison's prestige was, if possible, even greater in Germany than in the United States, and that my identification as 'ein Schuler von Harrison' always meant open sesame. . . . A young German embryologist, unaware that Harrison was paying Berlin a brief visit, suddenly found himself in the dazzling presence of the great man. On approaching the table where Mangold's group lunched daily, he became rigid as if in tetanic shock, then bowed, wall-eyed and speechless, deeply from the waist. What wonder that many Europeans visiting Yale for the first time were shocked at the easy air of informality that his students sometimes assumed in their relations with 'the Chief.'"[12]

By the time Hutchinson arrived at Yale, Harrison had less contact than previously with undergraduates. An anonymous account by a student who had taken Harrison's undergraduate embryology course credited him with fine undergraduate teaching: "Until his attention was dominated by research, administration and graduate teaching, the course in Embryology was given by Prof. Harrison. During [these] years . . . undergraduate interest in Embryology reached its peak. . . . It was the good fortune of the writer to . . . study his methods of bringing facts to the student in an enlightening way. His was not the obscure method which presents an excess of detail in which the student is hopelessly lost but a helpful and suggestive method by which the student could find *his own way* through the numerous facts and theories" (emphasis in the original).[13]

Whether Hutchinson's teaching methods were modeled after Harrison's, or, more likely, Hutchinson had arrived at similar methods from his Cambridge experiences, it is clear that they were in sympathy about both the aims and methods of teaching, methods unusual then and, at the undergraduate level, probably still so. Hutchinson taught the embryology course at Yale himself in his early years. One imagines that he continued Harrison's method of having the student find his own way.

Hutchinson credited Harrison with providing what he most needed in these early years at Yale: encouragement and appreciation. Nearly sixty years later, on receiving the Kyoto Award, the nearest equivalent for an ecologist of the Nobel Prize, he told the audience that Harrison, "a very great embryologist, chairman of my Yale department, found something in me that appealed to him, though his scientific interests were very far from mine. Basically it is because of Harrison that I am standing before you now."[14]

Harrison wrote a recommendation for Hutchinson in July 1930, less than two years after his arrival at Yale: "He has conducted an undergraduate course in Embryology with both lectures and laboratory work, has participated in the instruction in the introductory course in General Biology, and this past year has offered a course of lectures on the Ecology of fresh water animals for advanced or graduate students. Mr. Hutchinson impresses me as a man of broad and thorough learning, as an indefatigable and skillful investigator, and as one who will surely succeed in his chosen line of work. He presents his reports with clearness and has the faculty of arousing unusual interest in his hearers, particularly on the part of advanced students."[15] In a later letter, Harrison recommended Hutchinson for a promotion at Yale. After discussing Hutchinson's intellectual leadership, he added, "I say these things all the more cordially because he has worked in a field [ecology] far removed from my own special interest—a field which I have learned to respect mainly through what I have seen of Hutchinson's work."[16]

Another View of the Young Hutchinson

Not everyone in the department agreed with these views of Hutchinson's teaching. J. S. Nicholas, embryologist and later Zoology Department chairman, wrote of Hutchinson: "He applied for a Fellowship and when we suddenly lost an Instructor we cabled him and got him for the job sight unseen. When he came the Department followed its usual routine and placed him in elementary teaching which was a terrible mistake for neither he nor the beginning student fit that picture. When his advancement came up [1930] I opposed it strenuously (Pete [Petrunkevitch] said viciously) in order to drive the whole situation into the open and have him assigned to more advanced courses."[17]

Nicholas was never a friend of Hutchinson's, although Nicholas claimed his friendship at the end of the letter cited above. Hutchinson credited him with enabling the experimental methods of embryology to be used on mammals, but he had no kind words for Nicholas himself and commented, "Later in life [he] became too much engrossed with worldly power to continue as an effective scientist, which distressed all who knew him at all well."[18] A former department colleague reported to me that this comment was not harsh enough.

Limnology Teaching and Research

Albert Eide Parr, whom Hutchinson credited with starting the buildup of zoology at Yale's Peabody Museum, was director of the recently built Bingham Oceanographic Laboratory, which was affiliated with the museum. In fall 1929 Parr had begun teaching oceanography at Yale. Hutchinson was asked to give an advanced course on freshwater biology (limnology) in the second term, the course Harrison referred to in his recommendation.

Hutchinson also gave a seminar to the department on his limnological work in South Africa. It was about the pans he (and Grace Pickford) had studied in the Transvaal. It was based on Thienemann's views; he sensed approval from Harrison, who had studied in Germany. Hutchinson, either in hindsight or perhaps at the time in anticipation, sensed that these views would not be in accord with those of such contemporary American limnologists as E. A. Birge or Chauncey Juday. Hutchinson later said that if any members of the North American limnological community had been present at his seminar they would have responded to him as a seventeenth-century Puritan congregation might have to a Dominican or Jesuit sermon. No limnologist was present, apart from Pickford; there were not even any ecologists among the Zoology Department senior faculty.

By this time Hutchinson's main interest was in limnology, but he had not given up on experimental physiology. He started to do the research for which he had originally applied for the Sterling Fellowship. He was trying to establish a conditioned reflex with a time lag in a cold-blooded (poikilothermic) animal. He had originally proposed using a tortoise, but in this laboratory devoted to salamander embryology, he probably used a spotted or tiger salamander. In any case he was unsuccessful.

He did, however, carry out an experimental investigation related to limnology. Magnesium salts were supposed to keep cladocerans, the group of crustaceans to which the common water flea belongs, out of Lake Tanganyika. He placed members of several cladoceran families in water six times more concentrated than that of Lake Tanganyika—and they continued to swim about! In a letter to his parents written in 1929 in which he discussed these results, there is a rather infamous remark about limnology, dubbing it "the dumping ground for inferior off-scourings of the profession of zoologists," a remark that he himself called grossly unfair fifty years later.

Hutchinson did indeed teach the undergraduate embryology course despite lack of much training in this field. That he should be assigned this course seems strange in a department of embryologists! Hutchinson wrote that he had kept up with Needham's work in chemical embryology since leaving Cambridge, and with a Miss Pelham helping him, he thought the students might have learned almost as much as he did.[19]

It was against Yale mores then to allow women into Yale undergraduate courses. Even in the 1940s, Margaret Wright, while a graduate student, was not allowed into an undergraduate physics course. Nevertheless, Hutchinson allowed Dorothy Benton to take embryology. She and her husband, who was a mathematics instructor, both became Hutchinson's friends. They shared his early interest in mathematical aspects of population biology and what was later to become the competitive exclusion principle in ecology. Hutchinson taught embryology until 1936 when he inherited Petrunkevitch's course in the natural history of animals and turned it into an ecology course.

It was the graduate course in freshwater biology or limnology that eventually attracted graduate students to Hutchinson, but that was not until the early 1930s, the beginning of fifty years of graduate students for Hutchinson. But before the graduate students arrived, another important event occurred in Hutchinson's life: the Yale North India Expedition.

CHAPTER SIX

Chief Biologist:
The Yale North India Expedition

In 1932 a scientific expedition was launched to northern India, including Kashmir and specifically Ladakh, under the auspices of Yale University. It was generally referred to as the Yale North India Expedition. The expedition leader was Professor Hellmut de Terra, a German geologist with considerable earlier experience in Central Asia. He was formerly at the University of Berlin but in 1931 was a research associate in geology at Yale University. De Terra wrote a "Memorandum for an Expedition to the Himalaya and Karakorum" and persuaded Yale to let him organize it. In his memorandum de Terra described his experience on a previous expedition. De Terra had realized during that earlier expedition that northern Kashmir included some of the highest and geologically and zoologically least explored mountain regions on earth. "Its formations have never before been systematically searched for paleontologic treasure; nor has its highly interesting animal and plant life ever been [scientifically] examined."[1] He pointed out that the Himalayan uplift was relatively young and that adjustments to Pleistocene glaciation should be evident in both plants and animals, probably including prehistoric man.

At that time these regions were all part of India, then in the British Empire. He proposed to visit the Salt Range, the Kashmir basin, the northwest Himalaya south of the upper Indus Valley in Ladakh, and the completely unsurveyed Eastern Karakorum Mountains and plateau on the boundary of Kashmir and Tibet where the mountain ranges were "absolutely unexplored." There were important scientific problems in their stratigraphic geology, including paleontology. "The region chosen," de Terra wrote, should yield "most interesting material of marine invertebrate fauna of Triassic, Jurassic, and Cretaceous formations, which would enable us for the first time to get some idea of this ancient ocean belt [the Tethys Sea], whose length must have surpassed the widths of the present Atlantic and Pacific oceans combined."[2] The Yale Peabody Museum would benefit also from collections of vertebrate fossils since the Siwalik formation of the Himalaya, according to de Terra, contained some of the world's richest vertebrate fauna, particularly early mammals. Structural geology and geographic exploration were to be undertaken as well. (Knowledge of the Indian subcontinent in 1931 of course preceded contemporary knowledge of plate tectonics.)

In addition to fossil animals, the extant fauna was to be studied. Rare and unusual mammals and birds were to be found in this region, some of them endemic, that is, found only in the Himalaya and Eastern Karakorum. De Terra listed a number of these, including a wild yak, a wild horse, the Tibetan antelope and gazelle, a black bear, and even the snow leopard, as well as wild goats and sheep. In addition, de Terra thought that the fauna of Tibetan (the word he used for this region of India) mountain lakes would have survived recent geological events. An investigation of the fauna of the Tibetan mountain lakes would therefore be of the greatest significance. It is quite possible that Ross Harrison read this line and thought of Hutchinson, who was already a student of lakes.

The original proposal included a section on anthropology; de Terra had hoped to acquire ethnological collections as well and learn about the ancient inhabitants of the region. "My discovery of a prehistoric cave in the western Karakorum," he wrote, "leads me to hope that similar finds may also be made in the eastern parts."[3] This part of the proposal was crossed out, however; apparently de Terra thought there would not be sufficient funds for an anthropologist.

The earliest extant letter that mentions the possibility of a Yale zoologist joining the expedition is dated April 14, 1931. It was written by George Parmly Day, Yale University treasurer, to Ross Harrison. The letter advised Harrison that the Prudential Committee of the Yale Corporation had voted to authorize a "sum of not to exceed $6000 to cover all expenses and equipment of a biologist representing the Osborn Laboratory, to be approved by the President to accompany Dr. de Terra's expedition."[4] The appropriation for a Yale zoologist was a result of a letter from Harrison, who wrote saying that a comparatively small sum from the university would procure material of great value to the laboratory. President Angell recommended this and the committee approved the funds.[5]

An April 17 letter from Harrison to Day mentions Hutchinson, obviously Harrison's choice, and requests $1,000 of the total allotment for him to purchase equipment that summer. Harrison wrote, "The expedition certainly promises very interesting and important results and I think that Hutchinson is very well qualified to carry on the work."[6] In his answer the following day, the treasurer wrote that he had advised Mr. Ostrander, the Yale cashier, of $1,000 needed for equipment purchase, but he also reminded Harrison that Hutchinson had to have formal approval by President Angell.[7]

Hutchinson's Expedition Plan

Approval was given, apparently after Hutchinson submitted a "Program of Research That Should Be Undertaken by a Freshwater Biologist Accompanying Dr. de Terra's Expedition to Kashmir." Hutchinson's program, four pages long, includes problems to be investigated; examination, after the expedition, of the biological material collected; and the equipment needed, with estimated costs. It is quite impressive to read what could be bought for $1,000 in 1928. He needed, among many items, a collapsible boat, Nansen bottles with thermometers, a dredge, a bottom sampler, a Hellige Comparator set, and dynamite (for collecting fish). Many of the items listed needed to be ordered from Europe unless they could be specially made locally. Hutchinson made the interesting claim that no limnological apparatus was being made commercially in the United States. However, E. A. Birge and Chancey Juday, Wisconsin limnologists, had published designs from which, Hutchinson suggested, some of the equipment might be made at Yale.[8]

The problems to be investigated in freshwater biology were divided into "Faunistic," "Ecological," and "Correlation with Geological Results." Hutchinson planned to collect fauna, including fish, mollusks, crustaceans, and aquatic insects. The ecological problems are of particular interest, as this plan probably presents Hutchinson's first explicitly ecological field research. He wrote that no ecological observations of life in a large high-altitude lake (14,000 to 18,000 feet) had ever been made. He planned a variety of measurements, including temperature at different depths, oxygen content, transparency, and chemicals in solution. He planned to study the diurnal migration of plankton and the zonation of bottom fauna and plankton. These were to be compared with lower-altitude lakes in the region.

He postulated that because these lakes at high elevation lacked outlets, salinity differences should be an important influence on the plankton communities present. In addition, structural modifications had been found in Central Asian lake fish by a previous investigator. Hutchinson proposed to investigate the biology of this group (Colitid fishes), if any were found. Variation found in characteristics of other groups, such as phyllopod crustaceans, in relation to their ecology would also be investigated.

Biogeography had always been of interest to Hutchinson. Since the Yale North India Expedition was primarily geological, it would be important and possible to correlate his faunistic data with the past geologic history of the lakes. In a very ambitious proposal he hoped to be able to find fossil forms from adjacent ancient lake beds to compare with present fauna. In this way it would be possible to determine conditions in these lakes in past times.

In an important paragraph Hutchinson explained that much of the material collected would have to be identified by specialists. He himself could do the aquatic water bugs, the phyllopod crustaceans, and perhaps the rotifers. He would need to spend some time examining the Kashmir specimens in the Indian Museum in Calcutta before returning home. As it turned out, finding the specialists, sending them the collections, and arranging for the publication of their results was a major undertaking. All of this was up to Hutchinson himself.

Expedition Reports from the Field

De Terra sent an authorized account of the expedition to Yale after three months in the field. He was distressed by an account of the expedition that appeared in the *New York Times* of April 10, 1932, which reported that the expedition's main purpose was "to explore the Himalayas and Tibet and search for the cradle of man." De Terra did not know how that report reached the press, certainly not through any interviews he had given. He thought it advisable to give out his own account and sent President Angell this "authorized account" for the press. In his report, de Terra wrote that Hutchinson's research, "a study of the animal life in the inland waters of these regions will throw light on the ancient geography of its river systems."[9]

Up to that time the expedition had spent some time in the northern Punjab, where G. E. Lewis, the paleontologist of the expedition, found fossil mammals. They collected invertebrate fossils as well, with the help of a member of the Indian Geological Survey. At the same time, Hutchinson was investigating lakes and swamps in the Kashmir Valley. His collections and ecological data "would undoubtedly throw light on the relations of this fauna to that of western Central Asia from which it was largely derived," wrote de Terra.

A new member had joined the expedition, an Indian topographer, Khan Sahib Afraz Gul. He was to map an unsurveyed part of the Karakorum Range north of the Himalaya. His work was paid for by an anonymous donor to the American Geographical Society.

When the expedition arrived in Kashmir, Hutchinson rented a houseboat and installed a biological laboratory in it, "a floating offspring of the Osborn Laboratory at Yale," according to de Terra, who related to Yale President Angell that he had found the first data on the youngest upfolding of the Himalaya and that "Mr. Hutchinson's glass tubes and tins begin to get filled with queer water animals."

They were to leave Srinagar, the capital of Kashmir, on May 15, 1932, to spend five months in what is now Ladakh but was then referred to as "Little Tibet." Just before their departure, from the houseboat in Srinagar, de Terra wrote to Angell, "It was a great pleasure for all of us to hear from you and to feel that you take such an interest in our work. We are just about to start for Western Tibet; 15 pony loads of provisions left two weeks ago and we shall follow with 35 additional ponies on the 16th. Hutchinson has wound up his work here."[10]

De Terra had found crude stone and bone implements, which he thought were the first evidence of prehistoric man, probably Paleolithic, in the Himalayas. Lewis, the paleontologist, had found a mastodon skull and other Tertiary mammal fossils. A member of the Indian Geological Survey had collected many fossils in the Salt Range, and these and Hutchinson's first batch of zoological material collected in Kashmir and the northern Punjab were to be shipped from Karachi to New York. They were waiting for the unseasonable rains to stop so that they could be off to Ladakh.

Hutchinson at Work

Hutchinson collected all kinds of insects, aquatic life, other animals and plants, even fossils during the North India Expedition. He was also called upon to skin large animals. He was the only one on the expedition who knew how to do this, a skill he had acquired during his Cambridge days. His own particular interest was in the ecology and biota, including people, of the very high altitudes of Ladakh. He studied lakes at over 17,000 feet and took many chemical and other measurements.

He made much of his own equipment en route, quite ingeniously, from whatever he could scrounge. He described his equipment and the experiments he conducted in the many letters he wrote to his wife, Grace Pickford. He proved himself very adept on this expedition at making and repairing his equipment. He described a limnological tool kit, including a large collection of pieces of different sorts of wire. Much of the equipment he had tried to bring with him had been delayed, lost, or damaged. The transport methods were primitive, mostly four-legged. Yet he was able to gather much environmental data and to conduct many experiments.

His work in Ladakh provided insights into biogeography and paleo-limnology as well as much new data on high-elevation lake ecology. Many of these lakes contained no fish; a crustacean, in some cases a shrimp, was the top predator. At several high-elevation lakes that he studied he took ultraviolet (UV) light intensity measurements. Hypotheses were characteristic of Hutchinson's method of doing science, even at this early date. His hypothesis in this study was that there was a correlation between high UV and the black color of the *Daphnia*, or water flea, a tiny but dominant crustacean. He wanted data from a range of elevations. At this highest-elevation lake the black Daphniid and a "copepod, scarlet red, are the only plankton animals one sees coming within 1 or 2 centimeters of the surface."[11]

He wrote to Grace his general conclusions about the high-altitude Ladakh lakes, so different in water chemistry from the South African pans they had worked on together. Hutchinson at twenty-nine had already studied lakes on three continents, both their physical chemistry and their biota. In Ladakh, he collected data that showed that the qualitative variation of plankton was correlated with the chemical composition of the lake. He described a bay where river and lake water mixed. It was incredibly full of plankton, he noted, with different planktonic crustaceans

15. Hutchinson and Rhoda de Terra, wife of expedition leader Hellmut de Terra, in "Tibetan" costume. Manuscripts and Archives, Yale University Library.

at different levels in the lake, which he had correlated with the UV light levels.

Nearly three months after their arrival in Ladakh, Hutchinson wrote a four-page letter to Ross Harrison about their travels. All the mail sent to and from Ladakh seems to have traveled with much delay, by runner. Hutchinson apologized for not writing sooner, but their runner had "arrived several days early, late at night, only to leave the next morning" without leaving time for replies to be written. Hutchinson described for Harrison human life in Ladakh, "the perfect model of a medieval country." The important lamas were like bishops in the medieval church. "Everywhere Llasa seems more important than Srinagar or London, and the Dalai Lama [more important] than the Maharajah of Kashmir or the Indian Government." The expedition members went across the frontier to Tibet into "an uninhabited part where we did not meet even nomads, but we have had some suggestion of being watched from the Tibet side."

Hutchinson had been working on a large lake, Pang-gong Tso, where he had collected much limnological data. It was interesting but difficult work. He wrote to Harrison: "The lake is very large for a small rubber boat that one has to propel oneself and which is stable but not very mobile. . . . I only had to work during one snow storm. The lake is certainly the most beautiful one I have ever seen, a very transparent blue; the fauna consists solely of Gammarus [a shrimp], a bright red copepod and a deep black Cladoceran [minute crustaceans]. The black color of the Cladocera here is characteristic and I cannot help thinking that it is correlated with the great amount of ultraviolet at these altitudes."[12]

He had also been working in a small lake at 17,000 feet, which was covered with ice that he thought permanent except at the edge. He was surprised to find a great many copepods. He had also made numerous observations on the altitude limits of many terrestrial invertebrates. He had by then amassed a very large collection of invertebrate and vertebrate animals, many probably new species, for which he knew no specific names except for two fish! Experts were needed.

What would be the fate of Hutchinson's collections? He preferred to have them sent directly to specialists, who would work up reports and keep the specimens if they wished. He conceded that some might think it strange that Yale material should be sent out all over the place, molluscs to Calcutta, lacewing flies to Norway. Stoneflies were to go to entomologist John Needham in Ithaca, caddis flies to London, and so on. The alternative was for Yale to keep type specimens (those for species new to science). This alternative did not seem reasonable to Hutchinson since the Peabody Museum did not have other Asian specimens of these groups. The Peabody curator, S. C. Ball, and Hutchinson had agreed that the latter could make arrangements that seemed scientifically desirable.

The problem was that de Terra had signed a form agreeing that all material collected by the expedition was to be the property of the museum. Hutchinson wrote to Harrison, "Since I was appointed as the representative of the [zoology] department, I think technically you would have the last say in the matter." The problem in Hutchinson's view was the extra travel; Hutchinson wished to distribute his collections to specialists in India and Europe without first transporting them to New Haven. This would avoid subjecting fragile material to the hazard of two additional ocean trips. Equally important, Hutchinson thought, these taxonomists, especially those at museums, would not tackle large collections unless they could keep most of the specimens. He would use American specialists as much as possible but feared that even Petrunkevitch, the spider expert at Yale, would not wish to identify all the spiders Hutchinson collected.

Some were of special interest since they came from the upper limit of ter-restrial life. Harrison conveyed this information to President Angell.[13]

While the question of where Hutchinson's specimens were to be sent was being debated at Yale, the Yale North India Expedition returned safely to Srinagar after five months in the Tibetan-Kashmir frontier re-gion. The expedition had, in de Terra's words, "covered a long and difficult trail across 6 passes above 18000 ft. and sometimes over a region which has hardly seen a white face. Our scientific results are most successful in every way and if you could see the pile of boxes which contain our harvest you would get an idea of what we amassed. I hope that both the Peabody Museum and the Osborn Zoological Laboratory will be pleased with the things which we will add to their treasures."

"After the hazardous trip and after staying for such a long time in high altitudes, a certain deterioration set in and we felt as if we would never redevelop into thinking human beings," continued Hellmut de Terra to Angell. But "this depressing feeling has however given way to new ener-gies so that we feel fit again for new tracks of travel and thought."[14] One wonders whether Hutchinson also felt this "deterioration." He was prob-ably the youngest of the regular expedition members; he was not yet thirty. The group would now divide up, de Terra wrote, and "I shall lose my good companion Dr. Hutchinson who is going via Calcutta southwards." He himself and his wife, Rhoda, who had participated in all the hardships and excitement of the trip, were staying in Kashmir until winter arrived and they could no longer camp and work in the mountains."

Another authorized report was also sent, presumably for the press. The expedition had traveled 1,300 miles across the "barren mountain ranges north of the Himalaya which are known as the Eastern Karakorum." They had had to move all their scientific apparatus and provisions on ponies and yaks over snow-covered passes above 18,500 feet. They had had to establish food depots along the route since much of it lay above the upper limit of human settlements; the highest permanent settlement was at 14,500 feet. "Contrary to rumors at times circulating in North Indian ba-zaars that all the transport animals had perished on the high storm-swept plateaus, the expedition encountered no serious losses."[15]

Almost five thousand square miles had been mapped in detail, thanks to the support of the American Geographical Society, which had spon-sored "one of India's best mountain topographers," Khan Sahib Afraz Gul. Invertebrate fossils had been collected from regions never before explored; these would "throw new light on the geological history of Central Asia." As to Hutchinson's contributions as expedition biologist: "Mr. G. E. Hutchinson . . . made a detailed examination of nine of the most elevated lakes yet studied chemically and biologically. Considerable attention was also paid to the life conditions of the terrestrial invertebrates inhabiting the highest regions visited and much new information was amassed as to the factors limiting the altitudinal distribution of animal life."[16] Further, an ethnological collection from Ladakh "of considerable size and unusual interest" was made for the Peabody Museum; it is likely that Hutchinson had a large part in amassing this collection as well. He certainly gathered much information on the inhabitants of Ladakh, which he related in his book, *The Clear Mirror*, published in 1936.

Hutchinson wrote to Harrison on October 8, 1932, saying that they had arrived back in Srinagar "in good spirits." The journey down from the mountains had been unexciting as they were all too tired to do much work. They had packed up and shipped all their collections. That took a week. Hutchinson was on his way to Calcutta to work at the Indian Museum and then he was going to the "Nilghiri Hills to see under what conditions the supposed relict Himalayan fauna is living."

Hutchinson wrote that he had now had time to read about the Swiss alpine lakes described by limnologist Otto Pesta. He found very great differences between the lakes he studied in the high altitudes of Ladakh and the Swiss alpine lakes. Oddly, the Ladakh lakes, one apparently ice-covered all year, were much more like the low-elevation eutrophic lakes of Europe and North America. He would return to New Haven in early January 1933.[17]

A Battle Won — Distribution of Specimens and Publication of Findings

Also in October Harrison wrote to Hutchinson, saying that the decision about the distribution of his specimens was beyond his authority and that it was taking some time for consultation. He did not tell Hutchinson how much he had done on his behalf in regard to this question.[18] President Angell himself made the final decision that Hutchinson did not need to send the specimens to Yale for committee inspection before sending them to European and Indian specialists. Angell did insist that publication based on the material be credited to the Yale North India Expedition.

He wanted Hutchinson's attention "specially called to the necessity of making this clear in correspondence and conversations" with the scientists who received the material.[19]

The decision about the distribution of specimens, to some extent over the head of the museum director Ball, was extremely beneficial to Hutchinson. He had Harrison's active support to thank. Collections were promptly sent to specialists, particularly in Europe. The further question as to financial arrangements for publication of the results under Yale auspices remained to be resolved. Hutchinson was worried about the problems of publication. Even before leaving Ladakh he wrote to Grace, "I foresee that my now polychrome beard will be white and India independent and 5 Russian expeditions achieved and published on Ladakh lakes before my stuff is finished."[20]

Hutchinson did, however, soon publish a number of papers on his findings, including "Limnological Studies at High Altitudes in Ladakh" in Nature in 1933 and, later, "Limnological Studies in Indian Tibet" and "Ecological Observations on the Fishes of Kashmir and Indian Tibet."[21] He compared the Ladakh lakes to Central European alpine lakes. The lake he studied most intensively in Ladakh, Yaye Tso at 4,686 meters (15,374 feet), proved very different from the European alpine lakes. The latter were oligotrophic, that is, nutrient-poor, with high oxygen in the hypolimnion (lower layer) and a poorly developed bottom fauna. Yaye Tso was a eutrophic, nutrient-rich lake with an oxygen deficit in the hypolimnion and a rich bottom fauna, also true of several other Ladakh lakes he studied. In Yaye Tso there were approximately five thousand organisms per square meter in the bottom fauna, largely chironomid (midge) larvae and tubifex worms. The lakes in semi-desert Ladakh also differed markedly from lakes in another semiarid region, the pans in South Africa that he had studied and published on together with Grace Pickford and Johanna Schuurman.

Further limnological publications on the North India expedition awaited taxonomic studies by Hutchinson himself and many other experts worldwide. Hutchinson worked tirelessly to distribute his specimens to experts on the various taxonomic groups all over the world and to encourage these workers to complete and publish their results. He arranged to have many of their papers published in the Memoirs of the Connecticut Academy of Arts and Sciences between 1934 and 1936, until his funding ran out.

Even John Nicholas, longtime Yale Zoology Department chairman, and Hutchinson's severest critic in his earlier tenure and promotion proceedings, was impressed. Nicholas recalled in a 1945 letter: "When his [Hutchinson's] advancement came up I opposed it strenuously. . . . Then the North India expedition came along and instead of retiring or folding up, Hutchinson became the backbone of the whole expedition. When he returned he was an entirely different person. He had matured and found himself. . . . Since then his gifts of intellect, wit and breadth of knowledge have become steadily more apparent. . . . His book on the Tibetan paintings [sic] was the first big evidence of the change in the man, and then followed the Northern Indian reports due in no small way to Hutchinson's editorial insistence that laggard systematists and others get their work finished."[22] Not all of Nicholas's observations are reliable, but his view of Hutchinson's role in getting the scientific results of the expedition published was correct. There are a tremendous number of letters to and from scientists all over the world to attest to his encouragement and persistence.[23]

Yale University did come through with some funds for publication, together with the Connecticut Academy of Arts and Sciences, of which Hutchinson was a member throughout his Yale career. An item in the academy's minutes, from April 1934, recorded a vote to reserve academy funds for three years and up to $3,000 for publication by the academy of the results of the Yale North India Expedition. The understanding was that Yale University would contribute $4,000; this was indeed recorded at the next academy meeting.[24] Volumes 8, 9, and 10 of the academy *Memoirs* were devoted to expedition publications. Hutchinson had to work hard for these publications: a 1933 letter to M. E. Mosely at the British Museum (Natural History) in London noted, "Since I last wrote I have had a discussion with the President of Yale University about the . . . publishing of the Yale North India Expedition. The situation seems considerably more hopeful than when I wrote before."[25]

Hutchinson wrote instructions to all those examining materials collected on the expedition. He urged prompt publications in "English, French, German, Italian, or Latin." The experts were to bear in mind that the main purpose of the biological collections was ecological, and they should thus consider the distribution of their groups at high altitudes, especially for specimens collected at over 5,000 meters.[26]

He wrote to Malcolm Cameron in England, to whom he had sent

beetles (family Staphlylinideae). One of the specimens was from the highest elevation for any beetle collection that Hutchinson knew of. He asked Cameron if he knew whether the Everest expeditions might have collected beetles at even higher elevations.[27]

Academy funds ran out before the expedition findings did, but Hutchinson continued to write letters and encouraged publication elsewhere. He wrote to limnologist August Thienemann, whose 1925 book he credited with, in large part, his own conversion to limnology. In this letter he discussed Ororotse Tso, the lake that, according to the people who lived in the nearby settlement of Phobrang and pastured their yaks near the lake, never lost its ice cover except at its margins.[28] He also wrote that he was unable to publish Thienemann's paper in the academy *Memoirs.* He had already had to cancel a "large" paper of his own and another excellent one on copepods for lack of further funds from Yale or from the Connecticut academy. These papers were published elsewhere. The "large" paper was probably Hutchinson's "Limnological Studies in Indian Tibet" which he published (in English) in a German journal. Hutchinson's Yale colleague and later Smithsonian secretary S. Dillon Ripley referred to this paper as a classic.[29]

Hutchinson's First Book, a Literary Venture

Much of the scientific material in Hutchinson's papers is presented in a very different form in his 1936 book, *The Clear Mirror: A Pattern of Life in Goa and Indian Tibet.* This book marked the beginning of Hutchinson's literary career. Few twentieth-century biologists have left a literary legacy, and very few wrote as well about biology as Hutchinson did.

Many of Hutchinson's essays for the *American Scientist,* starting in the 1940s, were read by biologists and nonbiologists alike, and were later collected into several books. But this 1936 book, about the North Indian expedition as well as his impressions of Goa, was his first truly literary venture. It was also the first to reveal his very diverse interests and his knowledge of archeology, religion and ritual, painting and sculpture, which had been developing throughout his earlier travels, especially in Italy. On the North India expedition he observed closely and photographed widely and was able to include cultural side trips.

It was his literary skill that most surprised his colleagues. The style is most unusual for almost any scientist of this period. He described a scene around a bandstand in Goa after a festival: "The sky suddenly turns pink with flecks of blue that darken and fuse toward the east. The trees flatten like shadows against this painted canvas sky. The band plays gaily; the crowd seems to walk faster. The girls have all come to one end of the path, and like a chorus, have grouped themselves together. In these few minutes after sunset, the square becomes so theatrical that it seems in-evitable that all these people should start to sing, transforming the canvas sky into the backdrop of a musical romance, in which the principal lovers are to emerge shortly from behind the flat dark trees."[30] The back cover of the reprint of this book by Leete's Island Books gives this literary evalua-tion: "In 1932 a young biologist traveled to Goa and Tibet as a member of the Yale North India Expedition. He recorded his observations with the eye of a trained scientist and the style and sensitivity of a novelist."[31]

Luckily for the progress of ecological science, Hutchinson never be-came a novelist, but this kind of imagery is also found in the most sci-entific writing in this same book. The word pictures in the chapter "The Lakes in the Desert" are equally striking, more so to me as an ecologist since they evoke a landscape, a special ecology:

> Where the valley of the Indus narrows, the country is sparsely settled, but north and south of the gorge, lying in a dusty land like a few remaining stones on a worn-out headdress, many lakes are scattered. They vary in colour within a restricted range of blues and greens, just as the turquoises that they resemble [which Ladakh women wear] differ one from another. Each shade of blue or green sums up in itself a structure and a history, for each lake is a small world, making its nature known to the larger world of the desert most clearly in its colour. . . . The meaning of their colours is best seen if they are compared with lakes in other countries where, among swamps and fertile fields, turquoises are forgotten and their colours re-placed by more earthy yellows and greens.[32]

Later we meet the inhabitants of these high-altitude lakes as the last snow of June falls "on the enameled surface, silently strikes it and rests for a moment, then disappears." There are no resident fish, but Hutchinson describes the three dominant crustaceans, one black, one red, one pale gray, all nearly microscopic: "In the blue water small black spots move lei-surely here and there, and still tinier bright red points can be seen skip-ping among them. . . . In the deeper water and along the shore, small pale shrimps scurry about. Higher up, the lakes are ice-bound even in summer. There the ice melts from below on the lake edge in July: Underneath it is sculptured into thousands of spikes and icicles, hanging into the green water of the lake like stalactites. . . . Below lies the greenish water, in which the tiny red crustacea swim, more numerous than in the turquoise Pangog Tso, for here, in this shallow lake [Ororotse Tso] the greenness of the water indicates that there is an abundant supply of food."[33]

Hutchinson did surface tension experiments in Goa with water striders, those long-legged water bugs that live right on the water surface: "In the heat of the day, as if propelled by the sun itself, water striders shoot about like balls of quicksilver on the surface of the water; later, in the cool evening, when the human inhabitants of the town come out again, the precarious progress of these insects on the boundary of sea and air will be so slow that they will seem like clockwork toys, whose machinery, wound up by the sun, has now almost run down."[34]

These word pictures are not simply literary; they contain a great deal of ecology. The reader learns why so many animals can live in these icy waters, even forty feet down at the bottom of the lake, whereas in the icy lakes of the European Alps, for example, few such aquatic creatures are able to exist. This book and Hutchinson's subsequent writing influenced many people, most of whom never met him, including scientists, both in his own and in very different fields.

The Clear Mirror was a literary success. But it may not have been viewed so positively by some of his Yale Zoology Department colleagues; much of the book has nothing to do with zoology. As noted above, however, he did publish scientific papers on his expedition results in appropriate journals, even before the book was published. Findings were also published in Indian and other foreign journals. But it is quite possible that more people read about the Yale North India Expedition in Hutchinson's 1936 book, and especially in its 1978 paperback reprint. By the latter date the author was a famous ecologist.

CHAPTER SEVEN

The First Crop:
W. T. Edmondson, Gordon Riley,
Ed Deevey, and Max Dunbar

A fter Hutchinson returned from the North India Expedition, Yale graduate students began to study with him. The first two were Gordon Riley and Edward Deevey. But even before these two began their studies with Hutchinson, W. T. (Tommy) Edmondson had come to work in Hutchinson's laboratory; he remained there throughout his undergraduate and graduate career. All three of these men became eminent scientists, best known in oceanography, ecology, and limnology, respectively.

W. T. (Tommy) Edmondson

Edmondson, who was later known for his leadership of a large group of limnologists at the University of Washington and for the spectacular cleanup of Seattle's Lake Washington, was a lifelong student of a phylum of microscopic animals, the rotifers. He made his first appearance at Yale as a high school freshman.

Tommy Edmondson grew up in southern Indiana. As a child he was given a little metal tub, which he sank in the backyard and filled with water. He tried putting crayfish in it, but they did not survive well. But there were little things moving in the water that fascinated him. Some small ponds lay along the nearby railroad tracks, and there he saw more tiny creatures, including a beautiful fairy shrimp. He was given a toy microscope for his twelfth birthday; with it he saw his first rotifer.[1] But unlike Hutchinson's boyhood experiences in Cambridge, there was no one to help Edmondson identify anything.[2]

Edmondson's father had died young. When his brother Frank went to Yale as an undergraduate, his mother moved the family to New Haven, Connecticut. Tommy went to a very large public high school with excellent teachers. In 1930 Hill House High School was, according to Edmondson, very highly organized.[3] Each department chair had either a doctorate or a master's degree, and many of the teachers took courses at Yale "just to keep up." As Edmondson recounted his story: "I went down the hall one day looking to see if somebody would let me use a microscope [to look at pond water]. I walked in and here was this wonderful Miss [Ruth] Ross." As a high school freshman, Tommy had not as yet had a biology course, but his brother Frank had found, even before they moved to New Haven, the best book for a "kid like him." It was Ward and Whipple's *Freshwater Biology*, "a big heavy book. My grandmother sewed me a carrying case for it." (Much later Edmondson was to edit the next edition of this book.)

Using Miss Ross's microscope after school, he settled on a complex but microscopic aquatic phylum, the rotifers, as his own domain. One cannot look at pond water or on aquatic vegetation, or even at dried herbarium specimens of mosses, without finding these fascinating tiny animals. Ward and Whipple had a whole chapter on this group, and Edmondson soon was able to recognize many of the local rotifer species.

Miss Ross was taking a histology course at Yale with Professor Roswell Coe. One day she took Tommy with her, and after the lecture he was introduced to L. L. Woodruff, who studied protozoa. "When he heard I was interested in rotifers he took me down the hall and introduced me to the youngest faculty member, Evelyn Hutchinson. He had just been collecting and he had come back with a wonderful colonial rotifer." Edmondson related that Hutchinson poured some water from a thermos bottle into a finger bowl and put it under a dissecting microscope. In the water was a rotifer, *Conochilus hippocrepis*, that Edmondson had read about in a book by Hudson and Gosse but had never seen, and here before him were hundreds of colonies.[4]

Hutchinson set Edmondson up in a corner of his own lab and said that Tommy could use his microscope. "I spent every minute I could there all during high school." Asked how much Hutchinson was able to help him, that is, whether Hutchinson knew the rotifers, Edmondson answered that Hutchinson knew the freshwater fauna very well. The following year, 1932, Hutchinson went on the Yale North India Expedition. Edmondson recounted that de Terra, "a big husky imposing man," had come into Hutchinson's lab many times to discuss the expedition they were about to embark on. Edmondson remained in Hutchinson's laboratory while Hutchinson was on the expedition.

When Hutchinson returned from the expedition, he invited Edmondson, still a high school boy, to work on the rotifers that he had brought back with him. Edmondson recalled: "In many ways Hutchinson wasn't the most organized person. . . . There I was working on this unique collection of rotifers. I'd write the name down in pencil on little sheets of yellow paper. I'd write at the head the name of the lake; then I'd go through the sample. I'd identify rotifers. Sometimes I'd make sketches. These got put in an envelope. I never, never in all my years with him used a notebook for that work. Later I caught on. Somewhere I hope there is an envelope with the yellow papers with my handwriting on them."[5] So far they have not been found in the Yale archives.

Edmondson had also been collecting in Linsley Pond and "every body of water" around New Haven. Although his original interest had been in identification and taxonomy of rotifers, he soon became interested in their ecology. Did he remember discussing that with Evelyn? "Oh sure," he answered. Hutchinson's early career had been similar to Edmondson's. Hutchinson seems to have been interested in ecological questions about his water bugs early on and throughout his career. That was not true of many of Hutchinson's biologist contemporaries, who had continued doing pure taxonomy of whatever group of organisms they worked on. Edmondson suggested that there might be a development one could trace from "the big paper he did with Pickford and Schuurman on the South African pans. . . . Everything is there and he is trying to relate it to the chemical characteristics of the situation."

Edmondson claimed, "As far as I know that was original. How much of that kind of thing he got from Elton I don't know." Hutchinson had read Elton's book, *Animal Ecology,* in 1928, after the pan fieldwork in South Africa had been completed but before the papers about it were written. Edmondson also mentioned S. A. Forbes's paper, "The Lake as a Microcosm." This paper, much cited as a classic (in hindsight), was rather obscurely published in 1887 and reprinted only in 1925, but then in a local journal.[6] It is doubtful that Hutchinson had read it at an early date, although later he quoted it on page 2 of the first volume of his *A Treatise on Limnology*.[7]

Edmondson rapidly became proficient at identifying rotifers. Others had done so before him. During the nineteenth century in England, serious amateurs, particularly tradesmen, studied some specialized aspect of natural history, something Hutchinson himself later wrote about.[8] Indeed, amateur British microscopists, even including one wine merchant, studied and wrote monographs about rotifers.

Edmondson recollected watching Hutchinson in the laboratory in the early 1930s: "As I was sitting at one end of the room looking at rotifers, Hutchinson would be running chemical analyses at the other end, calculating the data, and expressing surprise or pleasure at some new finding."[9] As to Hutchinson's North India collection Edmondson said, "I just assumed he would write up the report and that both our names were going to be on it: 'Hutchinson and Edmondson.'" Hutchinson disagreed about the order of their names. He said to Tommy, "No, you identified the rotifers."[10] This paper was published in 1934 authored by "Edmondson and Hutchinson."[11]

Edmondson declared that Hutchinson "was very sharing. He didn't hog the spotlight. He just didn't seem to think that way. He acknowledged people that helped him." This apparently applied even to high school students. Tommy Edmondson also took Hutchinson's invertebrate zoology course at Yale as a high school student, examinations and all.

When asked about Ross Harrison, still head of the Zoology Department, Edmondson called him "a grandfatherly man, very kind." Tommy used to work in the lab after dinner. The front door of the building would be locked. He would ring the bell, and Harrison's technician would come down from the third floor to let him in. "One day she came into Evelyn's office and said to me, 'Professor Harrison would like to talk to you.'" He went into Harrison's office. "Harrison said, 'I understand that you do a lot of work here at night.' He then handed me a key."[12]

Hutchinson said in 1990 that Harrison had very much encouraged him in his early research at Yale to explore various fields, from embryology to physiology and ecology, instead of staying with just one topic or method, a more usual research path for a young zoology professor. Hutchinson credited Harrison with allowing him this sort of freedom of exploration.[13] Hutchinson, in turn, encouraged Edmondson's interest in the ecology and development of rotifers in addition to their taxonomy.

16. W. T. (Tommy)
Edmondson and
G. Evelyn Hutchinson.
Courtesy of W. T.
Edmondson.

From whom did Edmondson take his Yale undergraduate embryology course? From Hutchinson: "It was a very good course. What was going on in the department was Harrison's experimental embryology. . . . This course that normally would just be looking at chicken slides . . . had a lot of experiments. Hutchinson taught it as a functional thing."

Another later Yale graduate student, Ray Rappaport, now a well-known developmental biologist, did his graduate work in embryology but was a student in one of Hutchinson's graduate seminars. He felt that he was influenced by Hutchinson even though his research was in a different field. Hutchinson, he said, was teaching students to be scientists early in their careers.[14] That seems to have been true even in the undergraduate courses Hutchinson taught early in his Yale career.

Tommy Edmondson added that professors in other courses would invite Hutchinson to give guest lectures. "He would go to give a lecture to the beginning biology course. . . . He'd come in and lecture on some aspect of ecology. . . . He brought the house down every time. He'd go on way beyond time. There would be lots of questions." But according to Edmondson, not everyone approved of Hutchinson's undergraduate teaching. "He didn't go over with everybody. He had a thick [English] accent; some people couldn't understand him. And he wasn't the sort of teacher who would lay things out, 'Now here is a list of stuff to memorize.'"[15]

While an undergraduate zoology student, Edmondson helped Hutchinson's first two graduate students, Gordon Riley and Edward Deevey, with their fieldwork. During this time, when he had space in Hutchinson's laboratory, Edmondson often met and eavesdropped on conversations with the many visitors, some of them famous foreign limnologists, who went out of their way to visit Hutchinson's laboratory.

After graduating from Yale in 1938, Edmondson spent a year at the University of Wisconsin. Hutchinson had written to the chairman of the Zoology Department at Wisconsin on his behalf: "Mr. W. T. Edmondson tells me that he is applying for an assistantship in your department. This was really done at my suggestion, as I am most anxious that he should have the opportunity of working at least for one year with Dr. [Chauncey] Juday." Hutchinson continued with a glowing description of Edmondson and his professional career to date. "Mr. Edmondson is one of the most remarkable students that I [have] come across in my teaching experience. . . . It would hardly be possible to find a more promising student or one better prepared to enter his chosen field." Hutchinson then provided a detailed description of Edmondson's research accomplishments as a high school and undergraduate student.[16]

Juday's limnology course proved especially important because a doctoral student in the course, Yvette Hardman, helped Edmondson to collect rotifers; they were married several years later. Yvette (Hardman) Edmondson finished her doctorate and later became the editor of the journal *Limnology and Oceanography*.

Edmondson returned to Yale as a graduate student in 1939 after a productive summer spent at Trout Lake in northern Wisconsin. Juday had provided him with a research assistantship. He worked on sessile (attached) rotifers and collected much ecological data. He continued his work on these rotifers for his doctoral dissertation at Yale.

Asked how Hutchinson directed his Ph.D. students, Edmondson answered, "He let you do your own thing. . . . I don't remember that he ever said 'Now here's your project' or anything like that. I came with some background . . . and I had some ideas about it. He would tell me something was good or bad or wouldn't work. One of my best ideas he was sure wouldn't work, but it did."[17]

This idea, as often happens in science, was the result of a chance observation. He found *Utricularia* (bladderwort), an aquatic flowering plant, covered with rotifers. These rotifers, except the smallest ones, had a dark band across the tube, such as seen and pictured in the 1880s by Hudson and Gosse.[18] The dark bands were pellets made while the water was muddy. Edmondson thought that by marking the tubes one could tell how fast the animals were growing; the clear part above the dark band would show the amount of growth between the time the mud settled and the time they were collected. First he was going to stir the mud up himself; then he thought of pouring in carmine dye.

The eggs held inside the tube would be the next generation of rotifers. He thought the egg count could be used for calculating birth rate, a method he later used successfully for research on reproduction of planktonic rotifers. "Marking rotifers with carmine to measure their growth, that's what I'm famous for," said Edmondson. He is actually more famous, at least among ecologists, for his research leading to the reversal of pollution of Lake Washington in Seattle.

Gordon A. Riley

Gordon Riley was actually Hutchinson's first graduate student. Riley died in 1985; therefore I was unable to interview him for this book. He, however, wrote an account of his professional life, "The Reminiscences of an Oceanographer," which is very revealing.[19] In addition, oceanographer and history of science professor Eric Mills interviewed Riley about his Yale and later professional experiences,[20] and Hutchinson published his own reminiscences about Riley.[21]

In his memoir Riley recalled growing up in the hill country of Missouri, far from the ocean. He was a biology major at Drury College in Springfield Missouri; the college had only one biology and one chemistry professor. Washington University in St. Louis gave Riley a tuition scholarship for graduate study in embryology and twenty dollars a month to live on. When he finished his master's degree, Riley wrote, he was ten pounds lighter. His advisor at Washington University, Caswell Grave, thought

17. Gordon A. Riley. Manuscripts and Archives,
Yale University Library.

that descriptive embryology was dead and that Riley should continue his graduate studies where experimental embryology was done. This was in 1934 or early 1935. Grave gave him the same advice Hogben had given Hutchinson in 1928: go to Yale to study with the great Ross Harrison.

Riley did go to Yale, originally to work with Harrison. He was offered a teaching assistantship that paid his tuition and sixty dollars a week. "I could live on that without losing weight," he wrote. Harrison asked Riley to come up with a thesis proposal. On his own, Riley tried some of the experimental grafting techniques, using hydra and flatworms, but he was not successful. No thesis topic appeared. As Riley told Eric Mills, Harrison had about fifteen graduate students and little time; Riley did not find the zoology courses inspiring. Then he took a second-semester limnology course taught by a young assistant professor, Evelyn Hutchinson: "One lecture was enough to make me begin to sit up straight and bright-eyed and to struggle to assimilate every word of his thick British accent. He was dynamic and obviously very bright, full of new ideas, and was dissecting the literature with keen and frequently witty comments. Within a week, I knew that limnology was where I wanted to be. Unsolved problems stuck up all over the place. One didn't have to search for some little unfilled chink. . . . Evelyn described his excitement about that meeting. It could hardly have equaled mine. In the course of a half hour, my future was channeled into a very different direction, and the floundering of the previous few months was ended."

Hutchinson wrote of Riley: "Gordon was my first graduate student. The excitement I felt when he asked, I think after a lecture in my limnology course, if he might work with me, is still a vivid memory." In Riley's words, "When I got into a course with a sprightly young fellow known as Evelyn Hutchinson I quickly decided to convert to limnology."

Contrary to Edmondson's comment, Hutchinson did, in fact, choose Riley's dissertation topic. Unlike Edmondson, Riley felt himself to be completely inexperienced in his new field. Hutchinson asked about Riley's chemistry background. It was excellent; he even knew colorimetric techniques. The copper cycle in Connecticut lakes was the thesis project Hutchinson promoted for Riley. The latter noted the appeal of doing something that had never been attempted before. He searched the literature for methods of measuring copper.

Copper had not been previously studied but Hutchinson thought it might be important because it was essential to crustaceans that contained hemocyanin in their blood. The two men looked for small lakes that two could manage; Hutchinson credited Riley with finding Linsley Pond, "the nearest more or less natural lake to New Haven." Edmondson had actually "discovered" it earlier and collected rotifers there.

Hutchinson was busy with undergraduate teaching during the week but Saturdays were free. He and Riley did their fieldwork together in the mornings, ate a hot lunch cooked by Hutchinson's wife, and did chemical analyses in the lab on Saturday afternoons. Riley analyzed several types of copper found in the lake, while Hutchinson determined levels of phosphorus and nitrogen, hoping to explain some of the earlier Wisconsin nutrient data.

Gordon Riley described all of this rather more colorfully in his memoir. As soon as Linsley Pond was ice-free, the two of them went out in the very same inflatable rubber boat that Hutchinson had used in Ladakh. The boat leaked. They got cold and wet but managed to get a few samples before the boat collapsed. Later they bought a better boat that they left at Linsley Pond; they rented boats on other lakes. There were perils to this work, from tippy canoes to bitter cold and thin ice, which Evelyn walked on. But Gordon related the positive aspects, too: "The best part of course was that Evelyn was collecting samples, too, so that most of the time we went together, and this was a rare opportunity for me to further my education . . . with a man whom I have always regarded as having the keenest and best informed mind of any scientist I have known. We talked at length about new papers that were coming out and about limnological problems. . . . Ed Deevey and I, his first students, had the best of it in the old days when he could afford to be generous with his time."

Tommy Edmondson, as an undergraduate, helped Gordon Riley with the field collections one summer when Hutchinson was in England. Evelyn loaned them his old Buick. It had dysfunctional brakes; this was apparently typical. Riley related:

> Evelyn lived in that magnificent, well-ordered mind, which was a good place for him to be, for he was surrounded by chaos. His clothes were shabby, his car decrepit. Every surface in his office was piled high with books and papers, although he could instantly locate anything he wanted. His performance in the lab was a disaster, frequently accompanied by crashes of glassware and a fervent . . . broad English—"Oh Blahst." One day he came to my lab to borrow some Nessler tubes—those long test tubes with optically ground glass . . . used in visual colorimetry. He stuck a half dozen in his jacket pocket. They went through a hole in his pocket and smashed on the floor.

Ruth Patrick and others told similar stories. This was only a few years after his return from the North India Expedition, yet for his lake work on the expedition he had managed to put together his own equipment and to repair it himself.

As for Riley's copper results, they were not spectacular. Copper was not as variable with depth as other nutrients, but it did vary significantly seasonally, showing striking increases in concentration in the autumn. Over forty years later Hutchinson wrote that much was still unknown about copper cycles.

For his freshwater research, Riley borrowed some techniques used previously in oceanography at the Plymouth Laboratory in England. One was a quantitative plant pigment analysis that H. W. Harvey had developed. The phytoplankton were filtered, the pigment extracted with acetone and measured against a color standard. This gave an estimate of total phytoplankton more easily and perhaps more accurately than the usual laborious cell counts. Riley found that carotenoid pigments in the freshwater phytoplankton masked the chlorophyll. He was able to separate the carotenoids from the chlorophyll. He would use this important method later in his postdoctoral phytoplankton studies.

While Riley was still a doctoral student, Albert Parr, director of the Bingham Oceanographic Laboratory, invited him to join an oceanographic cruise—two weeks on the ship *Atlantis* off the mouth of the Mississippi River in March 1937. Riley was to do the plant pigments and nutrient chemistry. Riley had never been to sea before and was worried both about his competence and about seasickness, but he finished the first draft of his Ph.D. dissertation and took the train to Mobile, Alabama, to meet Parr.

Albert Eide Parr was a Norwegian trained in Bergen as a fisheries biologist. He came to the United States and found a job at the New York Aquarium "sweeping the floor and feeding the fishes." Fortune was with him, however. He met Harry Payne Bingham, a wealthy yachtsman, and persuaded him to set up a deep-sea fish collection program using Bingham's yacht. Parr also persuaded him to set up the Bingham Oceanographic Foundation with a small laboratory in New Haven, connected to Yale's Peabody Museum.[22]

Later Parr was able to use the Woods Hole ship, the *Atlantis,* in preference to a Bingham Laboratory one for winter research cruises. It was heavily used by oceanographers from several universities in the summer, but there was little resident staff at Woods Hole the rest of the year. On Riley's 1937 cruise were fellow seamen Yngve Olsen and Martin Burkenroad. As to seasickness, to which Olsen was prone, Riley turned out to be "a fortunate one who has never served as an unwilling intermediary between terrestrial and marine food webs." As to Riley's research: "The deck officer put the collecting bottles on the wire and took them off, and a seaman manned the hydro winch. . . . All this was a far cry from Linsley Pond [with Hutchinson]—one bottle on the wire, lowered by a hand winch to successive depths."

Riley worked on the samples in the lab below. There he did the chlorophyll analyses, including extracting the carotenoids with ether. The limnological methods he had pioneered worked fine for oceanography as well as for limnology. He loved the deep-sea trawling: it was "the first time I had seen the strange deep-sea life except as illustrations or faded preserved specimens. They were marvelous." Riley was being transformed into an oceanographer. When they returned to Mobile he had enough data for a paper showing phosphate enrichment and an associated increase in chlorophyll indicating enhanced productivity around the mouth of the Mississippi.

After the cruise, Riley finished his dissertation and with his Ph.D. in hand, he needed to find a job. He traveled to a job interview in Iowa, at a small denominational college where he would be the zoology half of the biology department. There would be no smoking, no drinking, and church every Sunday. The salary was $1,200 a year or perhaps "$1,800 if the corn crop were good." He decided instead to accept a postdoctoral fellowship at Yale to do what he wanted to do: limnological research with none of these restrictions, also at $1,200. So back he went to Yale and to Linsley Pond, where Hutchinson was still doing nutrient chemistry. Riley would do plankton studies, including chlorophyll analyses and photosynthesis and respiration determinations using light and dark bottles suspended in Linsley Pond.

Eric Mills in his interview asked Riley whether the light and dark bottle experiments, a method now very widely used, were his idea or Hutchinson's. Riley couldn't remember. By 1937 both of them would have known about this method. It had first been used by Gaarder and Gran in 1916 in the Oslo Fjord (first published in English in 1927). The method was improved by Marshall and Orr in Scotland, who published their results on a Scottish loch in 1928. Penelope Jenkin had also used it extensively at the Plymouth Laboratory between 1932 and 1935. Although Jenkin's work was not published until 1937, Hutchinson would surely have known of it. He and Jenkin, his Cambridge University classmate, corresponded and saw each other at meetings. Riley wrote that the decision to apply the method to Linsley Pond was the natural thing to do, whoever first suggested it.

Riley used the plant pigment methods he had initiated, particularly the carotenoid separation technique. He also used multiple regression to analyze the results. He had learned statistical techniques in Oscar Richard's course at Yale, but no one had previously applied them to limnology. In his memoir Riley related what he thought had been Hutchinson's most important influence: "I am deeply indebted to Evelyn for introducing me to his own scientific philosophy about ecology, which was ahead of its time and has permeated all my plankton work. He maintained that populations needed to be studied in terms of dynamic processes — rates of production and consumption and the way these are affected by ecological factors."[23]

Unfortunately, Linsley Pond was not owned by Yale. Many people had cottages around it. It was ironic that the householders used copper sulfate to control the algal bloom on Linsley Pond. Riley's experiments on copper were abruptly ended, a "bitter disappointment."[24] It was, however, probably the first light and dark bottle work done in North America, and the statistical approach was certainly new. Too new for the editors of *Ecological Monographs*; the well-known Wisconsin limnologist Juday was editor at that time. The paper came back, Riley wrote, with "an enormous number of nitpicking comments and criticisms that were almost as long as the paper. I had to do a great deal of revision, eliminate some things I liked best."[25] The journal ended up publishing the data but not the statistical analysis.

Albert Parr saw the value of the latter, however, and published it in his own new *Journal of Marine Research*, even though the research had been done in freshwater Linsley Pond. Parr, moreover, saw Gordon Riley's potential and offered him a staff position at the Bingham Laboratory. To quote Hutchinson, "From now onward Gordon was an oceanographer." He did much pioneering work in oceanography, a young science especially in North America. Riley was a colleague of Hutchinson's for a long time after their joint Linsley Pond adventures and then went to Dalhousie University in Halifax, Nova Scotia, where he developed his own research group. He was not the only Hutchinson graduate student to move from fresh to salt water, where Hutchinson himself almost literally did not set foot. Several others of Hutchinson's graduate students later joined Riley in Halifax.

Edward S. Deevey, Jr.

Hutchinson's second graduate student was Ed Deevey. He and Riley started their graduate work the same year, but Deevey received his Ph.D. in 1938, a year after Riley. Deevey came from Albany, New York, where he had spent a year at the New York State College for Teachers (now the State University of New York at Albany). He transferred to Yale and completed his undergraduate work in 1934.[26]

Deevey had majored in botany at Yale and originally intended to continue in botany as a graduate student. He was doing some research at Linsley Pond. Hutchinson was also working there, and Deevey was giving him a ride to Linsley Pond several afternoons a week. Deevey's proposed research involved the history of the lake and of human impact on the lake environment. He felt that a lot of information could come from lake cores. No one in the Yale Botany Department was interested in this. Hutchinson, however, was really excited about the lake cores and urged Deevey to come and work with him in the Zoology Department. He did so.[27]

Deevey's breadth of interests and knowledge, even as a graduate student, ranged from botany and zoology through geology, paleolimnology, and paleoanthropology. He wished to unite these fields in his research. The use of pollen cores at Linsley Pond was Deevey's idea.[28] Paleolimnology had been pioneered by the Swede G. Lundquist in the 1920s. Before Deevey's study there had been terrestrial research using pollen analysis by Paul Sears and others in North America. P. W. Bowman had used pollen analysis in a Quebec peat bog in 1931, and both Sears and V. Auer had studied the vegetational history of the area surrounding Caribou Bog in Nova Scotia. Thus the statement by Edmondson in his National Academy of Sciences memoir that Deevey made the "first pollen stratigraphy for northeastern North America" is not literally true.[29] Nevertheless, no comprehensive study comparable to Deevey's doctoral research had previously been attempted.

In addition to the pollen analyses and their interpretation, Deevey also carried out two related studies. One was on the current limnology of Linsley Pond and several other lakes in Connecticut and New York, with particular attention to their bottom faunas. The other was not current limnology, but zoological paleolimnology; Deevey studied the fossil fauna in the sediments of two Connecticut lakes throughout their post-glacial history. Using lake deposits for all these studies was in itself innovative.

Most of the previous pollen analysis had been carried out using peat deposits from bogs. Deevey's major purpose was to study the developmental history of Connecticut lakes.

Deevey also examined many other topics, from the presence of human cultures during the post-glacial period in the Northeast, a continuing interest of his, to cyclomorphosis of fossil crustaceans. Cyclomorphosis, which is the presence of very different forms of the same species, particularly in *Daphnia* (water fleas) and in a related genus, *Bosmina*, was found in Deevey's fossil assemblages. It had been studied earlier in Europe and was analyzed much further by John Brooks, one of Hutchinson's later students. Interestingly, Deevey, like Riley, used techniques learned from Oscar Richard in some of his pollen analyses.

In his overall pollen analyses, Deevey got help from George E. Nichols, a plant ecologist and chairman of the Botany Department at Yale, at that time separate from zoology. Hutchinson noted Nichols's presence at Yale during his early years there but wrote that though botany and zoology shared the freshman course and a common library, "There was a great gulf artificially fixed between the two disciplines; this could not be crossed during working hours."[30] Nevertheless, Deevey crossed it. It was Nichols who loaned Deevey a Davis peat sampler with which to take his cores, and it was Nichols who discussed forest types and ecological classification in relation to his pollen findings with Deevey.[31]

Deevey amassed a great deal of original data from what were certainly at that time considered disparate fields in order to solve a particular problem. He raised as many questions as he answered. Some of his ideas have since proved incorrect, but he carefully pointed out in his dissertation where more research was needed and what methods might be used to answer questions that he discussed but for which he had no definitive answers.

And what of Hutchinson himself? He was clearly excited about the paleolimnology and the comparison between the present and fossil microfauna. He was out there at Linsley Pond at the same time as Deevey, as was Gordon Riley. Many measurements had already been made by these two: the oxygen cycle and cycles for nitrates, phosphates, and silicates, as well as for copper, were being worked out by Hutchinson and Riley.

It was clear that in the 1930s Linsley Pond was a eutrophic lake with poor visibility and high alkalinity. Deevey found more phytoplankton than zooplankton, the latter being dominated by *Bosmina*, a cladoceran, and by rotifers. The bottom fauna was dominated by two types of midge lar-

vae, *Chironomus* and *Corethra* (or *Chaoborus*). Deevey did his own temperature readings, depth soundings, dissolved oxygen, and alkalinity
measurements for Linsley and eleven other lakes, largely in Connecticut. He also studied Queechey Lake, Canada Lake, and Lake George in
New York, with help from Emmeline Moore, then New York State's chief
aquatic biologist. Hutchinson no doubt gave advice on methods and was
certainly knowledgeable about zooplankton. Riley cited the help both he
and Deevey got as students in an era when Hutchinson was a field companion and had ample time for them. Much of the material in Deevey's
dissertation and his early publications on pollen analysis and paleolimnology cry out (in hindsight) for an independent method for determining
chronology. Such a method, carbon 14 dating, became available only after
World War II.

Hutchinson and Deevey were among the first to utilize radioisotopes in their ecological studies: Hutchinson used them to study mineral
cycling in lakes (see chapter 9), Deevey to provide post-Pleistocene dating
for his cores. After Willard Libby had used carbon 14 for dating archeological samples, Deevey got a Rockefeller Foundation grant to establish
Yale's Geochronometric Laboratory in 1951. Thus the ages of cores, previously only relative, became actual; it became possible to correlate the age
of cores from Europe and North America, and to use lake sediment pollen
cores in order to date climate change as well as to better understand the
limnological history of lakes. Hutchinson was very much involved in all
of this, as was R. F. Flint of the Yale Geology Department. Margaret B.
Davis, Deevey's postdoctoral student, was later an authority in this field.
As to the establishment of carbon 14 research, Hutchinson wrote that he,
Flint, and Deevey were all close to Libby and helped each other in getting
carbon 14 on its feet. He thought Deevey largely responsible for the speed
of the spread of the use of carbon 14.[32]

While he was working on his dissertation, Deevey analyzed pollen
from interglacial beds that Hutchinson had collected in the Pang-gong
Valley of northern India. He published a stratigraphic paper from this
core, including the climatic interpretation of these data, probably the first
such data from this geographical region.[33] Deevey went on to study paleolimnology in many other places, including Guatemala and San Salvador.
He always kept in close touch with Hutchinson through many fascinating letters. He went to the Rice Institute in Texas as instructor in Biology
from 1939 to 1943. During the rest of World War II, Deevey was one of a

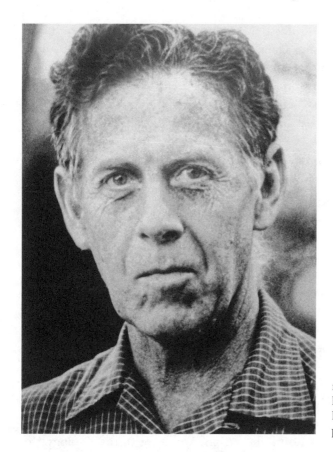

18. Edward S. Deevey.
Photograph by
Brian Deevey, with
permission.

group of Hutchinson's former students working, as civilians, at Woods
Hole on war-related marine projects. In 1944, however, he was invited by
limnologist Arthur Hasler, the successor to Birge and Juday at the University of Wisconsin, to set up a future program for limnology at Wisconsin.
Deevey wrote a four-page proposal for such a program, which he also sent
to Hutchinson.

Deevey had reservations about accepting this position. Hutchinson
concurred, noting that his career would "be seriously hindered by any ob-
stacle that prevented your developing your position as the leading American bio-climato-geographical historian of the past million years."[34] That
was before carbon 14 dating came in use, although after Hutchinson's
radiophosphorus studies of lake metabolism. By the end of the war, a Yale
position became available; Deevey never went to Wisconsin. In 1946 he
returned to Yale as a member of the zoology faculty.

In 1937, Assistant Professor Hutchinson had written a confidential statement concerning Deevey stating, "Of the graduate students I have met here in the past nine years, few have impressed me as much as Mr. Deevey has, both personally and intellectually." He went on to discuss Deevey's interests in subjects outside biology, including botany, geology, and American archeology, and how he was able to use these interests and appropriate methods to integrate the results into his doctoral research. "He has been working under my direction most of this time [Deevey's three and a half years as a graduate student] and I have been in very close association with him."[35]

Maxwell (Max) Dunbar

There was one other student who worked closely with Hutchinson during this period, Max Dunbar. Dunbar was not a graduate student at this time but an advanced undergraduate. He spent the academic year 1937–38 at Yale on a Henry Fellowship, which enabled students from Oxford and Cambridge Universities to study at Harvard or Yale. Dunbar, a Scot from Edinburgh, had been doing undergraduate work in zoology at Oxford University but was interested in becoming a physical anthropologist; his fellowship would enable him to study anthropology at Yale. When he arrived in 1937 he spent three weeks with the anthropologists, but he realized that this was a "youthful aberration" and went to the Zoology Department to talk to Hutchinson. "I simply walked into his office one day . . . and told him what I was all about. He was interested and said, 'Come and work here.'" Dunbar wanted to work on osmotic effects in marine or brackish water animals. He decided on Vorticellid protozoans, one-celled ciliate organisms. They attach to algae or other substrates with a stalk that expands and contracts like a Slinky toy. One can observe their reactions under a microscope. Dunbar wanted to test their reactions to metallic ions such as sodium, calcium, and potassium in seawater, "to see what effects isotonic solutions of these salts, chlorides of these metals, had on the activity of the contractile vacuole, which kept on receiving water and pushing it out."[36]

Hutchinson took him down to the oyster station just southwest of New Haven, run by a Russian named Loosanoff who gave Dunbar some Vorticellids from the bottom of one of his tanks. On discovering what Dunbar wanted to do Loosanoff said, "Now that's a good idea, but don't make the mistake of one of my students in the past. If he had two animals and one died, he said, '50 percent died.'"

Before coming to Yale Dunbar had spent the summer in Greenland, working on the mechanism that maintained upwelling currents in places with active glaciers. A fellow student at Oxford went with him. They chose northeast Disko Bay and got there with the help of some Danes who lent them a motorboat. They spent three months doing plankton surveys at the glacier base and in the whole area. Dunbar arrived at Yale with these data, not yet written up, and showed them to Hutchinson. He said:

> Nobody had looked at these [upwellings] before but they were very important ecologically because they yanked plankton up from the deep water, even big crustacean plankton. There was a constant journey back and forth on the part of sea birds and also of hundreds of seals into these areas. At that time the man who ran the Peabody Museum was Albert Eide Parr from Norway, a physical oceanographer. There were very few physical oceanographers in the world at that time. We had to do our own. He was just about to publish the first volume of the *Journal of Marine Research*. Hutchinson showed him my [Greenland] work. He grabbed this stuff of mine and got it all beautifully drawn out, all the figures done by one of the museum staff and it appeared in the first volume of *JMR*.

Clearly, Hutchinson was interested in marine research very early. When I asked Max about this, he replied, "Hutchinson was interested in everything." Dunbar gave up any further thoughts of pursuing physical anthropology. "He put me in an office right next to his, and I stayed there all year." They conversed on many subjects, from ecology to folk music. (Max Dunbar later in life recorded Scottish ballads for Folkways.)

The fellows were intended to "cement the bonds of friendship between the U.S. and the U.K., which we did," said Dunbar. Another Henry Fellow from Cambridge University was at Yale that year, working with zoology professor John Nicholas on colchicine, "a new thing to play with at that time," said Dunbar. Nicholas wanted this student, named Wallace, to work with him in the summer when the fellowship terms were up, but Dunbar had already inveigled Wallace to come up with him to Glacier Bay in Alaska. The Henry Fellowship, according to Dunbar, provided ample funds for travel. Dunbar felt there was friction between Hutchinson and Nicholas, perhaps exacerbated by the fact that Dunbar, who was working with Hutchinson, was taking Wallace to Alaska instead of leaving him to work on colchicine with Nicholas.

Dunbar and Wallace did go off together. They were on their way to Glacier Bay in a single-engine plane when "the ceiling came down to zero." They landed in an Indian village, and a Scandinavian with a salmon trawler took them to Glacier Bay and picked them up two and a half months later. They lived on an island in Glacier Bay and did plankton studies.

Thereafter Max Dunbar returned to Oxford and finished his undergraduate degree, then went to McGill University in Montreal for a Ph.D. He had always wanted to work in the Canadian Arctic, which was probably more feasible from a Canadian university. His doctoral work was completed during World War II. He served in the Canadian Cavalry for a year while at McGill. "I was suddenly yanked out of there and sent to Greenland—as consul—at the age of 26 and after only two years in Canada." He explained that the previous consul, a career man, had spent a most unhappy winter in Godthaab, Greenland. This consul told External Affairs that they should send scientists, not career diplomats. "Scientists would be completely understood by the Danes [who controlled Greenland], and would have something to do besides send dispatches . . . and watch for saboteurs."

Indeed, Dunbar did carry out significant research in Greenland. He wrote Hutchinson a long letter from Godthaab in 1943 about plankton production at the head of the fjord, about size differences in two coexisting subspecies of zooplankton, and about his methods of salinity measurement. He also wrote about a young sea eagle he had weaned who could recognize humans by the color of their pants. Dunbar obviously felt that Hutchinson was interested in everything. Dunbar did three tours of duty as consul and went on to a distinguished career of Arctic research while a professor at McGill.

Charles Elton, whose ecology book had influenced Hutchinson before he even arrived at Yale, was at Oxford in the 1930s during Dunbar's undergraduate years there. Elton, according to Dunbar, got along very well with people at Oxford, was beginning to develop the Bureau of Animal Population, and was becoming a well-known scientist. I (the author) told Dunbar that I had met the legendary Charles Elton in 1967, when I was a demonstrator at the Botany School in Oxford. He participated in one of our student ecology field trips. Dunbar replied, "Someone had seen Elton and Hutchinson, the two people responsible for most of modern ecology, plotting together on the Broad [Broad Street in Oxford]."

19. Larry B. Slobodkin while a Yale graduate student.
Courtesy of Yale University Office of Public Information.
Manuscripts and Archives, Yale University Library.

One wonders what new ideas in ecology they were "plotting." When
Hutchinson, many years later, was headlined in a series of newspaper
articles as the "Father of Ecology," he insisted that that title rightly belonged to Charles Darwin. The title of "Father of Modern Ecology," he
felt, belonged to Charles Elton. Elton pioneered new ways of looking at
the interrelationships of animals, an enterprise that Hutchinson and his
many students took much further. Dunbar said he thought Hutchinson
was the first to recognize the significance of size differences within different groups, including copepods, minute crustaceans. "He wrote a magnificent paper called 'Copepodology for the Ornithologist.' . . . Elton was
on to the same idea but he didn't write about it as Hutchinson did."

Hutchinson defined for Dunbar the importance of different functions
in the Zoology Department at Yale. "He put the professors last. . . . He
always put the graduate students at the top. They were the most important people in any department." In the 1950s when another cohort of exceptional graduate students arrived at Yale, including Larry Slobodkin,
Hutchinson credited a group of them, by name and in print, with their
role in the development of one of his most important ideas.

20. G. Evelyn Hutchinson with glassware in his Yale laboratory, 1939.
Courtesy of Yale University Office of Public Information.
Manuscripts and Archives, Yale University Library.

Max Dunbar and Ed Deevey were among a small group of students that Hutchinson took on field trips to the woodlands around New Haven "to chat about what he was seeing, like birds and spiders." Asked whether Hutchinson knew about birds, "Hutch knew about everything," replied Dunbar. Such was the folklore. In truth Hutchinson continually added new areas of knowledge, and his old knowledge could always be altered or updated. Hutchinson's graduate students had told Dunbar that every September when they came back to Yale from their summer research sites, Hutchinson would have a new hobby, such as learning Chinese pictographs. The year Dunbar arrived Hutchinson was studying early music, music before Palestrina.

In contrast, Dunbar told a story about a graduate student who came in and asked about a paper Hutchinson had written. "Hutch talked about it for quite a while and the graduate student said, 'That's not what you wrote in this paper.' Hutch said, 'No, that's what I was thinking six years ago.'"

CHAPTER EIGHT

The Old and the New Limnology

Hutchinson's Early Interest in
Aquatic Ecology Reprised

Hutchinson's interest in the fresh water world started by age five or six. By then he was already collecting organisms in ponds and ditches near his home in Cambridge and making aquaria. Later, at Gresham's School in Holt, he was an accomplished student of water bugs. Collecting could be done only during his free time on Sunday afternoons: "Throughout Sunday we had to be dressed in our best clothes; striped trousers, black waistcoat and jacket, stiff turn-down collar with black tie. . . . Thus attired with pond-net, killing-bottle and jars for live specimens I set out each week for some favorite locality, which ordinarily was on a private estate. . . . To me it was amply worthwhile to run the risk of being caught by a gamekeeper."[1] One Sunday he not only escaped the gamekeeper but found the rare pond skater (water strider) *Gerris lateralis asper*, previously thought to be almost confined to Scotland. He published a paper about it, his second paper, in 1919 at the age of sixteen.[2] His publications in limnology were to span nine decades.

During his secondary school years, Hutchinson's family inherited a holiday house in the north of England. On visits there Evelyn had a whole new region to explore, including a series of shallow lakes or tarns in the Pennine Range. When he visited these lakes he found them teeming with a water bug new to him, in the genus *Arctocorisa*, a northern and mountain genus.[3] In a mountain lake in Ladakh he was to find a related species not only new to him, but also new to science.[4]

Hutchinson wrote that while he was still at Gresham's School his interests were becoming more ecological. He was observing the aquatic environments as well as their fauna. He realized that organisms had chemical environments. He practiced titrations. "I should have to be an adequate chemist if I were to become a good field zoologist."[5] Later Hutchinson became a more than adequate chemist. One of the fields in which he was to be well known early in his professional career was biogeochemistry. Limnology and biogeochemistry were interwoven in his first twenty years of research.

At Cambridge University Hutchinson studied a very broad range of zoological subjects, several of which, like embryology, were far removed from limnology. His limnological interests continued, however. He listened to lectures in J. T. Saunders' hydrobiology course together with Penelope Jenkin and collected aquatic insects at Wicken Fen with Grace Pickford.

The Wicken Fen expeditions led to two Hutchinson publications on true bugs (Hemiptera-Heteroptera). These include aquatic ecology as well as taxonomy. Special habitats were noted for the aquatic and semi-aquatic species living in or on the water or in the moss or mud of the water's edge. He discussed predaceous, plant-eating, and parasitic bugs and their hosts. Careful observation was obviously an integral part of his collecting. He observed that sweeping with a net was too indiscriminate a technique to throw much light on habitats or food plants.

While at Cambridge he was also able to go to the Hebrides, an expedition that he described to this author nearly seventy years later as though it had taken place the previous summer. He published the results of this trip in the *Scottish Naturalist*.[6] His Cambridge entomology instructor, F. Balfour Brown, "a committed Scot with a passion for water beetles," made this trip possible.[7]

Hutchinson's primary mentor in his study of water bugs was E. A. Butler, a retired schoolmaster and the leading authority on the true bugs in Britain. In 1923 Butler published a major tome on the Hemiptera-Heteroptera, including major aspects of their biology from reproduction to behavior. He included and credited many observations Hutchinson had sent him, particularly about the aquatic groups, the Corixidae or water boatman (previously known as boatflies) and the Notonectidae or backswimmers.[8]

Butler wrote to Hutchinson just before he went to the Hebrides, suggesting that he "should look well for all water-bugs." Butler continued, writing what was an ecological guide for Hutchinson to various genera and species in many freshwater habitats, including *Sphagnum* (peat) moss, and also in saltwater habitats, such as salt marshes and stranded seaweed. Butler clearly knew the ecology of his bugs and encouraged Hutchinson to think ecologically. Perhaps he was a literary influence as well. He wrote of one water bug "putting one in mind of Shakespeare's men whose heads are in their breasts" if one tries to imagine oneself a *Corixa*.[9] By the time Hutchinson arrived in South Africa he himself had become a mentor on this group of aquatic insects to workers on several continents.

When Hutchinson left Cambridge after his graduation in 1925, he was not planning to pursue either entomology or limnology. But his research on octopus physiology at Naples was unsuccessful. Nor did he accept a teaching position in South Africa with freshwater ecology research in mind.

South Africa's appeal to Hutchinson was its completely new fauna, not only the huge elephants, lions, and ostriches but also the strange small terrestrial creatures of the phylum Onychophora (velvet worms), which he did indeed find and write about. In South Africa he did his first work that could truly be labeled limnology (see chapter 4). Lancelot Hogben sent him and Grace Pickford, by then his wife, off to work on the pans or vleis, the shallow depressions characteristic of dry regions, that generally had flat bottoms and no outlets but varied greatly in size. Here, to judge from the resulting papers, their study was of much more than the water bugs or even the lake biota. Their pan study included many other aspects of lakes, especially water chemistry, and environmental physics, formation, and classification. It was Hutchinson's idea, and perhaps Pickford's as well, to include these other aspects of lake study, in addition to the organisms. European limnologists were already doing so, but at the time of the pan research Hutchinson had not yet read Thienemann's pioneering limnology book.

Hutchinson and Pickford also used quantitative methods. Their 1932 publication has "hydrobiology" in its title and "an ecological study" in its subtitle.[10] Comparative limnology, with some emphasis on process, seems to have been the end result. Plankton were sampled in sixty-five localities; statistical sampling methods were presented. They carefully followed, described, and graphed seasonal succession for different pans. This paper came out several years after the earlier one by the same authors, "The In-land Waters of South Africa,"[11] because many of the groups of organisms collected needed to be identified by experts. But even the earlier paper involved innovative limnology. Hutchinson did read Thienemann's book before leaving for Yale in 1928 and declared, as noted earlier, that he him-self had become a limnologist.

The Early History of Limnology

Lake studies originated very early in human history; fish symbols are found in many cultures. There were even early attempts to study applied limnology; Romans cultivated fish in lakes and managed fishponds—and wrote about them. But most would agree that F. A. Forel (1841–1912) was actually the founder of limnology as a scientific field. His Swiss lake inves-tigations were underway in the 1860s; he published an instruction manual for lake studies as early as 1869. He started to study the limnology of Lake Geneva (Leman) in 1872 while living in the Swiss village of Morges; his monograph appeared in 1892.[12] In 1901 he published the first limnology textbook, his *Handbuch*. Even before that, limnological commissions for Switzerland and an international commission were set up in 1887 and 1890, respectively, both due to Forel's efforts. As Hans-Joachim Elster pointed out: "The rapid upsurge of limnology in the first half of the [nine-teenth] century was mainly due to the fact that, particularly in lakes, the reciprocal effects of animate and inanimate parts of the environment are easily measured." Henson introduced the word "plankton" in 1887 and used quantitative methods to capture plankton.[13]

It was probably the work of August Thienemann that was most impor-tant in early twentieth-century European limnology (see chapter 4). His book that so influenced Hutchinson, *Die Binnengewasser Mitteleuropas*, was published in 1925, but Thienemann had written important papers on limnology in the first decade of the century. By 1915 he had recognized different bottom fauna correlated with oxygen concentration in differ-ent types of lakes. When the international limnology society (SIL) was founded in 1922, Thienemann's work was important in working out a lake classification scheme that would include the ecosystem as a whole. Thienemann's scheme, based primarily on bottom fauna, was combined with the trophic types of E. Naumann, based on chemistry and primary production. In the 1930s W. Rodhe proposed a multidimensional system.

Hutchinson certainly developed his own ideas along these lines, developing his knowledge of biogeochemistry and of zooplankton, and later of energy transfer among trophic levels in lakes. These were all quantitative studies, and all before the use of computers.

"That such a system can be elaborated into a multidimensional one is self-evident in this age of computers," wrote Elster. He added that correlations found from such analyses are no substitute for experiments.[14] Such experiments were carried out at Yale by Hutchinson and his graduate students and by faculty and students of Dalhousie University in Nova Scotia.

The trophic-dynamic aspects of lake ecosystems were studied in the early forties by Hutchinson's postdoctoral student Ray Lindeman. These and the P^{32} studies of Hutchinson and the carbon 14 studies of the late forties expanded into radioecology and systems ecology in the fifties (see chapter 9) with pioneering work by Hutchinson student H. T. Odum and many others.

Almost all the limnology research from the first part of the century was published in German. Hutchinson's facility with foreign languages surely helped him to absorb these limnological ideas of European scientists and to develop his own quite early in his career.

As Eric Mills, a Hutchinson graduate student (Ph.D. 1964) who also worked closely with Gordon Riley, has clearly shown in his book on biological oceanography, it is difficult to separate the history of limnology from that of oceanography; some of the techniques were passed back and forth between the two fields, and many players, including several of Hutchinson's students, moved back and forth from freshwater to marine research as well. Gordon Riley's early career showed several such moves.[15]

Paul Welch stated in his *Limnology*, written in 1935 when Hutchinson was publishing some of his first papers in this field, that limnology as a distinct field had existed for less than forty years. He pointed out its very early antecedents, beginning with Leeuwenhoek's use of the microscope in the seventeenth century to examine "invisible" aquatic organisms. A Dane, Otto Friedrich Muller, wrote a classification of these tiny aquatic organisms, both freshwater and marine in 1786, entitled "Animalcula Infusoria Fluviatilia et Marina." Oceanography, originally called "marine zoology," really took off in the nineteenth century, when techniques later useful in limnology were invented.[16]

Places as diverse as the English Lake District and the midwestern

American lakes became centers of limnology in the early twentieth century. The American Charles Kofoid visited European biological stations in 1908 and 1909. There were European freshwater laboratories in Schleswig-Holstein in 1890, and by 1908 also in Denmark, Sweden, and Hungary, the latter on Lake Balaton. The European stations were often directly supported by local or state funds, sometimes in connection with universities, but that was rarely true in Britain. John Murray organized a Scottish freshwater lake survey in 1897–99 with private money; there was also a private station in Norfolk in 1901.[17]

W. F. Pearsall returned to the Lake District of northern England after World War I. He noted a progression of lakes with differences in pollution, plankton, algae, and rooted vegetation as well as in fish. University teaching of "freshwater biology" started at Cambridge with J. T. Saunders. E. B. Worthington was in his first class. Hutchinson and Penelope Jenkin were also his undergraduate students. Saunders together with Pearsall and algologist F. E. Finch wanted to start a freshwater biology station. They found Lake Windermere the most suitable location. Wray Castle on the lakeshore resembled a medieval castle but was actually built in the 1840s; it then belonged to the National Trust but soon became the home of the Freshwater Biological Association of the British Empire, the FBA ("of the British Empire" was later dropped).

Penelope Jenkin became the Windermere station's first researcher. A team of five workers subsequently studied the ecology of inland waters. The insect specialist was T. T. Macan, one of Hutchinson's correspondents. Another researcher was Clifford Mortimer, who later became a professor at the University of Wisconsin. An Austrian woman, Marie Rosenberg, studied algae, and Winifred Frost carried out fish studies. They not only worked at Wray Castle but also lived there unless they were married. Mortimer was attempting to obtain cores of bottom deposits. Jenkin's father, V. M. Jenkin, designed better devices for obtaining the cores. Winifred Pennington (Tutin) started British paleolimnology there; she later became an FRS.[18]

Pennington's cores taken at Windermere must have interested Hutchinson. One of his first graduate students, Ed Deevey, did similar studies in Connecticut (see chapter 7). The top 20 cm of Pennington's cores contained the diatom *Asterionella*, rare in earlier deposits. Its presence indicated human influence, probably water-borne sewage in the lake deposits. The railroad arrived there in 1847 and with it tourism in the town of Windermere. E. B. Worthington became director of the institute in the "great sham medieval castle," as he called it, in 1937.

During World War II the staff was dispersed from Lake Windermere to various places around the world. Macan worked on malarial control in the Far East. W. F. China, another Hutchinson correspondent, came from the British Museum to head the FBA. From their original program, devoted largely to pure science to find out how "plants and animals were born, lived and died below the surface" of freshwater lakes, FBA workers turned to applied research on food fishes during the war, especially on perch and eels, including the latter's fascinating life history. By the forties, according to Worthington, ecological principles had been developed at Windermere, including the importance of limiting factors and the transfer of energy in food chains of many interacting species, as well as in paleolimnology.[19] All these subjects were also of interest to Hutchinson in the 1940s; Hutchinson was able to make a visit to the station with Guggenheim funding.

North American Limnology Studies

In the United States, Louis Agassiz and his students carried out the first scientific study on the Great Lakes.[20] They started out from Sault Sainte Marie in 1848 in two canoes and one Mackinaw boat to study the north shore of Lake Superior. Louis Agassiz studied the fishes and Joseph Leconte the water beetles. Some limnological measurements were made, including water temperatures and transparency. By the early 1870s Stimson had studied the deep-water fauna of Lake Michigan, and Smith and Verill had conducted deep-water dredgings of Lake Michigan. The first freshwater lab, Allis Lake Laboratory in Milwaukee, Wisconsin, was started in 1886. It was privately funded and had a short life.

Four different midwestern freshwater biological stations were founded in the 1890s, all longer-lived than the Allis Lake Laboratory: Gull Lake, Minnesota, 1893; Illinois River, Illinois, 1894; Turkey Lake, Indiana, 1895; and Flathead Lake, University of Montana, 1898. Kofoid, who as noted, traveled in Europe to visit the biological stations he wrote about in 1910, did significant research on the Illinois River.[21]

The Birge-Juday Era of Limnology

The University of Wisconsin became an important early center of American limnology. In 1875 E. A. Birge arrived there, marking the beginning of the Birge-Juday era of limnology. Birge was publishing on the Wisconsin inland lake fauna by 1878, and by 1895 he was publishing limnological work on the plankton crustaceans of Lake Mendota, including their vertical distribution. Chancey Juday did not begin his long Wisconsin research career until the early 1900s; he and Birge were to publish more than 400 limnological papers in a research collaboration of nearly forty years. Limnologist David Frey wrote: "The studies of Birge and Juday, although they are largely what is known today as descriptive limnology, are of interest not merely for their limnological descriptions of Wisconsin lakes but also for their significant contribution to our understanding of limnological processes in general."[22]

Another early Wisconsin limnologist, C. D. Marsh, may have started on the limnology of Wisconsin even before Birge. He was a professor at Ripon College from 1883 and studied the plankton, especially copepods, of nearby Green Lake. He also studied the daily migration of the planktonic crustacea, using a vertical tow net (dredge) that could be closed at any depth. He took temperature measurements but did not have a good thermometer, or at least one that could be operated from his sailboat.

What was the state of limnology in America at the end of this "nearly forty years"? The International Association of Limnology (SIL), formerly the International Association for Theoretical and Applied Limnology, of which Hutchinson was to be a future president, was in existence by 1922. It had held six international congresses by 1932, none of them in the United States. There was no such American society until 1934, when a founding committee was formed for an independent society of hydrobiology, including Paul Welch as chair, Juday, and four other members. It had its first actual meeting in 1936 at the AAAS meeting in St. Louis, with Juday as president. All who sent in one dollar for dues would become charter members of the Limnological Society of America. Hutchinson was one of these. (Oceanography was not added to the organization's title until 1948.)

Two 1938 letters indicated that the field was not well recognized near the end of the Birge-Juday era. Hutchinson had sent a paper on his limnology research to the *Proceedings of the National Academy of Sciences.*

He had omitted the "subject heading" and explained that he had looked up earlier articles on limnology and found them listed under various headings, including "zoology," "chemistry," and "physics." He didn't think his paper fitted any of these: "No animal is mentioned throughout, nor do I think it can truly be called chemistry, as the chemistry investigations were merely a tool in elucidating the water movements. . . . You have no category of limnology." An editor of the *Proceedings,* the eminent E. B. Wilson of Harvard, wrote back, "Every time we get a paper on oceanography or limnology we are stumped." They did not fit into the academy's hierarchy of sciences. This was probably also true of limnology at other institutions at that time. "Physics will suit us perfectly," Wilson decided.[23] The paper was presumably the one on oxidation-reduction potentials of lake water published in 1939.[24]

Birge and Juday and their students had also invented new limnological techniques and equipment. In 1931 when Hutchinson was about to go on the North India Expedition, he wrote to Juday about equipment for limnology, most of which had to be obtained from Europe. Juday wrote him about a hand centrifuge, plankton filters, and other equipment and advised him to make a special case for his reversing thermometers so that he could carry them over his shoulder to transport them without breakage on a pony caravan. One wonders if Juday had ever been on a pony caravan.

As we have seen, W. T. Edmondson, Hutchinson's undergraduate and graduate student, spent a year with Juday. Hutchinson arranged this, and the research worked out very well for Edmondson, who afterward returned to Yale to complete his graduate work. Other Hutchinson interactions with Birge and Juday were not so positive, particularly in their reviews of his students' papers, both those of his doctoral student Gordon Riley, as recounted above, and his postdoctoral student Raymond Lindeman.

Hutchinson's first years at Yale were taken up with heavy teaching in introductory biology and embryology and with analysis and publication of the data accumulated while in South Africa. His research and publications between his 1925 Cambridge graduation and the 1932 Yale North India Expedition cover a surprising variety of subjects: the branchial gland of the cephalopods, the Onychophora of South Africa, biological aspects of psychoanalytic theory, the blood of the hagfish, and the 570-million-year-old Burgess Shale and the fossils of strange multicellular creatures of the Cambrian ocean. As pointed out by Stephen Jay Gould in his book on the Cambrian Burgess Shale fossils, *Wonderful Life: The Burgess Shale and the Nature of History,* Charles D. Walcott, their discoverer, got the analysis of these early and bizarre organisms nearly all wrong; Hutchinson in an early paper got one of them right. Gould wrote:

> G. Evelyn Hutchinson, who described the strange *Aysheaia* and the equally enigmatic *Opabinia* . . . (getting one basically right and the other equally wrong) . . . later became the world's greatest ecologist and my own intellectual guru.[25]

Much later, starting in the 1970s, three paleontologists at Cambridge University, Harry Wittington, Simon Conway Morris, and Derek Briggs, currently the director of the Yale Peabody Museum, studied these enigmatic fossils intensively and elucidated many of them, as Gould described in his book, and they did in important papers—which Hutchinson almost surely read.

Many of Hutchinson's early limnological papers concerned the water bugs of South Africa and elsewhere. Yet in addition to his innovative South African research with Pickford and Schuurman, he wrote another paper indicating new directions. Entitled "Experimental Studies in Ecology. I. The Magnesium Tolerance of Daphniidae [aquatic submicroscopic crustaceans] and Its Ecological Significance," it was published in a German hydrobiology journal.[26] In fact, many of Hutchinson's early papers were published in European or South African journals. American limnologists knew little about Hutchinson during his early period at Yale, and experimental limnology was largely unknown in the United States.

Hutchinson's wife, Grace Pickford, by now with a Yale doctorate but not a Yale academic appointment, was able to do some additional limnological research at Mountain Lake in Virginia. Hutchinson did not come on that research trip; Pickford did the data collection with some help from Petrunkevitch, her mentor at Yale. Hutchinson did much of the analysis and writing.[27] It was their last joint paper and Pickford's last in limnology.

Limnology became Hutchinson's major concern in the thirties. Although he collected a great variety of organisms on the Yale North India Expedition (see chapter 6), in 1931 in his plan for his expedition research he described himself as a freshwater biologist and was recommended as such by his chairman, Ross Harrison. It is clear, however, from Hutchinson's detailed letters to Grace Pickford from the expedition that in spite of the lack of specialized equipment and the presence of very difficult weather conditions, his research was not simply the collection of aquatic organisms; it was real quantitative limnology. His first expedition paper came out in *Nature* in 1933, the year following the expedition.[28] It was entitled, "Limnological Studies at High Altitudes in Ladakh." In 1937 he published a long paper, "Limnological Studies in Indian Tibet."[29] An additional long paper from the expeditions discussed the fishes of Kashmir and Indian Tibet, one of Hutchinson's few ventures into the world of fish. It is significant that this latter paper was published in a leading American ecological journal, *Ecological Monographs*.[30]

During the summer of 1933, following his return from the North India Expedition, Hutchinson traveled to Nevada to obtain the divorce he and Grace Pickford had agreed to before he left on the expedition. Hutchinson used his short stay in Nevada to do a study of its arid lakes, comparable in some respects to the pans he had studied in South Africa. He published "A Contribution to the Limnology of Arid Regions," a much-quoted tome of nearly ninety pages, in 1937.[31] Such lakes are now of particular interest in limnology, an area of research surely encouraged by Hutchinson's early work. This paper, in fact, became an important one in enhancing Hutchinson's reputation as a limnologist.

By the late thirties Hutchinson was teaching a graduate course in limnology and was acquiring graduate students of his own. Local limnology was booming, especially in the Connecticut lake most closely associated with Hutchinson and his students, Linsley Pond. There ensued many studies of many aspects of small lakes, from chemical stratification to oxygen deficits, productivity, and the ecological significance of the oxidation-reduction potential of lake waters. These were made by Hutchinson together with Ed Deevey and research assistant Anne Wollack, both of whose names appear with Hutchinson's on early papers. Paleolimnological research had begun as well; the results of the chemical analysis of a lake core by Hutchinson and Anne Wollack was published in 1940.[32] Hutchinson published another important limnological paper the following year, on the mechanisms of intermediary metabolism in stratified lakes, this one in an ecological journal.[33] Limnology and aquatic ecology were by then considered one field by Hutchinson. By this time the limnological and paleolimnological studies of Deevey were well underway as was the lake biogeochemical research of Gordon Riley. Lake biogeochemistry became a major focus of the work of Hutchinson and his students in the 1940s.

From Birge and Juday to Modern Limnology

As the founders of the Wisconsin school of limnology, Birge and Juday left a great deal of data and an important legacy, continued by Arthur Hasler, who was Juday's student. Professor Birge became heavily involved in University of Wisconsin administration and had no doctoral students. Juday, on the other hand, began his career with only the title of biologist with the Wisconsin Geological and Natural History Survey and did not officially become a professor until 1931 at age sixty. Nevertheless, he supervised thirteen Wisconsin doctoral students between 1928 and 1940. Hasler and several of his outstanding students, including David Frey and Gene Likens, later did much innovative and important research.

By the early 1940s, Birge and Juday had become the old guard and were critical of new methods, particularly of the use of theory and statistical analysis in limnology. Juday specifically objected to Hutchinson's more hypothesis-testing and mathematical approach to limnology.[34] In a letter to his graduate student Robert Pennak in 1941 Juday wrote, "Deevey tells me that H. [G. Evelyn Hutchinson] is writing a book on Limnology and it is to be chiefly mathematical. So you can look forward to the worst." In a 1942 letter to Pennak, Juday wrote, "In a short time I shall expect them [Deevey and Hutchinson] to tell all about a lake thermally and chemically just by sticking one, perhaps two, fingers into the water, then go into a mathematical trance and figure out all its biological characteristics." Elsewhere he compared their work to that of a spiritualist: "It will not be necessary to visit a lake at all in order to get its complete chemical, physical and biological history."

Both Juday and Paul Welch, a well-known University of Michigan limnologist, wrote negative reviews of Lindeman's now famous 1942 limnology paper, "The Trophic-Dynamic Aspect of Ecology," which marks the beginning of ecosystem ecology.[35] Lindeman was Hutchinson's postdoctoral student, and Hutchinson had influenced his theoretical views and the notation used in the paper. Hutchinson wrote to Thomas Park, the editor of *Ecology*, in Lindeman's defense and to answer the criticisms of Juday and Welch.

21. Raymond L. Lindeman,
one of Hutchinson's
postdoctoral students.

Juday had written, in part: "Lakes are rank individualists and are very stubborn about fitting into mathematical formulae and artificial schemes proposed by man." He wanted Lindeman to go out and study many more lakes. Welch wrote: "Limnology is not yet ready for generalization of this kind. . . . What limnology needs now most of all is research of the type which yields actual significant data rather than postulations and theoretical treatments."

Hutchinson, who had espoused the testing of hypotheses, including the use of mathematics and statistics, a view also adopted by his early student Gordon Riley, fundamentally disagreed with Welch. He wrote to Thomas Park: "Far from agreeing with Referee 2 as to what limnology needs, I feel that a number of far-reaching hypotheses that can be tested by actual data and which, if confirmed would become significant generalizations, are far more valuable than an unending number of marks on paper indicating that a quantity of rather unrelated observations has been made. . . . At times I have felt quite desperate about the number of opportunities that have been missed in the Middle Western regions for obtaining data confirming or disproving the hypotheses that have been forced on us by our little lake."

Robert Cook tracked down these reviews and responses, which he published in *Science*.[36] They have also been quoted by historians of the Birge-Juday school. These authors rightly defend Birge's and Juday's contributions in terms of bringing new instruments and approaches to observing lakes and collecting the first precise and accurate data from the lakes they studied: "Limnology was such a new science and there were still so few limnologists in the world that the research conducted by Birge and Juday and the scientists and students working with them must still be considered pioneering and a major contribution to the understanding of limnological processes." But they also cite "G. Evelyn Hutchinson, who with his students would lead the way in theoretical ecosystems ecology and limnology."[37]

A Treatise on Limnology

Hutchinson had been contemplating as early as the 1940s a large book on limnology. Originally it was to be called "The Study of Lakes," as he wrote to L. G. M. Baas Becking. This book was to "include all aspects of the subject; physical, chemical, and biologic. At the present time most of the chemistry, about half the biology and the chapters on optics and temperatures are complete."[38]

The original idea for the book evolved into the four-volume *Treatise on Limnology*. The first volume, *Geography, Physics, and Chemistry*, became a bible for limnologists and limnology students, still in use today. In the preface to this volume, Hutchinson wrote:

> The aim of this book is to give as complete an account as possible of the events characteristically occurring in lakes. The author, by training a biologist, is by inclination a naturalist who has tried to examine the whole sequence of geological, physical, chemical, and biological events that operate together in a lake basin and are dependent one on another. The book is addressed to all who are professionally concerned with limnology but also to biologists who may wish to know something of the physiochemical environment, mode of life and evolutionary significance of such fresh-water organisms as they may study . . . [also] to geologists who are desirous of learning something of modern lakes in order that they may better interpret the record of inland waters in past times.

Hutchinson also addressed this volume to oceanographers who wanted to compare the results of their own science to what has been learned about lakes and other inland waters.

Volume 1 is thus a lengthy, more than 1,000 pages, account of geographical and physiochemical limnology. Hutchinson did not get to limnobiology until volume 2, *Introduction to Lake Biology and the Limnoplankton.* Volume 3, published in 1975, is *Limnological Botany.* Volume 4, *The Zoobenthos,* was published posthumously. Yvette Edmondson, who served as editor, completed this volume with the help of Hutchinson's devoted friends, students, and colleagues.[39]

Volume 1 of the *Treatise* includes some of Hutchinson's and his early students' biogeochemical research. With the publication of volume 2 on lake biology, particularly plankton, in 1967, Hutchinson was widely recognized in Europe as well as in the United States as America's foremost limnologist. A great many ideas and the research of his graduate students are included in this volume, which discusses numerous ecological problems. These range from seasonal succession to feeding relationships, population dynamics, and distribution. Cyclomorphosis, or change in form of planktonic species in relation to environmental factors, is of special interest; 144 pages are devoted to this topic. Hutchinson's student John L. (Johnny) Brooks had completed his Ph.D. in 1946 on cyclomorphosis in *Daphnia;* much of his work is included in volume 2.[40] Brooks later became a Yale zoology professor.

Volume 3, *Limnological Botany,* contains less of the work of Hutchinson and his students. Hutchinson rarely worked with plants, but he does cite his own work in this volume on three species of water milfoil (*Myriophyllum*) in the long chapter on the chemical ecology of freshwater plants.[41] In addition, the research of Ursula Cowgill, a longtime research associate at Yale with Hutchinson, is included. She worked on the hydrogeochemistry of Linsley Pond in the 1970s and on lake cores from many different regions of the world.[42] This author's (Nancy Slack's) early postdoctoral research with Dale H. Vitt on the ecology of peat mosses (*Sphagnum*) in relation to bog lakes is also discussed in this volume.[43]

Ruth Patrick was an early and continuing research associate and good friend of Hutchinson's; her research is included in both volumes 2 and 3, in chapters on the algal benthos. She studied diatoms at the Philadelphia Academy of Sciences and had a long and fruitful research career. Several of her papers, on diatoms of both lakes and streams, including an early one (1943) on the diatoms of Linsley Pond, appeared in the sections on phytoplankton in volume 2. Patrick was especially interested in the diversity and ecology of diatoms and published an important paper (volume 3) on diatom community ecology using slides as "islands" in 1967.[44] When I interviewed Ruth Patrick in her eighties, she had just come in from a collecting trip in a nearby river.

Hutchinson included both discussions of ecological problems and a great deal of quantitative data and statistical analysis in these volumes. The growth of American limnology (and aquatic ecology) since 1940 is amazing, and of course the books include research from other countries as well. The bibliographies are immense; that of volume 2, for example, is eighty-three pages long. Hutchinson helpfully included in the bibliographies the pages in his books where he discussed each paper. He also starred items to indicate the papers he considered of "exceptional limnological importance."[45] Robert G. Wetzel, another well-known limnologist, wrote two papers that Hutchinson starred and discussed in volume 3. All the papers referred to above are starred except the Hutchinson reference; he did not star any of his own papers.

Volume 4 of the *Treatise, The Zoobenthos,* was published in 1993, after Hutchinson's death. When he died in 1991 he had finished all the major writing, but the illustrations, figure captions, the index, and other aspects of the book were left unfinished. In 1988 he had asked Yvette Edmondson to help with the completion of the book. This was a heroic task. Many people helped, even this author, who was a visiting scholar at Yale during 1990–91. In her foreword, Yvette Edmondson especially thanked Anna Aschenbach, Hutchinson's able and devoted editorial assistant since 1967. Hutchinson's former students W. T. Edmondson, Alan J. Kohn, and John Edwards gave professional help on particular groups. Saran Twombly, a professor at the University of Rhode Island, whom Hutchinson always called his post-ultimate student (since she did her Ph.D. at Yale after his official retirement), did many drawings, as did Marion Kohn. Others, among them Jon Moore, Victoria Mason, and Robert Wetzel, helped. Wetzel told this author that he had read all of Hutchinson's papers. Several Yale graduate students also participated—all of them organized by Anna Aschenbach and Yvette Edmondson.[46]

Hutchinson himself explained that he knew he would be unable to complete his projected volumes 4 and 5 as planned and wrote in his preface to volume 4: "For several years my late wife [Margaret] was dying of Alzheimer's disease. It proved practical to write on what I really knew best . . . Hemiptera."[47] After he had started working on these and other insects he realized that a single author could not do a book on the animal benthos (bottom-dwelling animals) as comprehensive as the limnological botany of volume 3. The zoobenthos or animals of the lake bottom include both attached and motile organisms belonging to many different groups. He was unable to do them all in the time he knew he was allotted and therefore chose to do some well and omit others.

He therefore recommended several other books covering the benthos, writing: "By far the most significant is the noble volume by Ramon Margalef, *Limnologia* (1983), which being written in very clear Spanish has the great advantage of enabling the foreign reader to acquire a reading knowledge of Spanish as he completes his limnological education."[48] Hutchinson was still, sixty years after his arrival at Yale, encouraging his students to learn as many languages as possible. He recommended another limnology book, this one in Italian, by Emilia Stella, "the most recent product of a feminist tradition [in Italian limnology] exemplified by Rina Monti and Livia Pirocchi Tonolli," the latter his longtime friend. He dedicated volume 4 to a number of his students and colleagues who had died, including Ed Deevey, Gordon Riley, Grace Pickford, and Vittorio and Livia Tonolli.

The benthos is home, in addition to animals, to aquatic plants attached to the lake bottom; these were covered in volume 3. The zoobenthos is extremely species-rich, in large part the result of the many different niches it encompasses. Hutchinson reminded the reader of his n-dimensional niche construction, applied here to the animals of the zoobenthos: "If we regard the niche of any organism as an abstract n-dimensional space in which each axis can have a low, medium or high value, a minimum number of three axes would give us a 3^3 or 27 niches, while six axes representing, for example the ranges of temperature, light intensity, some aspect of food, pH, electrolyte content and some physical property of the substratum, would permit the development of 3^6 or 729 inhabited niches."[49]

Hutchinson had earlier pointed out that in addition to mud and other benthic substrates there were also complete assemblages of organisms that lived in sand (the psammon, or sand fauna) and another rich assemblage of species that live in *Sphagnum* mosses. In volume 4 of the *Treatise* he discusses many sessile (attached) or subsessile benthos animals, including a long section on gastropod molluscs. Much of the book is devoted to insects, including those that are aquatic only in their juvenile stages, such as dragonflies, and those that have aquatic adults, such as Hemiptera-Heteroptera (water bugs) and Coleoptera (water beetles). A good deal of this latter chapter is devoted to the ecology of these aquatic insects. As with the first volumes there are many indices—to authors, lakes, and rivers (mentioned in the text) and to species of organisms. There are as well, eighty-one pages of references, including thirty-three references to works by Hutchinson, from his very first paper of 1918 to a 1984 paper.[50] At the end of his preface Hutchinson listed a great many people who had helped him "each in a special way." This includes a paragraph about the influence of his father, "Arthur Hutchinson, OBE, FRS" who was characterized by one of Arthur's students as the "best teacher of mineralogy in Britain in the early half of the twentieth century." Hutchinson clearly wanted to be remembered in a similar way. He wrote here: "At the end of my career of over seventy years of publication, I particularly want to emphasize that, in any effective university, you find mature scientists playing an enormous role as mentors of younger ones, which provides a major or even the most important aspects of their educations."[51]

Hutchinson was only twenty-five and hardly a mature scientist when he began teaching at Yale, but from an early date he looked at his role as mentor, especially of graduate students, as one of his most important. Many others, including this author, who were not his students at Yale benefited greatly from his writings. As Robert Wetzel wrote in his text, *Limnology*, "Some of the most conspicuous influences are to be found in the writings of others, as, for example, all of us have been influenced to some degree by the perceptiveness of G. E. Hutchinson."[52] Wetzel referred to eighteen of Hutchinson's "writings" in his text.

Hutchinson continued to do research in limnology for many years, research often with colleagues and involving lakes in many countries. Apart from Italy, Hutchinson did not himself work in these countries. Some of this foreign research involved Americans, former students, such as Ed Deevey, and research associates, especially Ursula Cowgill. Among the lakes studied were Lake Patzcuaro in Mexico, Lake Petenzil in Guatemala, Lake Zeribar in Iran, and later Lake Huleh in Israel.

Limnological Work in Italy

Most of these studies were done from lake cores. In 1964 Cowgill and Hutchinson published a paper titled "Cultural Eutrophication in Lago di Monterosi during Roman Antiquity."[53] There were more studies of this lake in Latium, Italy, in succeeding years. Hutchinson actually did go to Italy and was a visitor at Vittorio Tonolli's laboratory, directed after his death in 1967 by his wife, Livia. The story of the Tonollis, well-known limnologists and Hutchinson's close friends, is fascinating.[54]

Vittorio Tonolli (1913–67) was born in Milan, and his original training was in medicine. During World War II, however, he got involved in antifascist activities. Through an old friend who had been parachuted into northern Italy by the Allies, a radio transmitter was set up in a partly bombed-out house that Tonolli owned. The Italian fascist and German authorities found out, and he was arrested by secret agents of the German Luftwaffe and was to be executed at the local cemetery. He had written a farewell letter to his mother, but somehow escaped execution and was hidden by friends, including Dr. Livia Pirocci at the Instituto Italiano de Idrobiologia (hydrobiology) in Pallanza for nine months.

By the time the war ended Tonolli, in hiding in the laboratory, had learned sufficient limnology to be appointed to a position there. He wrote fifty papers on many aspects of limnology. He married Livia Pirocci and together they developed the Instituto Italiano di Idrobiologia at Pallanza into an international center. As Hutchinson wrote, "Hydrobiologists from all over the world have worked in great happiness and have had the opportunity to appreciate the finest gifts of the Italian and human character."

Hutchinson was one of those hydrobiologists and carried out limnological work while visiting there. He and Livia Tonolli, who became director of the laboratory after Vittorio's early death, remained close friends. Many of Hutchinson's papers, largely with Ursula Cowgill, concern the Lago di Monterosi in Latium, Italy. Hutchinson put to good use the Italian he learned in his early days at Naples; he both gave a talk in Italian at a meeting of the international limnology society (SIL) and wrote a paper in Italian about his niche theory.[55]

Hutchinson continued to collaborate with Ursula Cowgill and to publish limnological papers. There were a whole series of papers on Lake Huleh; one by Hutchinson and Cowgill was published in 1972. Sharon Ohlhorst and Hutchinson wrote others in 1977 and, with collaborators,

22. Hutchinson at Wicken Fen, 1991. Photograph by David Furth, with permission.

in 1982. The latter part of Hutchinson's career from the late 1950s onward, however, was primarily devoted to population and theoretical
ecology, for which he is now best known (see chapters 14, 17, and 18).
He continued to use aquatic insects as examples, notably in his famous
paper "Homage to Santa Rosalia" (1959). He continued to think about
the population biology and community ecology of lake organisms, particularly the smallest ones. Another of his best-known theoretical papers
is entitled "The Paradox of the Plankton" (1962).[56] The aquatic habitat
and its biota continued to intrigue him to the end of his life. Entomologist David Furth took him back to his beloved Wicken Fen in March,
1991, shortly before his death.

Hutchinson did a great deal to enlarge the field of limnology, particularly in its ecological and biogeochemical aspects, on which he conducted
nearly sixty years of research. He championed the use of modern statisti-

cal and mathematical methods in limnology. He is remembered by European and American limnologists at the international meetings of SIL and the yearly meetings of the American Society of Limnology and Oceanography (ASLO). At the latter, present-day limnologists win the Hutchinson Medal for their outstanding research.

CHAPTER NINE

Radioisotopes: Radiation Ecology and Systems Ecology

G. Evelyn Hutchinson and his graduate student and research assistant Vaughan Bowen are usually credited with doing the first field experiments using radioisotopes as tracers. In doing so they invented a new major field of ecology, later referred to as radiation ecology. This chapter explores how this technique was first used in the United States and the later development of isotope research in ecology. The cultural context of this story is the interest in and fears about atomic energy after World War II, on one hand, and the opportunities that radioisotopes provided for a new type of "big science," on the other.

Hutchinson's Early Isotope Experiments

When radioisotopes first became available for scientific research just after World War II, Hutchinson was able to obtain the radioactive isotope P^{32} incorporated into phosphoric acid. He and Vaughan Bowen converted it into sodium phosphate and released this into Hutchinson's longtime study area, Linsley Pond. In this first experiment they released twenty-four portions along two transect lines in order to disperse it uniformly in the surface waters.

A week later they returned and collected the water in bottles from four depth layers of the lake. They evaporated the water, precipitated the phosphate, and with the precipitates on filter paper, measured the radio-activity. There was only one Geiger counter available, and they had some difficulties getting it to work. Nevertheless, they decided their data were statistically significant and accounted for 74 percent of the P^{32} intro-duced. Much of the rest of it they found had been taken up by aquatic plants in the littoral zone—the shallow water or shoreline zone of the lake.[1]

The following summer Hutchinson and Bowen did a similar experi-ment. By that time P^{32} was available in larger quantities from the Oak Ridge National Laboratory. They also had more sensitive counting equip-ment. They ran the experiment for four weeks to get better measures of uptake rates, collecting data each week. They concluded from the first experiments that the phosphates had moved both down and up again between layers of the lake. In the second experiment they found that the phytoplankton, the photosynthetic microscopic organisms, took up the radiophosphorus first. This happened in the epilimnion, which is the lighted upper portion of the lake. It was later taken up by zooplankton that fed on the phytoplankton and then became sediments of dead phyto-plankton and the feces of the zooplankton. These sediments moved the phosphorus into the lower layers of the lake, the hypolimnion. Hutchin-son and Bowen also concluded that the phosphorus is rather rapidly re-generated from these lake sediments into the free water above, with a re-placement of phosphorus in the upper layers every three weeks.[2]

The first paper was published in the *Proceedings of the National Academy of Sciences* in 1947 and is titled "A Direct Demonstration of the Phosphorus Cycle in a Small Lake." Hutchinson was not a member of the National Academy until 1950; the paper was probably submitted via Ross Harrison, Yale's famous experimental embryologist who was a member. The second paper, "A Quantitative Radiochemical Study of the Phospho-rus Cycle in Linsley Pond," was published in the journal *Ecology* in 1950. It is a better paper scientifically, but it is the earlier 1947 paper that is widely cited as the first use of radioisotopes in ecology.

Two immediate questions arise. First, what were the antecedents for this research? Why did Hutchinson want to do these experiments? And second, what became of this research in terms of Hutchinson or his graduate students? A brief answer to this second question is that no more radioisotope experiments were done in Linsley Pond by Hutchinson,

Bowen, or other graduate students. Neither Hutchinson nor Bowen is still alive, but there is correspondence between them starting in 1942 and continuing over the next ten years. Bowen's Ph.D. work involved the biogeochemical cycling of manganese, work he started before World War II and finished when he came back to Yale after the war. It was his research assistantship that most probably explains why he was the one in the boat with Hutchinson during these Linsley Pond experiments. Bowen later had a joint appointment at Brookhaven National Laboratory and at Yale and became an expert on the use of radioisotopes in biology, about which he taught a course at Yale. It did not, however, involve ecological field experiments.

Why did Hutchinson do these experiments? Field experiments were not exactly the rage during this period in limnology, the study of lakes. Ecology was still largely a descriptive, though often quantitative, science. Hutchinson, however, had been brought up as a student at Cambridge University both in the natural history tradition and in the experimental tradition in zoology. In terms of the natural history tradition, his original interest and area of great expertise was in aquatic insects; this knowledge he later used in some of his most important theoretical work in population ecology. Many of his students followed in this tradition, became experts on some particular group of organisms, as we have seen with Edmondson, and did innovative research with them.

Hutchinson had carried out experimental work both in England and at Yale, and also on the North India Expedition. These researches attempted to answer particular questions. For example, way back in 1929 he had asked, "Do high concentrations of magnesium salts keep the so-called water fleas, microscopic crustaceans (*Daphnia* species), out of Lake Tanganyika?" He tried various concentrations and eventually had them swimming about happily in even higher concentrations of magnesium than in the lake itself, thus disproving his hypothesis. These were, of course, laboratory experiments, but the experimental approach was a natural one for him.[3]

In the earlier of the two radiophosphorus papers he explained the reason for the Linsley Pond experiments. He and others had observed both in the United States and in England that there were several pulses of plankton per summer, although it would seem that the phosphorus should have been depleted by the time the later pulses occurred. Hutchinson's earlier studies in Linsley Pond led to his hypothesis that phosphorus was continually liberated from the lake sediments and moved into the upper illuminated layers, and then essentially recycled by the phyto- and zoo-plankton. He published a theoretical paper suggesting this hypothesis in 1941. In the 1947 paper he wrote: "The possibility of obtaining relatively large amounts of the radioactive isotope of phosphorus, P^{32}, permits the hypothesis to be tested."

Interestingly, similar experiments were being conducted at Dalhousie University in Halifax, Nova Scotia, at almost the same time, the late 1940s. These experiments were not done by two men in a small boat, but by a team of nine or ten, with much better equipment, under the direction of F. Ronald Hayes. Hayes had recently switched from experiments in chemical embryology to lake experiments, that is, limnology. Daniel Livingstone, later an important paleolimnologist, a student of African lakes, and a professor at Duke University, was at that time an undergraduate student at Dalhousie. He was Hayes's summer research assistant and took part in these experiments together with Hayes's graduate students. Dan Livingstone said in a recent interview that there were at least ten people, including two professors, graduate students, and technicians, involved. In his words: "For the first few hectic days you run these things in an exponential time sampling series. Initially you are really going hammer and tongs twenty-four hours a day with every hand you can get." They put the P^{32} vials into the lower level of the lake, the hypolimnion, and cracked the vials open there with a dynamite cap.[4]

It was all clearly more high tech than Hutchinson's and Bowen's experiments. Livingstone said that Hayes knew *how* to do these experiments but didn't know why. Hutchinson knew *why* but didn't really know how; his first experiment was relatively crude.

Hayes's group did not intend to study the cycling of radiophosphorus through the ecosystem. They simply were doing a fertilization experiment. They found that fertilizing the lake with the amount of phosphorus they were using had no longtime effect. You have to "add phosphorus continuously for thirty years to change the nutrient economy of the lake," Livingstone said. By adding the radioactive label they did find out where the phosphorus went—into the phytoplankton and zooplankton and fish—but they were not thinking of that beforehand. "It came as a surprise to find how little of the phosphorus was actually in solution in the lake after fertilization." Hayes was not a theorist. Nevertheless, with a good protocol, good equipment, and so many hands, they got interesting if unexpected results using P^{32} as a radioactive tracer. These experiments were done completely independently of those by Hutchinson. Before Hayes published his results, however, he sought advice from Hutchinson. Hayes's paper was not published until after the two Hutchinson-Bowen papers.

It seems worthwhile to try to track the ideas for isotope tracer experiments before 1946—before these two sets of experiments in Connecticut and Nova Scotia. When did Hutchinson or any of his students first think about them? I recently found some materials in the Yale archives bearing on these questions. In 1944, Edward Deevey, the early Hutchinson graduate student who had gone on to Rice University and then to Woods Hole during World War II, was recruited by Arthur Hasler for a possible position at the University of Wisconsin. He was asked by Hasler to propose a research program for limnology or aquatic ecology. In it Deevey wrote the following about food cycle dynamics: "This is the central problem of all ecological research. . . . It should be observed that I am building almost entirely on the structure laid out by the late Raymond Lindeman in his paper 'The trophic-dynamic aspect of ecology,' which seems to me a prospectus of ecology for years to come."[5]

Lindeman's 1942 paper, as discussed previously, has been widely cited as the beginning of ecosystem ecology. Lindeman had done his graduate work in Minnesota, but when he wrote the paper he was Hutchinson's postdoctoral student. Bob Cook wrote in 1977 in *Science* about Hutchinson's influence on getting Lindeman's paper published in *Ecology* in spite of the negative reviews of well-known limnologists to whom hypotheses of almost any kind were anathema (see chapter 8).[6] Joel Hagen and I have also demonstrated Hutchinson's influence, which was acknowledged by Lindeman in the paper.[7]

Deevey's proposal to Hasler continued with several suggestions for following up Lindeman's work. One suggestion was "the possibility of using tracers, such as radioactive phosphorus and heavy nitrogen in exploring the metabolism of the plankton community." Before sending his proposal to Hasler, Deevey sent it to Hutchinson to review and got an immediate answer. Hutchinson sent several "major comments." One reads: "The radiophosphorus method of studying lake metabolism will probably turn out to be by far the best. It should be emphasized to the extent of enquiring if they have a cyclotron with a willing physicist attached who could make about 100–2000 milliliters of radiophosphorus in a single operation. We just could not get results with 20 ml in Linsley. Heavy N would, I think, be impractical unless a new method of making it in vast amounts were available."[8]

Clearly, Hutchinson had been thinking about this for some time. As it turned out, Ed Deevey did not go to Wisconsin or do any of this work. He instead returned to Yale, as did many Hutchinson graduate students, forming what I have called elsewhere the "Hutchinson research group" of professors and graduate and postdoctoral students. Deevey went back to the field in which he was an important innovator, paleoecology. He also became involved with the use of isotopes, namely carbon 14, in radiocarbon dating. Carbon 14 has also been used in ecology in many ways, particularly to measure productivity, but here we are talking about radiocarbon dating, which was developed in large part by Willard Libby after World War II. R. F. Flint, a geologist at Yale, and Evelyn Hutchinson were prime movers in obtaining $42,000 from the Rockefeller Foundation for the establishment and support of a geochronometric laboratory at Yale for carbon 14 dating, of which Deevey became the director; Hutchinson was on the board of directors.[9]

What was the meaning of Hutchinson's comment to Deevey about not getting results in Linsley Pond? Hutchinson, I discovered, made an earlier attempt to use P^{32}. One of the participants in that early experiment, the late W. T. Edmondson, described it in an interview with the author. In 1941 it was Tommy Edmondson who was the graduate student in the boat with Hutchinson, though it was not *his* thesis topic either. As he described their experiment, they got some P^{32} from E. C. Pollard, a physicist at Yale, who made it in his cyclotron. The cyclotron, however, broke down, and they only got half as much as they needed. But they went out anyway in a small rowboat and "spread it out in dollops." They later took samples at different depths and found that it was distributed in the water column, but there wasn't enough to be significant.[10]

The experiment was not tried again until 1946. Probably the cyclotron was needed for more important things during the war—but radioisotope tracer experiments clearly had their beginnings before the United States was involved in World War II.

Two New Fields: Radioecology and
Ecosystem (Systems) Ecology

As Robert McIntosh wrote in his book *The Background of Ecology*, "The atomic age had brought with it both the use of radioactive materials in studies of natural environments [and here he quoted the first Hutchinson and Bowen experiment] and the threat of adverse effects of radiation on individual organisms and ecological systems."[11] Thus Hutchinson's radioactive isotope experiments led directly to radiation ecology. In government research radiation ecology was originally called "health physics"; at least in some quarters there was concern about the adverse effects of radiation on the human organism. There was, at any rate, considerable interest by both the United States government and by the United Nations in the biological effects of radiation and the applications of radioactive tracers in biology and medicine. In 1948 Hutchinson received a letter from H. H. Goldsmith, a United Nations senior consultant who was compiling an international bibliography on the scientific aspects of atomic energy. Leading scientists throughout the world were being asked for additions to the bibliography. Most of the work had nothing to do with ecology. Hutchinson sent three references: his own radiophosphorus work with Vaughan Bowen, Bernice Wheeler's work on iodine metabolism of *Drosophila* using I^{131} published in 1947, and a third paper on the use of isotopic iodine in biogeochemistry.[12]

Also in 1948, the director of the Brookhaven National Laboratory, Philip Morse, sent to Yale a tentative program for nuclear research. The Brookhaven Laboratory was created by the Atomic Energy Commission (AEC) to carry out such research. Its Biology Department was to study the "mechanisms by which radiation produces its effects on living systems" and also "basic biological problems in which tracer techniques can be profitably employed." The latter, however, were to be largely physiological or biochemical, for example, the biosynthesis of tagged materials such as plant sugars. Hutchinson was a consultant to the Brookhaven Laboratory, and Bowen and other Yale graduate students did some of their research there.[13]

It was only later that "health physics," primarily interested in finding out how radioactive substances affected human and other organisms, was turned into "radiation ecology" by Stanley Auerbach and others at the Atomic Energy Commission. This has been well documented by Stephen

Bocking and others—and by Auerbach himself in a 1965 article entitled "Radionuclide Cycling: Current Status and Future Needs."[14]

Stanley Auerbach was in the Radiation Ecology Section of what was still the Health Physics Division at Oak Ridge. To quote Auerbach's 1965 article, "In the early 1950s the conceptual framework of modern ecosystem ecology was established." He too referred to Lindeman's paper: "In the middle 1950s, ecological research was undertaken in a significant and cohesive fashion within the Atomic Energy Commission." Radionucleotide contamination was being studied in an ecological context, the purpose of this research being to "provide radiation protection to man." In terms of future needs, Auerbach was recommending extending experiments using tracers to "measure the complete cycle in an ecosystem." In particular, Auerbach wrote, there was a need for more experiments on simple Arctic ecosystems because of concern about the significance of Sr^{90} and Cs^{137} in these ecosystems. Thus in the mid-1960s radioactive tracer experiments were going strong but in a different context from Hutchinson's original experiments and with much government funding by the Atomic Energy Commission.

Others have written about radiation ecology and the rise of ecology. Eugene Odum, well-known ecologist and author of the most popular ecology textbook of the era, wrote in 1964 that "the exploitation of atomic energy was responsible for ecology's rise to a front-line position." Peter Taylor, historian of biology, also wrote insightfully about the transformation of American ecology.[15]

The AEC certainly affected the direction of ecology after World War II. The research it funded was a result of the fallout from nuclear bomb tests and of concern about the radiosensitivity of different organisms. Isotope tracer experiments were done in both aquatic and terrestrial sites starting in the mid-fifties. Arthur Hasler, Gene Likens, and others did aquatic isotope studies using several different isotopes in lakes, but most of this tracer research was under AEC auspices. By the mid-sixties really elegant tracer experiments involved large teams and whole ecosystems.

John Wolfe, head of the Division of Biology and Medicine at the AEC from 1955, recalled at the National Symposium on Radioecology in 1967 (the first such was in 1961): "It was more than a decade ago that ecology came of interest to the AEC. In 1955 Gene Odum was doing work at the Savannah River Plant, Lauren Donaldson was investigating radiation effects on fish in the Columbia River, and a small group at the Nevada

Test Site was tracking fallout from weapons tests." All of this work was directly related to nuclear fallout.[16]

Tracking of Radioisotopes through Ecosystems

Later the AEC funded similar research to solve basic ecological problems. Beatty noted that the tracking of radioisotopes through food chains could "be seen on a much broader scale of studies Hutchinson had begun, taken up by Howard (H. T.) Odum," who had been Hutchinson's graduate student. His Ph.D. was on strontium cycling but, ironically, before strontium 90 was a concern in relation to nuclear weapon tests. H. T. Odum later worked on large-scale projects under AEC contracts, for example, irradiating a whole rain forest ecosystem in Puerto Rico.

Frank Golley wrote, "The study of complex problems . . . does not move forward entirely through the activities of creative individuals like Howard Odum."[17] H. T. Odum's older brother, Eugene (Gene), did much of the earlier AEC tracer research. In 1958 a Gene Odum team studied uptake of P^{32} and primary productivity in marine algae, and two years later another Odum team studied population density of an underground ant species by tagging with P^{32}.

In the mid-1960s Stanley Auerbach put together a team that included Jerry Olson, one of the originators of systems ecology, as well as George Van Dyne and Bernard Patten. Olson published a radioactive tracer project done with cesium isotopes for an entire tulip tree (*Liriodendron tulipifera*) forest. It was published in *Health Physics*. Tulip trees were tagged with Cs^{137}. It was found that transfer to the soil was greater by way of tree-root turnover and by way of rain leaching and leaf fall from the tree canopy. A relatively small fraction of the radioactive cesium was recycled through other plants and animals. Some transfer coefficients could be determined directly from measuring the rates of these processes, but others had to be determined by computer simulation. "Ecological complexity," Jerry Olson concluded, "does demand more careful analysis" and thus more experiments.[18]

Olson wrote articles about the advances in radiation ecology in the 1964 and 1966 volumes of the journal of *Nuclear Safety*. In the 1966 paper he referred to much prior information in this field, brought together at the first National Symposium on Radioecology in 1961, and to plans for a second national symposium in 1967. He noted the trend of using analog or digital computer simulations to model ecosystem structure as well as the actual experimentally introduced tracers—radiophosphorus, carbon, and cesium in terrestrial, freshwater, and marine ecosystems, including the Columbia River and even the Pacific Ocean. He quoted dozens of such experiments and computer simulations.[19]

Concern arose about the safety of radioisotopes in food chains. In another paper in the same volume of *Nuclear Safety* (volume 8), Martin Witkamp worried about the "biological concentrations of isotopes of cesium and strontium in Arctic food chains." The fallout was found to move from lichens to caribou and domesticated reindeer and from there to Eskimos and Lapps. Studies were carried out both in Alaska and in Lapland. These people had high body concentrations of these isotopes related not to higher fallout levels but to the nature of ecological food chains in the Arctic.[20]

Lichens in particular accumulate radionucleotides. Eville Gorham had reported as early as 1959 that lichens even in England contained three times more radionucleotides per gram than higher plants and predicted a high intake by animals feeding on them. This was later found to be the case for both mosses and lichens in both Alaska and northern Finland. More experiments were recommended for the Arctic, too—particularly by Auerbach, despite the possible dangers of adding yet *more* radioactive isotopes to the environment, aquatic or terrestrial.

Why Did Hutchinson Leave These Fields?

Both radiation ecology and ecosystems ecology became major and very well-funded fields. Hutchinson, together with his students, was instrumental in founding both fields in the United States. But Hutchinson did not take part in any of these large AEC programs, nor did any of his later graduate students. Why not?

One answer is that his interests had turned to other areas of ecology by the mid-fifties, particularly to population ecology, where he, together with his graduate students, particularly Robert MacArthur, made important theoretical advances. Another possible answer is that he viewed his major role as innovator. Hutchinson, together with a new crop of graduate students, thought up ideas and hypotheses, as with his n-dimensional niche theory (see chapter 14). Many others came along to test these hypotheses, sometimes with more elegant experiments than his. He was extremely busy in the 1950s and 1960s writing books and papers on a great variety of subjects, mostly but not entirely within ecology, and with training another decade of graduate students, more than four decades of them in all. Dan Livingstone, Yale graduate student and Duke University professor, and a correspondent of both Hutchinson's and myself, concurred with these ideas.

But I believe there were also other reasons. To understand these we have to look at the political and social context of the later isotope research. Frank Golley wrote in his *History of the Ecosystem Concept in Ecology*: "Many American ecologists were unconcerned that their studies were closely linked to military activities. . . . They welcomed the opportunities military research made available."[21] Golley worked on this type of research himself.

Did Hutchinson disapprove? Hutchinson was not apolitical. I have documented that he turned down the President's National Medal of Science when it was offered to him because he disapproved of the Nixon administration during the Watergate era. But that was a later age. In the fifties Hutchinson actually sat on the AEC board that awarded postdoctoral fellowships to students. Hutchinson was, however, uneasy about government-funded big science. Big science done by large groups with major funding and all the facilities of the AEC laboratories had become normative science by the late 1950s. But as an individual creator of ideas, as Tomas Sodestrom once said, Hutchinson did not have to do this kind of mainstream science. He did not think such big projects were suitable for Yale, nor did they accord with his views on training graduate students.

As I have argued elsewhere, Hutchinson did head a research group or perhaps a research "school," though an atypical one, at Yale.[22] He never advocated only one program or one methodology. He did occasionally suggest Ph.D. research topics, but even then the student had to work out his or her own hypothesis and methodology. Hutchinson, Lindeman, Deevey, and Bowen were all interested in theoretical questions in relation to ecosystem function, questions that led to innovative science.

Did the later work in relation to weapons fallout also lead to innovative science? Certainly much good science was done and much was ultimately learned about nutrient and other types of cycling in a variety of ecosystems. Systems ecology may have eventually become bogged down in too much modeling—too much computer simulation with too little real data, something Olson seemed to be warning other AEC ecologists about in the mid-sixties.

When did concern about using radioisotopes in natural systems arise? Was it as a result of Rachel Carson's 1962 *Silent Spring,* along with concern about pesticides? Neither Hutchinson nor Hayes seemed to have any concern about it in the forties or early fifties. It is clear that attitudes and research have both changed. I attended the American Society of Limnology and Oceanography (ASLO) 2000 conference, where I organized a historical symposium on limnology and oceanography in which I participated with Keith Benson, Ron Rainger, and others. A look at the program for that meeting and of more recent ASLO meetings, especially at the research of younger members, indicates that ecosystem ecology using isotope tracer experiments is still alive and well but that radioactive isotopes are no longer added to our aquatic and other ecosystems for research purposes; today, because of concerns about environmental damage, stable rather than radioactive isotopes are widely used in this valuable type of ecological research.

Hutchinson never lost his interest in the function of aquatic ecosystems and in their preservation. But he moved on in research and writing to population biology (chapter 14) and in addition pursued his environmental concerns (chapter 15).

CHAPTER TEN

Biogeochemistry:
From Linsley Pond
to the Guano Islands

The image of Hutchinson as a young high school boy in the laboratory perfecting his chemical titrations before allowing himself to go out on a Sunday collecting trip relates to much of his work as a scientist. Among future ecologists in both Britain and the United States, he was exceptionally well trained at an early age in physics, chemistry, and geology. From his earliest work he saw applications of these fields to the lives of the organisms that he studied. There is a great deal of water chemistry in his early studies of the fauna of South African pans. When he came to Yale and started his studies of Linsley Pond he became interested in the chemistry of the lake sediments as well as of the living organisms. He extended this interest to the chemical elements found in marine animals and to the cycling of elements in the biosphere.

In the 1930s Hutchinson discovered the writings of V. I. Vernadsky, the Russian scientist who is usually credited with inventing the words "biogeochemistry" and "biosphere."[1] Hutchinson later wrote:

> Though I never visited the USSR, I have studied the works of some of your investigators so intensely that I feel as if I had been one of their students. I started my career as an animal ecologist and as a student of limnology. I soon became interested in the enormously important investigations of V. I. Vernadsky. By using some of the same techniques as are employed in mineralogy [Vernadsky's original field as well as that of Hutchinson's father] to which he applied great imagination and skill,

23. Vernadsky and Goldschmidt, founders of biogeochemistry,
photograph given to G. Evelyn Hutchinson by Vernadsky.
Manuscripts and Archives, Yale University Library.

Vernadsky introduced a new way of looking at biological science which
he called biogeochemistry in which organisms are described in the same
sort of way as a mineral might be considered. Ordinarily this approach
had been discouraged, but in Vernadsky's hands it began to show aspects
of organisms not otherwise appreciated. The chemical composition of
species of some genera for instance might be found to have new characteristics when looked at in this way.[2]

Vernadsky's pupil, A. P. Vinogradov, had examined a great many marine
organisms, often finding similarities in the elementary chemical composition of members of one genus. Hutchinson was instrumental in having
the work of both Vernadsky and Vinogradov translated into English and
published.

Several letters attest to Hutchinson's early interest in biogeochemistry. Shortly after his return from the Yale North India Expedition in
1933 he wrote a letter to Cecilia Payne, later Payne-Gaposchkin, a Harvard astronomer who had been a student with him at Cambridge University. He had looked over the spectrographic analysis of the material from

his lake residues from the high-elevation Himalayan lakes and wrote to her about the deficiencies of current methods of spectrographic analysis. He wondered if any of Payne's spectrographic friends had any new techniques. He wrote that he was anxious to know what range of elements might occur in these residues. He thought the most likely elements were lithium, rubidium, cesium, strontium, manganese, and copper. In the case of copper, he needed to know the exact quantities involved. Copper is of major importance in the life of many crustaceans since this group of animals uses copper-containing hemocyanin instead of hemoglobin in its blood. Many crustaceans, such as shrimp, occurred in the North India lakes that he had studied. He also sent to Payne a copy of the first part of Vernadsky's article on biogeochemistry, which he thought might be of interest to her.[3]

The same month he wrote to Dr. Hugh Ramage in Norwich, England, about the same samples. He had obtained them by evaporating lake water in glass beakers "in the cleanest conditions possible in a camp, in order that a spectrographic analysis of these residues could be made." He wrote that "no one has ever attempted this method of studying the chemistry of lakes" and thought it important since many freshwater animals require copper and perhaps other rare elements. It was possible, he thought, that the existence of very small amounts of such essential elements might be a limiting factor in determining the species present in these lakes. Hutchinson hoped that Ramage might undertake this analysis as it was impossible with the apparatus he had available at Yale to attempt it himself.[4] Ramage agreed to do this, requiring only 0.033 grams of each of Hutchinson's six 0.1-gram samples. Ramage also suggested an analytical chemist in London who could do further analysis with a more sensitive arc and spark method. Hutchinson's idea was indeed a new undertaking in ecological lake research.

By 1935 Hutchinson was requiring the students in his graduate freshwater ecology course to read Vernadsky's *Biosphere*, which had been published in French. (Hutchinson's graduate students were expected to read foreign languages.) He had earlier written to Yale's Sterling Library purchasing department asking that they acquire this very important work for the library.

In the 1920s and early 1930s the majority of Hutchinson's publications related to insects. By 1930, however, he was looking at chemical-biological interactions. He published papers on the chemical ecology of

Lake Tanganyika, and he also carried out an early experimental laboratory study on the magnesium tolerance of water fleas (*Daphnia*).[5]

Hutchinson's coauthored papers from the 1930s and thereafter indicate his interest in the biogeochemistry of lakes and his collaboration with colleagues, graduate students, and research associates. He and Ann Wollack published chemical analyses they had done on a core from the sediment of nearby Linsley Pond. He had already started working on intermediary metabolism in Connecticut lakes and on nutrient cycling. In short, he had carried out research in essentially all the areas of biogeochemistry discussed in Eville Gorham's 1991 review of the subject, except for the origin of life.[6]

Hutchinson's work was apparently well enough known by 1940 that he was asked to review, in the journal *Ecology*, the recent book, *Bio-Ecology*, by Frederic Clements and Victor Shelford, the reigning plant and animal ecologists, respectively. Robert McIntosh tersely summarized Clements's and Shelford's efforts to unify plant and animal ecology, two quite separately developed fields, in *Bio-Ecology*: "This effort failed to achieve notable success, for reasons detailed by Hutchinson." Hutchinson's review is highly critical, particularly of the authors' omission of "the language of mathematics" in dealing with competition and population biology, but also of their failure to discuss the metabolic aspects of ecology and particularly the "neglect of the biogeochemical approach." The authors had insisted that "the community and the environment must be separated and should not be considered as forming part of the same ecological unit."[7]

In Hutchinson's view a new type of ecology incorporating biogeochemistry had already arrived. In his review he cited not his own work, but that of Schroder and Noddack, in which the photosynthetic efficiencies of major vegetation types permitted an estimate of the earth's carbon metabolism. The first general ecology textbook incorporating these ideas, Eugene Odum's *Fundamentals of Ecology*, did not appear until 1953. When it did appear it incorporated ideas and research by Hutchinson and by his students, Raymond Lindeman and H. T. Odum, especially the ecosystem approach to ecology.[8]

The conceptual advances in terms of the global significance of biogeochemical cycles had a long history, according to Gorham, who wrote, in the introduction to his review of biogeochemistry, "The impetus for this article . . . owes much to the inspiration of G. E. Hutchinson and his academic descendants." Among antecedents, Gorham credited Halley (in the seventeenth century) for his work on the hydrologic cycle, Priestly's research (in the eighteenth century) on the interactive roles of both plants and animals in global gaseous metabolism, and especially Dumas's work (in the mid-nineteenth century) on the global significance of carbon and nitrogen biogeochemical cycles powered by the sun.[9]

By the 1940s Hutchinson had started to put much of his own energy and that of his students into the biogeochemical aspects of limnology, including nutrient cycling. Some of this work, such as Riley's on copper, has been discussed in chapter 7. Hutchinson himself studied nitrogen and phosphorus cycles and in 1943 published on the vitamin thiamin in aquatic organisms and lakes. In this paper he reported that free thiamin did exist in lake waters but that it was found mainly in suspension in the cells of the phytoplankton. The results suggested that thiamin synthesized in lakes by green plants is rather rapidly destroyed. He later, together with Jane Setlow, studied a more stable vitamin, niacin.[10]

The Biogeochemistry Project for the Museum of Natural History

In the 1940s Hutchinson was also involved in a major project in biogeochemistry for the American Museum of Natural History. It would provide a decade of support for his research and for that of his graduate students. He published an exhaustive study of the biogeochemistry of aluminum in 1943.

A 1945 letter explains how this project came about. Hutchinson wrote to a friend in Dublin, Edward Conway, that he had set out to write a comprehensive treatise on biogeochemistry, "more by accident than design." One of the first parts he completed was on the biogeochemistry of aluminum. He explained: "I was living close to large amounts of *Lycopodium* [club moss] plants, and started analysing them to check the old reports of large amounts of aluminum in them. When I had got the review done . . . I saw that what I wanted to do lay outside the powers of one man. My friend Albert Parr, Director of the American Museum of Natural History [formerly director of the Bingham Laboratory at Yale], was approached by one of his trustees for suggestions as to disposal of a large fund donated to further synthesis of scientific knowledge. He has put this at my disposal to further the scheme."[11]

Hutchinson was to direct a three-year survey of existing knowledge in the field of biogeochemistry, the interrelation of biology and geochemistry. Further research was intended to determine the effects of chemical elements on the lives of organisms. Hutchinson wrote to Conway, "Though the fundamental purpose of this fund is to increase human welfare, I can interpret this as broadly as can be desired, though the donors probably would like man to occur occasionally as one of the organisms involved." The American Museum of Natural History was to publish the resulting monographs. Hutchinson held the title of "consultant in biogeochemistry" at the museum during the duration of the project. He made several good friends there, including Margaret Mead and Ernst Mayr.[12]

Hutchinson encouraged some of his graduate students to undertake parts of the project and wrote to many scientists asking them to contribute to the research. Graduate student Vaughan T. Bowen worked on manganese and organisms that accumulate it. He also worked with Hutchinson on phosphorus, including the first published use of radioisotopes to study mineral cycling in lakes (see chapter 9). H. T. Odum did his doctoral dissertation on the biogeochemistry of strontium before this element became notorious in relation to fallout of radioactive strontium 90 (Sr^{90}) from atomic bomb tests. Hutchinson himself worked on lithium and rubidium and later titanium.

The project continued for much more than three years. Researchers outside Yale also participated in the project. W. A. Albrecht from the University of Minnesota, for example, was working on a monograph on the distribution of calcium as a factor in vertebrate nutrition.[13] Dr. Eva Low, Hutchinson's assistant, was preparing a volume on the halogens, including iodine and bromine. The resulting monographs were generally long and expensive to publish and did not come out very rapidly.

Hutchinson published a number of other biogeochemistry papers, including "The Problems of Oceanic Geochemistry," one of his few excursions into the marine environment, and a paper on aluminum in soil, plants and animals.[14] He got quite involved in the ecology and even the taxonomy of land plants for the only time in his career, investigating the biogeochemistry and systematics of club mosses (species of *Lycopodium* and related genera). His students collected these for him in the United States and in England. He found varying aluminum contents in different species that correlated with the species' morphology, such as leaf type, and with the alkaloids they contained. Much of this was published in the

1945 aluminum review, but his former students were continuing to collect *Lycopodium* specimens for him as late as 1949. By that time he was doing the last of the editing for his series; Dr. Didier Bertrand of the Pasteur Institute in Paris was completing a monograph on the biogeochemistry of vanadium.

Hutchinson's Treatise on Guano

Hutchinson himself wrote the largest volume for the biogeochemistry series, with the tongue-in-cheek title *The Biogeochemistry of Vertebrate Excretion.* In the preface of this book he thanked the funders of the survey, Robert Earl McConnell and George Monroe Moffett, and also "the imagination and foresight of Albert Eide Parr," director of the museum.[15]

The title of this 554-page tome did indeed engender many jokes, but it is in fact a book on guano. Hutchinson pointed out that the contact area between biology and geochemistry is important in terms of human efforts to cope with the environment. (For Hutchinson's environmental work, see chapter 15.) This particular volume involves not only the birds and bats that have produced the guano, but past and present human use of guano, with its high phosphorus and nitrogen content, for fertilizer.

Phosphorus, nitrogen, and other elements are not only concentrated in the bodies of living organisms and in their remains when they die, but these elements are excreted from living animals and sometimes accumulate instead of undergoing decomposition by bacteria and fungi. Accumulations of seabird excreta are called guano. This word is derived from *huanu*, the word for dung in Quechua, an indigenous language of Peru, on whose coast many of these deposits have been found. By the 1700s the Spanish colonists in Peru were already using guano to fertilize their vineyards.

Much of the first part of Hutchinson's book discusses the seabird islands and the exploitation of guano deposits along the coasts of Peru and northern Chile. Major guano deposits are formed by seabirds that feed over a large area but return to a limited site to reproduce; these sites are almost all on islands, where there is protection from mammalian predators. A dry climate is necessary so that the deposits are not washed into the sea by rain (one means of recycling the guano elements). Ocean currents, and especially upwelling, result in high quantities of nutrients in the water and are essential to guano formation. The birds feed largely on anchovies, which in turn feed on plankton; large plankton populations occur in such nutrient-rich portions of the oceans.

Hutchinson collected and interpreted data on guano and on a variety of interrelated subjects from all over the world, rather than visiting these places himself. In fact, after the North India Expedition, Hutchinson did little field research outside of the United States and Europe, although his students traveled far and wide obtaining lake cores from Guatemala, Israel, and elsewhere to be analyzed in his laboratory.

The guano volume is in a sense Darwinian in Hutchinson's use of so many different sources of information. He explained that because nearly all the important guano deposits were quarried during the nineteenth century, he had tried to collect a very scattered literature, including travel accounts, trade pamphlets, agricultural journals, political documents, and even travel books. Hutchinson wrote to a variety of people for information, including ship captains engaged in the ongoing guano trade.

Hutchinson's presentation of mathematical equations in his theoretical models is unlike Darwin, but his sources of information certainly are similar, particularly much of the correspondence. Some of the most interesting sources are letters from and whole books by travelers to the regions where guano islands are found. These include travel reports from as early as the twelfth century, as well as reports by Captain James Cook, Alexander von Humboldt, and those of Darwin himself. There were periodicals and government papers as well. *The Friend* was a monthly journal devoted to temperance, seamen, and marine and general intelligence published out of Honolulu in the 1870s. Hutchinson used reports on ships loading at guano islands. British parliamentary *Sessional Papers* from the 1850s were also searched for correspondence between Liverpool merchants and the British Foreign Office about newly discovered guano islands and about the importation of guano from Peru, Africa, and the coast of Arabia.

Many ornithologists, professional and amateur, contributed data to Hutchinson's book. Baron Lionel Rothschild, for example, wrote a history of the avifauna of the Hawaiian Islands to 1900. Many of the guano companies had ornithologists on their staffs.

More than 950 printed references are listed in the book, almost all of which Hutchinson read or at least saw; the few that he was unable to locate are noted. Of those references, more than 300 are from the nineteenth century or earlier. There are references in Spanish, French, German, Italian, and Portuguese. There are also letters to him in several of these languages. Darwin's *Geology* from the *Beagle* voyage is discussed, as well as voyages of earlier travelers who mentioned guano, for example: Captain Cook's 1776–80 voyage, Cieza de León's of 1532 to 1550, and even a twelfth-century voyage of Edrisi.[16] This research, Hutchinson noted, presented problems familiar to the historian, not only in terms of finding sources but in terms of their commercial bias and sometimes questions of their authenticity. It was certainly a departure from most of his previous work. He wrote: "The adventures of whaling captains, the political significance of the skull of the last male Tasmanian, the sidelong glances of the Peruvian *tapadas*, the black and white butterfly that lured Saint Rose of Lima into the Dominican Order, and other curiously colored tales have played their part in relieving the tedium inherent in the study of much seemingly unpromising material."[17] They relieve the tedium for the reader as well. In fact, this is a most interesting work, though it is probably the least read of Hutchinson's books.

Hutchinson thanked an incredible number of people in diverse fields for their help, from Robert Cushman Murphy for seabird data and old photographs to Sr. Carlos H. Benitez, general manager of Guanos y Fertilizantes de Mexico, S.A. One very interesting aspect of the book is archeology, for which he thanked Professor George Kubler of Yale, to whom he dedicated the book. Kubler's contribution permitted a new approach to the chronology of guano deposits as well as interpretations of many of the artifacts found in the deposits. Research assistants Ruth Jaffe and Jane Setlow, who worked on the niacin study, helped with this book, and several others, including Georgiana Deevey, prepared figures for it.

The Guano Islands of Chile and Peru

Hutchinson discovered the first account of the use of guano, along the coast of what is now Chile. It was written in 1554: "Farther on are the rich valleys of Tarapac. Out of the sea, in the neighborhood of these valleys, rise some islands much frequented by seals. The natives go to them in balsas and bring a great variety of the dung of birds from the rocks, to apply to their crops of maize, and they find it so efficacious that the land, which formerly was sterile, becomes very rich and fruitful. If they cease to use this manure they reap little maize. Indeed the people could not be supported if the birds landing on the rocks round these islands did not leave that which is afterward to become an article of trade between the natives."[18]

Hutchinson quoted another report from 1609, in Spanish. (Not only did he expect his graduate students to master foreign languages, but apparently his readers as well.) This quote, from Garcilaso de la Vega, reported that there was no other fertilizer available than that of the seabirds on islands great and small all along the coast of Peru. The deposits of guano were piled so high that from afar the mountains of guano looked like peaks of some sierra. He also wrote that the Incas exploited the guano and strictly administered the islands. Each one was assigned as a source of supply to an Inca province, or if a large island, to two or three provinces. Yet another early observer, Espinosa, wrote in 1618 that frigates loaded soil called guano from barren islands and brought it to Peruvian ports where farmers bought it to fertilize their crops and Indians carried it on their llamas for the same purpose. Its origin was not known; some said God put it there for this purpose, but others thought it was seabird excrement. (References for all these writers quoted by Hutchinson can be found in the bibliography of *The Biogeochemistry of Vertebrate Excretion*.)

The major guano birds on the Peruvian islands were a cormorant, a booby, and a pelican species. A Peruvian penguin once occurred there, too, abundant in records from the 1850s. It is no longer present on these islands because harvesting of the guano meant that the penguin could no longer burrow in it. The penguin was also eaten by fishermen and used for hats!

Alexander von Humboldt knew about the guano islands in 1806 but apparently never visited them during his travels in South America. He wrote about flamingos and herons rather than cormorants, boobies, and pelicans. He counted fifty native boats, each with about fifty metric tons of guano on board.

Modern guano collection began on these islands in 1909. The scope of this endeavor and of the amount of guano available is hard to imagine. For example, on one of the Lobos Islands 2,500,000 pelicans produced 10,000 tons of guano a year.[19] The total yield on the Guanape group was nearly half a million metric tons. G. E. Hutchinson quoted T. J. Hutchinson in his 1873 book, *Two Years in Peru*: "The rock of the islands is only 30 meters high, while the height to the top of the guano was 72 meters." On the Macabi Islands (Peru) the presence of 6,527,000 tons of guano was reported in 1863; the larger of the two islands is only 160 by 100 meters. Ninety percent of the islands' surface was covered by "guanays," the cormorant species, *Phalacrocorax bougainvillei*. By 1874, 400,000 tons of guano had already been collected and exported from these islands.

The Macabi Islands are also the site of a large number of archeological finds that lie under the guano indicating repeated visits by pre-Columbian people. The visits by indigenous people continued into the mid-nineteenth century. One report said that Indians filled their canoes with cormorants, boobies, and eggs; they collected 24,000 birds and 12,000 birds' eggs annually.

Correlations of Guano Deposits with El Niño Occurrences

One of the most innovative aspects of Hutchinson's book is his attempt to correlate El Niño occurrences with other factors, for which he collected data from many sources. This research was certainly one of the first attempts to examine these relationships. As noted above, the upwelling of the cold coastal current and the resulting high productivity in the waters off the coast of Peru where these birds fished, plus the flat islands and the arid conditions made these bird populations ideal guano producers. The deposits contained at least 16 percent phosphorus and from 2 to 12 percent nitrogen. However, Hutchinson found that there were warm currents (El Niño) in approximately seven-year cycles. These warmed the water, reducing its plankton productivity, its anchovy populations, and thus the bird populations.[20]

When Charles Darwin visited the Galapagos Islands on his *Beagle* voyage in 1835 he did not see the guanay, the cormorant that was the major guano producer. This might not have been an oversight but the result of an avian catastrophe in 1834. Elsewhere at Pabellon de Pico the whole promontory was covered with birds in the 1830s, but in about 1848 a plague attacked them and they fell into the water, dying by the millions.

There was a very wet season reported in 1891; accounts of that year include a great die-off of marine organisms and the presence of tropical animals on the temperate Peru coast, indicating warmer water than usual. In 1899 a ship commander reported passing 300 miles of dead pelicans off the northern coast of Peru. Hutchinson collected a wide variety of data on bird populations and water temperatures from 1790 to 1940 (figure 17 in his book). He also found data on the annual guano deposits from 1909 through 1943 from the Compania Administradora del Guano of Peru and graphed it with the bird catastrophe data (his figure 18).[21] The guano deposit curve simulates a bird population curve, based on the assumption that one adult bird produces about sixteen kilos of guano a year. In addition, the Compania Administradora del Guano kept food-fish data, particularly for shoals of *Eugraulis* (anchovy) reported by the so-called Guardians of the Islands.

Hutchinson concluded that the data summarized for individual years demonstrated that the bird "catastrophes" occurred when a large quantity of warm water moved from the northwest toward the coast. His graphs suggest a periodicity of seven years. He thought that additional hydrographic factors might be involved: the relation between bird populations and El Niños of 1911, 1917–18, 1925–26, and 1939–41, though probably not that of 1932, are well confirmed by the data.

Hutchinson also collected a qualitative record of the changes in bird species of the various guano islands over the previous two hundred years — and even earlier, since a pre-Inca ceramic of an island with booby and cormorant nests was found on the northern coast of Peru. Artifacts such as these have helped to solve the problem of the age of the guano. Artifacts were found through at least eighteen meters of the guano on the Chincha and Macabi islands, which thus must be within the span of human settlement of Peru. Dating was also done from the rate of deposition: a colonial artifact under six meters of guano was probably from the sixteenth century, and an object from the earlier Michica Period, representing a captive with a rope around his neck, was found at nineteen meters. These data are a result of George A. Kubler's work. A tentative date for the start of guano deposition was the first or perhaps second millennium BC.

Why was the Peruvian coast not suitable for guano accumulation more than two thousand years ago? Hutchinson thought that the most reasonable explanation was climate change: the abnormal years with warm water and violent rains may once have occurred much more frequently, so that neither bird colonization nor guano deposition could occur, that is, conditions now found north of the Lobos Islands were once found as far south as the Chinchas. This may have only been a temporary phase, but even so, previous guano deposits would have been washed away by the rains. The seven-year cycle, Hutchinson concluded, was a real phenomenon but not of great antiquity. Carbon 14 dating of guano deposits was possible by the time the book was published; Hutchinson suggested this in a footnote.

Chemical Analysis of Guano and
Its Relation to Seabirds

A whole section of Hutchinson's book is devoted to the chemistry of the guano, including the relationship of guano production to the birds' food supply. Hutchinson analyzed recent guano samples collected for him not only for nitrogen and phosphorus but also for boron, fluorine, alkaline salts, and other contents. He pointed out that although seals and sea lions might be present, they are not contributors to the guano because of their different nitrogen metabolism. Mammals, unlike birds, excrete excess nitrogen in liquid form as urea, which is mainly lost to the sea.

Birds excrete more solid uric acid which is very high in nitrogen. An analysis of the guano birds' main food fish, *Eugraulis* (anchovy), gave a ratio of nitrogen to phosphorus of about 10 to 1; a study of fresh pelican droppings gave over 11 to 1. Luis Gammara Dulanto actually did an experiment with captive cormorants in 1941 and showed that sixteen tons of fish produce one ton of guano. He concluded that as far as the guano birds–anchovy–plankton ecosystem was concerned, no commercial fishery could set up an artificial fertilizer industry based on anchovies without competing with the birds or depleting the fish population.

Records of guano production from the rest of the world take up most of the remainder of the book. These areas include the coast of Baja California; the Pacific islands; islands of the southwest African coast and those of the Indian Ocean; the Atlantic, and the Caribbean. Climatology, oceanography, and ornithology are also included for each region.

Of special interest is Hutchinson's discussion of the Galapagos Islands, visited by Darwin on the voyage of the *Beagle*. The Galapagos are five hundred miles off the coast of Ecuador in the Pacific Ocean, and the climate is semi-arid, except for the El Niño years. In addition to the giant tortoises, land and marine iguanas, and one penguin species (on the equator), nineteen species of seabirds are found there. One of these is the guano-forming masked booby (*Sula dactylatra*), which still nests there. Rumors from the 1850s of immense guano deposits in the Galapagos led to negotiations between Ecuador and New Orleans, arousing suspicions in England, France, and Spain and much excitement. But in spite of climatological and biological conditions that would give an expectation of guano islets, there apparently were never any significant deposits. Eventually it became apparent that no guano deposits existed, and no treaty between Ecuador and the United States was ever signed.

The guano deposits on the coast of Lower (Baja) California had been little studied. As was typical in his search for relevant data for the book, Hutchinson wrote to the company Guanos y Fertilizantes de Mexico, saying that he had made inquiries through the Los Angeles Chamber of Commerce and had been given its name. He explained his monograph and wrote: "For Lower California . . . deposits, no analyses seem to have been published. The information that is needed for each locality is the quantity of guano present before exploitation began, the species of birds that produced it, and chemical analyses indicating, at least, the quantity of phosphate and nitrogen present. . . . Data obtained prior to 1900 are more significant from my point of view than modern reports."[22]

Hutchinson did indeed receive information from this company, a government-sponsored organization that had all these Mexican guano islands under its control and was now trying to carry on a rational exploitation of these islands. Their general manager, Carlos H. Benitez, knew a great deal about the bird species present as well as the annual yield of guano and its chemical analysis. Hutchinson also received information from George Beermaker of the Union Fertilizer Company of Los Angeles and from his son. There is, in fact, much extant correspondence about requests for information for this book. Hutchinson had even put an advertisement in *American Neptune* in 1946, asking for information on guano birds that might be found in old whaling logs. One man, Gale Blosser, replied.[23]

In addition to information, Hutchinson received actual guano. Colonel J. N. Tomlinson sent guano from Ascension Island in the South Atlantic via James Chapin of the American Museum of Natural History. He wanted to know about the biological origins of these deposits. Hutchinson replied that he was delighted to have the material because he had a young man who was working on the X-ray mineralogy of guano phosphates. He wrote also that the Ascension deposits were formed by an ancient bird colony, though probably post-Pleistocene. Hutchinson added, "The still bigger deposits such as Nauru, which are certainly Pleistocene, are in my opinion due to a great increase in vertical oceanic circulation as glacial conditions gave place to interglacial, and the source of cold water at the Poles disappeared, so that nutrients locked up in the deeps of the ocean became available to the plankton and so to the animals dependent upon them."[24] This was partly speculation, of course, but it indicates Hutchinson's ecosystem approach and his long-term interest in climate change and its biological effects.

Correspondent Douglas Carroll wrote to Hutchinson about an island in the Pacific where he had seen colonies of bats: "Not just millions but undoubtedly billions . . . who darken the sky in the middle of afternoon as though night had fallen . . . when passing in their daily forages for food, always returning to their home roost to sleep heads down. . . . I could take you to this island which is a present day guano deposit in the formation . . . if your foundation has appropriations set aside for such purposes." Hutchinson replied, providing much information for Mr. Carroll, but saying that although he would greatly appreciate knowing where Carroll's bat island was, he feared he had no funds to be able to visit it.[25]

The southwest African guano coast is a region of major upwelling of the cold water of the Benguela Current, rich in phosphate and plankton, including huge populations of diatoms. The major guano birds of these coastal islands are a cormorant, the Cape gannet, and the Cape penguin, *Spheniscus demersus*. A verse about the penguins and their guano on Ichabo Island was quoted by an 1845 writer:

> Where penguins have lived since the flood or before
> And raised up a hill there a mile high or more.[26]

K. J. Andersson reported fishing cormorants passing overhead for two or three hours in 1872. A 1933 observer claimed to have counted 100,000 of these birds passing over.

It is not only the avian inhabitants of these South African islands that were of interest. Two human mummies had been found buried under the guano; Hutchinson attempted to use them to estimate the chronology of guano deposition. One of them, probably found in 1721 on Ichabo Island, had been publicly exhibited at Havre de Grace, Maryland, in the nineteenth century. In his typical fashion, Hutchinson asked the local chamber of commerce to put a letter in the local newspaper, but this failed to elicit any further information. The second mummy had been discovered in 1844 under 18 feet of guano, with an inscription from 1689. Hutchinson calculated the rate of deposition to be between 2.3 and 3.6 centimeters per year, the former rate comparable to that calculated for the Peruvian coastal islands. The chemical composition of the guano from these South African islands had been published in many government reports. There is evidence of the ancient existence of extensive bird colonies on former islands which are now part of the African mainland. On islands off at least three of the world's coasts—Peru, Baja California, and southwest Africa—Hutchinson noted similar geographical, oceanographic, and climatic conditions as well as three comparable guano bird species.

The Pacific Atolls and Other Islands

A large section of the book presents all the information Hutchinson collected on the Pacific atolls, very little of which can be included here. An example of his method of integrating data he collected from many sources is shown here.[27] For a section of the Pacific Ocean from longitude 140 degrees west to 180 degrees west (roughly between the Marquesas and the Marshall Islands), on both sides of the equator, he showed all of the following: rainfall, the guano islands, the quantity of plankton, the number of plankton-feeding sperm whales, the circulation of currents, the location of phosphate deposits, and the distribution of silicate and phosphate in the water. The sperm whale kills from nineteenth-century logbooks are immense. Many were killed in a high plankton zone that had been found on a cruise of the ship *Carnegie* in 1828–29.

Exploitation of guano from atolls in the Pacific Ocean began in the nineteenth century. Hutchinson tried to collect records of the quantity of guano before that time. Some of the guano deposits were apparently apocryphal. A number of atolls were registered with the United States Treasury as guano islands under the Guano Act of 1856 and published in the *New York Tribune* on March 5, 1858. There was strong commercial inducement to register islands as having guano where it later proved not to exist; in fact, some of the islands were nonexistent.

For a section on the Hawaiian Leeward Islands, Hutchinson collected much information on the birds of these islands, not all of them guano birds. A flightless rail (*Porzanula palmeri*) once lived on Laysan but is now extinct. It had been successfully introduced on Midway Island but was exterminated there by rats in 1944. Hutchinson wrote: "The catalogue

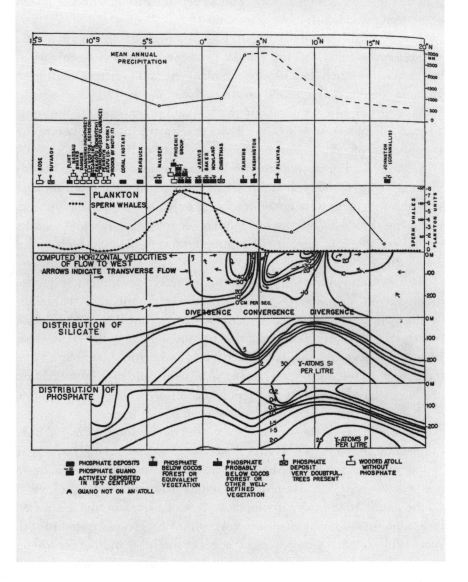

24. Section of the Pacific from longitudes 140 degrees to 180 degrees west showing distribution of rainfall, guano islands, plankton, sperm whales recorded in nineteenth-century logbooks, circulation, silicate, and phosphate. Reprinted from Hutchinson (1950b), figure 46.

188   BIOGEOCHEMISTRY

of the Rallidae is a pathetic cemetery filled with crosses commemorating flightless forms that have developed in different directions, considered of generic value, on oceanic islands, only to become extinct with the advent of modern man and his inevitable companions."[28]

Guano was first exported from the Hawaiian Islands in 1853. One of the islands, Nihoa or Bird Island, contained twenty different vascular plants and a rich flora of nitrophilous lichens, many species of both groups endemic, that is, found nowhere else in the world.

Hutchinson also wrote about the sad fate of Laysan, an atoll only 3 by 1.6 kilometers, which was called a true earthly paradise before the introduction of rabbits. Before the introduction of this alien herbivore, H. Schauinsland spent three months on this tiny island; he published his findings in 1899. Twenty-six vascular plant species were known from Laysan. Many of these were endemics found only on this one small island. Following the rabbit infestation, Laysan became a sandy waste, and its endemic plants and terrestrial birds, including the Laysan honeyeater, were extinct.

Hutchinson wrote about many other guano islands, including those of Australasia-Australia, New Zealand, and even the subantarctic islands, as well as those of the western Indian Ocean. The latter include Aldabra, for which Hutchinson provided a description and a map. It is really four islands plus several smaller ones, separated by narrow channels, all in a large lagoon. Hutchinson was later to be instrumental in the preservation of Aldabra (see chapter 15), home of giant tortoises and other rare biota. A student brought Hutchinson a tortoise shell from there.

To the north of the Indian Ocean are the guano-producing Bahrain Islands in the Persian Gulf. Edrisi described them as early as 1154 as "numerous islands inhabited by birds whose dung was transported to Basra and sold at a high price as a fertilizer for date palms, vines and gardens."[29] Of the South Atlantic Ocean islands, only Ascension and St. Helena had large amounts of guano. Beatson, governor of St. Helena Island in 1811, successfully applied guano to sugar beets at the suggestion of Sir Joseph Banks. Almost five thousand tons of good phosphatic guano was exported from St. Helena to Britain between 1855 and 1860.

Not many of the Caribbean Islands produced guano but the islands off the coast of Venezuela did. Here the North Equatorial Current produces upwelling. Hutchinson's map shows high phosphorus near Curaçao,

25. Hutchinson and giant Aldabra tortoise given to him and placed
in his Yale office by A. J. Kohn and Willard Hartman.
Photograph by Phyllis Crowley, with permission.

Bonaire, and Aruba. Hutchinson never visited the many islands he wrote about, with one exception: he and his third wife, Anne, visited Aruba in his old age. The original guano birds there were boobies, frigate birds, and terns. There was guano on the Cayman Islands also; it was shipped to New York from 1884 to 1890. At the time Hutchinson's book was written there was still guano on these islands; it was being shipped to the island of Jamaica.

In South America, even Patagonia and the Falkland Islands, sites of high plankton productivity, produce guano in spite of the cold temperatures and high rainfall. Several penguin species, including the rock hopper and gentoo, produce guano in the Falklands. And off the northeastern coast of Brazil, St. Paul's rocks have large colonies of boobies and noddy terns. Darwin visited there on the *Beagle* and reported that the brilliantly white rock resembled seals due partly to "the dung of a vast multitude of sea-fowl [boobies] and partly to a coating of a hard glossy substance with a pearly luster . . . intimately united to the surface of the rocks."[30]

The Biogeochemical Cycling of Phosphorus

How important is guano deposition in the general phosphorus economy of the earth? This was a question Hutchinson set out to answer with all the data in his book. He concluded from his numerous calculations that the guano birds of the world had little important effect on the global geochemical cycle of phosphorus. He calculated the guano production rate along the arid coastlines of South America, South Africa, and Baja California to be 185,000 tons a year or 880 metric tons of phosphorus from the sea. He calculated, however, that much more phosphorus was drained from land surfaces into the sea: 13.7 to 17.4 million tons of phosphorus enter the sea each year.

How much guano has been deposited on land in post-Pleistocene times? Hutchinson calculated 2.8 million tons, with the highest amounts coming from the Chincha Islands and the Mejillones of the Peruvian and Chilean coasts. Less than half of the earth's guano deposits were still growing in the early nineteenth century; at least half were the result of ancient but post-Pleistocene bird colonies. He concluded that Pleistocene deposits were much greater than those of post-Pleistocene times; guano-depositing climatic conditions were better in the Pleistocene.

What is the relationship of climatic and oceanographic factors to guano production? Guano accumulation is generally limited to arid conditions, in the Pacific to the equatorial dry belt south of the Doldrums. By contrast, where there are over 1,000 millimeters (about 40 inches) of rain a year there are no large guano deposits, none at all over 1,500 millimeters. All over the world guano deposits are related to ocean currents and upwelling. The many guano deposits were formed by birds fishing in water of "notorious" fertility characterized by intense upwelling.[31]

Population and Community Ecology of Guano Birds and Bats

Although *The Biogeochemistry of Vertebrate Excretion* constituted a major part of the series on biogeochemistry that Hutchinson wrote and/ or edited, one could argue that by the time of its writing population and community ecology had become his major research interests. The water bugs, his preferred freshwater research organisms, play no part in these marine ecosystems. He seems, however, to have had much interest in the marine birds that do play a major role and in their population biology.

What makes a good guano bird? Such a bird—be it cormorant, booby, pelican, tern, or penguin—is generally highly social, breeds in large colonies, deposits a large percentage of its excreta at the nest site (not in the sea), does not burrow excessively, and eats fish rather than shrimp or other invertebrates. Hutchinson thought that specific or subspecific behavioral and physiological differences in these birds, for example, age of breeding or whether the birds were territorial or colonial, could affect their guano production. He pointed out that the genetic aspects of bird behavior were important and little known. He noted a possible example of innate differences of behavior between two subspecies of the booby, *Sula leucogaster,* one being territorial (and producing little guano), the second a gregarious guano producer.

Hutchinson's interest in population and community ecology is evident throughout the book. For example, he discussed the importance of the size of breeding bird colonies: in sooty tern colonies which continually breed on the same oceanic island, a catastrophe can cause a decline in breeding population below a critical minimum, leading to local extinction. Population size has also been found to affect courtship behavior and other socio-sexual phenomena in some seabird species.

Community interactions involving guano birds are fascinating. For example, the endemic pipit, *Anthus steindachneri,* on the Antipodes Islands, New Zealand, feeds on flies breeding in penguin droppings and thus is dependent on the penguin colony. Plant-seabird interactions also abound. For example, large bird colonies promote rich local vegetation in the Arctic. Hutchinson reported a study by B. Lynge who found an extraordinary development of the lichen and moss flora in the Norwegian Arctic, including 208 lichen species. He thought the fertilizer from the large bird colonies responsible, especially for nitrophilous lichen species.[32] Some vascular plants also benefited; a saxifrage, *Saxifraga cernua,* was growing in "great luxuriance." Vascular plants elsewhere have been found to benefit from the guano, for example, an endemic composite, *Senecio atipodus,* on the Antipodes Islands, New Zealand, was found only where fertilized by the nesting giant petrel. Although plant species' populations are generally thought to be increased by bird-dropping fertilization, Hutchinson pointed out that certain plant species may not require the high phosphorus or nitrogen content but may simply be tolerant of them; other more widespread but less tolerant species would be prevented from colonizing these guano sites. Thus, the presence and large populations of unusual species may be the result of lessened competition. Guano birds can have negative effects on the flora also; large penguin populations can turn vegetated areas into bare mud, whereupon the island is often abandoned by the penguins.

Commenting on the interactions of birds, vegetation, and climate, especially rainfall, Hutchinson wrote: "A fascinating field for further research on the interaction of the precipitation, bird colonies and vegetation on tropical islands is evidently open to anyone in a position to take advantage of it. In any such work it is very desirable to bear in mind certain of the theoretical aspects of competition between species."[33]

Here Hutchinson was concerned about bird populations. He referred to Volterra and Gause and discussed their equations. Hutchinson also referred to the only one of his papers in the bibliography, a 1947 paper from before the better-known Robert MacArthur era: "A Note on the Theory of Competition between Two Social Species" that had been published in *Ecology.*[34] He commented in relation to that paper that although the outcome of competition between seabird species will often depend on environmental factors, in many other cases the initial sizes of the colonizing populations may be equally important. Hutchinson pointed to probable evidence for this type of competition from the Peruvian and other guano bird islands. He also discussed some theoretical aspects of island environmental gradients, their effects on competition coefficients, and the resulting zonation of the biota. In the cases discussed, the biota are guano birds, and their zonation is important in relation to guano production. But for Hutchinson, of course, the population biology is important in itself; it was to become a major area of his research (see chapter 14).

The effects of large bird colonies on fish catch is of particular importance to fishermen. Hutchinson concluded that the guano bird colonies actually increase fish productivity near the islands by biogeochemical concentration. Nutrients that would otherwise stay in fish and be distributed more widely in the seas instead are taken up by fish-eating birds and moved back into the ocean near the colonies, in the absence of complete aridity.

Bat Guano and Comparative Biogeochemistry

A fascinating section of the book provides data on guano deposits produced by bats in caves, including some caves within the United States. Most bats are insectivorous, but some are fruit-eating or even fish-eating; the food determines the type of guano. Guano of a frugivorous bat from a Jamaican cave was found in 1865, and even the guano of a Mexican blood-sucking bat has been analyzed.

Reports of cave guano go back to the early 1800s. Bat guano from the Bahamas was imported to England in the second half of that century. Texas and New Mexico have many bat caves. Guano from the most famous American cave, Carlsbad Caverns, has been exploited commercially; 100,000 tons of guano have been removed. Cave paleontology and archeology are of interest there. In another New Mexico cave the skeleton of a Pleistocene ground sloth was found, and in a Nevada cave dung of extinct ground sloths has been excavated, mapped, and analyzed. Eighty percent of the vegetable matter was the remains of yucca. There was clear evidence of human habitation below the sloth dung.

Biogeochemistry, especially the comparative geochemistry of bat and bird guano deposits, was important to Hutchinson. Bat guano and bird guano differ significantly since birds excrete uric acid, but bats urea. Also the bat guano, at least of insect-eating bats, contains chitin, high in nitrogen, from arthropod skeletons. Both types of guano contain sulfate and phosphate, alkalis and alkaline earths. When bird and bat guano decompose they both leave an insoluble residue, largely calcium phosphate. But many minerals are formed, which in both cases interact with inorganic materials in the environment.

Bacterial action also plays a part. For example, oxalates are characteristic minerals of nitrogenous guano. Oxalic acid is not found in bat guano since it is formed by bacterial decomposition of uric acid. In bat guano the last fraction of the original nitrogen to remain in the guano is in the form of nitrate, a function of bacterial metabolism, which oxidizes ammonia to nitrates. Nitrates are the most stable form of combined nitrogen in the biosphere, the form most needed by organisms other than bacteria. Another nitrogen-containing compound, saltpeter or niter (KNO_3), is a valuable product found in bat caves.

Many guano deposits have been analyzed and a great many minerals found, some of them rare, most of them unfamiliar to non-mineralogists. For example, stercorite, a phosphate mineral, is found only in bird but not bat guano; it contains sodium of marine origin. Calcium phosphate, magnesium phosphate, and aluminum and ferric phosphate minerals have been found in the guano analyses. Hutchinson, who had previously written an exhaustive monograph of the biogeochemistry of aluminum, was particularly interested in the origin of the aluminum minerals. These are formed, he concluded, by phosphatization of igneous and other aluminosilicate rocks in tropical and subtropical regions with seasonal rainfall.

"Previously Unrecognized Conclusions"

Near the end of the book Hutchinson gave a summary of eighteen previously unrecognized conclusions. Perhaps the most important in terms of the biogeochemistry of guano is that "the rate of removal of phosphorus from sea to land by all the seabirds of the world is . . . certainly not more than a few percent of the rate of entry of phosphorus from land to sea by chemical erosion."[35]

Biogeochemistry in Hutchinson's mind was a wide-ranging subject, and in the course of researching the literature relating to the biogeochemistry of guano he discovered an enormous number of interactions, including those among geology, present and past climate, ocean currents, geography, plants, nitrophilous lichens, seabird behavior and population biology, bats, and other biota, such as oilbirds that produce cave guano. In addition, there are many indications of human interference, from the extinction of species on guano islands to stockholder reports from guano-exploiting companies. Archeology, paleontology, and various methods of dating are also important in relation to guano.

The book, published in 1950, got a favorable nearly three-page review in *Ecology* by Robert Cushman Murphy, who wrote: "The diverse nature of the evidence—climatological, oceanographic, archeological, biological, chemical and mineralogical—and the inductions drawn, endow this work with fascination that pertains to all good sleuthing."[36]

Murphy emphasized Hutchinson's ecological study of the contemporary guano birds, nearly all of which (cormorants, pelicans, boobies) belong to the Pelecaniformes, his calculations of fish intake and annual guano production, and his conclusion that the birds convert no more fish into guano than permits the persistence of the whole ecosystem. Murphy asserted, as did Hutchinson himself, that this is not the case with the fish-meal industry, which should never be allowed to replace the natural fertilizer-producing industry of Peru.

Murphy also pointed out Hutchinson's ecosystem deduction that the birds increase the potential fish catch by biogeochemical concentration, that is, by increasing the fertility of nearby ocean waters. Murphy concluded that the "wealth of scientific findings" poses as many questions as it answers and will be a rich source for future doctorate theses "in half a dozen different disciplines."

Hutchinson received a copy of a letter to Parr from the chief geologist of the Bureau of Mineral Resources, Geology, and Geophysics of Australia who wrote saying that the volume on guano deposits would be of great interest to their geologists and mineralogists. Most pleasing to Hutchinson, however, may have been the letter from the ornithologist of the Compania Administradora del Guano of Lima who had Hutchinson's book "sobre la Geobioquimica del Escremento de los Vertebrados," which seemed to him "sencillamente maravilloso" (simply marvelous).[37]

Not all of the readers of *The Biogeochemistry of Vertebrate Excretion* were so enthusiastic. There were many jokes about the title. One Yale graduate student thought Hutchinson had wasted good research time writing so much about "excretions." A variety of scientists, including this author, found the book fascinating; it is doubtful that it had many lay readers. Hutchinson must have enjoyed researching and writing it. Its emphasis on the effects of El Niño and other aspects of climate change would interest many readers today.

Hutchinson, Catherine Skinner, and Biogeochemistry at Yale

Hutchinson taught Biogeochemistry at Yale together with Dr. Catherine Skinner and sometimes Dr. James Walker. Skinner was doing research at the Yale Medical School at the time they taught the course. She had been one of only four undergraduate geology majors at Mt. Holyoke College and afterward the only woman graduate student in geology at Harvard. After her master's at Harvard, where she learned X-ray diffraction techniques, she did her doctorate at the University of Adelaide in Australia. There were no other women graduate students in geology at Adelaide either. She carried out field research on the precipitation of calcium carbonates in lagoons in South Australia and was the discoverer of low-temperature precipitate dolomite. This type of dolomite proved to be present all over the world, including the Red Sea's low-lying lagoons. After Catherine Skinner finished her Ph.D. she returned to the United States where her husband, Brian Skinner, worked at the U.S. Geological Survey in Washington. Preston Cloud, who in 1959 was in charge of a branch of the Geological Survey, knew that Catherine Skinner had worked on carbonate deposition and offered her a job at the survey. She instead took a position at the National Institutes of Health (NIH) working on the structure of proteins. Her work in Australia had been on X-ray diffraction of low-temperature precipitates that were noncrystalline. There were few people at the time with this kind of experience in X-ray diffraction. She worked from 1961 to 1966 on the structure of a number of different proteins, such as chymotrypsin, growing crystals of it and trying to get "first class diffraction patterns so you could do the structure, which was at that time the cutting edge because no one had good enough crystals."[38]

In 1966 Brian Skinner took a position at Yale. His wife Catherine did not receive a Yale position, but she did obtain a big grant from the NIH Dental Institute and set up a laboratory in the Yale Medical School in the Orthopedics Department. There she gave lectures to medical residents on basic mineral materials and did research on bone.

Hutchinson met Catherine Skinner soon after her arrival at Yale. She lectured to the Connecticut Academy of Arts and Sciences on collagen and bone; Hutchinson, long an academy member, introduced her lecture. They found they had a number of common interests; she too was, as she put it, "an interdisciplinary person." Hutchinson was at that time anxious to teach a course in biogeochemistry involving everything from the chemical contents of living tissues to atmospheric chemistry. According to Skinner, it was Hutchinson who brought faculty with expertise in biogeochemistry into the biology program. Skinner and Hutchinson, together with an atmospheric physicist, Jim Walker, coordinated a graduate student course on biogeochemistry for three years in the late 1960s.

"We took the periodic table and went through the strange biological uses of particular elements," Skinner reported.

During this same time period, Ursula Cowgill was doing biogeochemical research in Hutchinson's laboratory at Yale. Cowgill was a research associate who carried out a great deal of research with Hutchinson involving the analysis of lake cores from all over the world. They published many joint papers. Skinner worked with Hutchinson and Cowgill on a project analyzing *Latimeria* scales. *Latimeria,* a coelacanth, is a so-called living fossil, a modern representative of a very ancient group of fish, the crossopterygians, which gave rise to amphibians. It had long been known as a fossil, but was discovered alive in the sea off Madagascar. As recounted by Keith Thomson, a *Latimeria* specimen had arrived frozen at Yale. Grace Pickford, Hutchinson's first wife, worked on it as well. A major discovery was that this ancient fish, like sharks but unlike all modern bony fish, retains urea in its blood.[39]

Hutchinson, a Well-Known Biogeochemist and Geologist

By the time Hutchinson's book and his work on biogeochemical cycling of aluminum and other elements had been published, Hutchinson was widely known in this field. He received inquiries from all over the world. Some of his ventures into biogeochemistry related to practical problems. For example, among his biogeochemical interests were the elemental constituents of marine invertebrates. A 1946 paper by Hutchinson, Jane Setlow, and John Brooks analyzed the elemental constituents of the starfish, *Asturias forbesi,* a common species in Long Island Sound that preyed on oyster beds there.[40]

Hutchinson's primary ongoing professional interest in biogeochemistry was in its relationship to limnology. His next major opus was the *Treatise on Limnology,* volume 1, published in 1957. Of the four volumes of the *Treatise* that were eventually published, the first three contain substantial material about the biogeochemistry of lakes (see chapter 8).

By the late 1940s when the last monographs for the biogeochemistry survey came out, Hutchinson was almost as well known in geological as in biological circles. He was by 1948 chairman of the Division of Geology and Geography's Committee on the Application of Biological Methods to Geological Problems of the National Research Council (NRC), one of the fifteen technical committees of this section of the NRC. He was one of the "non-survey" members of a Geological Survey Committee reviewing the work of the Geochemical Prospecting Unit. He was also involved with the geophysical laboratory of the Carnegie Institution. This laboratory had formerly been at Yale.

Hutchinson had much correspondence with members of the U.S. Geological Survey (USGS). His most important correspondent, at least in terms of the future of Hutchinsonian ecology in America, was E. H. Bradley. In 1948 Bradley and the USGS made a very determined effort to hire Hutchinson as a biogeochemist. To lure him away from Yale, the Geological Survey offered him unlimited research opportunities and funds to work on whatever he wished—and even to do all the paperwork for him. Bradley offered him a position as a "full-time member of our research staff with the charter written by yourself, to spend 12 months of the year, year after year, in laboratory or field research along whatever lines you think need to be pursued. Our responsibility would be to provide you with what facilities you would need and wish, and to shield you from red tape and chores."[41]

The salary was high for 1948: $10,000 a year. Hutchinson answered, "Of various offers that I have had in the past years it is the only one that I feel tempted by." His main objection to their very generous offer was his commitment to present and future Yale graduate students: "The chief thing that I am doubtful about is that I have had, as you know, very considerable success in turning out young investigators of quite outstanding ability. These men, notably G. A. Riley, E. S. Deevey, and V. T. Bowen are rapidly becoming leaders in their fields. The prospect of not being able to train such people is the most cogent . . . argument against your otherwise very attractive offer."[42]

Hutchinson chose to remain at Yale. There he was to have many more graduate students, both men and women, and with them the opportunity to develop new ideas and to invent more aspects of modern ecology. Although Hutchinson was not usually thought of as a geologist, at least by those outside the USGS, he continued to use his interest and knowledge of biogeochemistry in much of his limnological and ecological work, including his ongoing research with Ursula Cowgill and in his graduate course taught with Catherine Skinner and Jim Walker.

CHAPTER ELEVEN

The Three Wives and Yemaiel

M an was not meant to live alone. What is true for an ordinary man is even truer for a research biologist," commented a fellow ecologist while discussing Evelyn Hutchinson.[1] Women were always important in Hutchinson's life—his wives and also other women who were his close friends. As one of his graduate students remarked, these women themselves were of unusual interest—and not just because of their link with him.

During his long life, Evelyn Hutchinson had three very different wives. The first two, Grace Pickford (who kept her own name) and Margaret Seal Hutchinson, were English; the third, Anne Twitty Goldsby Hutchinson, was American. Grace Pickford, Evelyn's fellow zoology student at Cambridge University, became a very well-known scientist in her own right, originally working with Evelyn but later in a very different field. Their early marriage ended in divorce by the time Evelyn was thirty, yet they remained colleagues at Yale until Grace's retirement. Margaret, his second wife, was not a scientist, but she shared Evelyn's interest in music and in art. She was his wife for fifty years; he cared for her during her last years when she suffered from Alzheimer's disease. Anne Twitty Goldsby Hutchinson was much younger than Hutchinson, the wife of his old age. She was black, of Haitian extraction, and a biologist. When Hutchinson first knew her she was in charge of freshman biology laboratories at Yale.

She also taught high school biology in New Haven and was interested in improving the teaching of science in the inner-city schools.

Evelyn always wanted children but never had any of his own. During World War II, Yemaiel Oved, the half-Jewish daughter of an old family friend of Evelyn's in England, came to live with the Hutchinsons in Connecticut when she was two years old. Evelyn was close to Yemaiel when she was a small child but was not able to continue a close relationship with her after she returned to her parents in England at age five. Evelyn saw her later in his life, when she was an adult; he thought of her as his daughter in his old age.

The people in Hutchinson's early world—those from his elementary and secondary school, his dons and friends at Cambridge, his colleagues and students when he first arrived at Yale—were nearly all male. Women undergraduate students in biology at Cambridge University, including Grace Pickford, were exceptions. Some of them were also exceptionally gifted. Perhaps because of his respect for these women and their work, Hutchinson was always supportive of women students in science, an unusual attitude for male professors at Yale in his early years there. One can document this attitude as early as the 1940s, when he answered a letter from a woman student inquiring about a graduate teaching assistantship. He wrote to her that they had never had a woman teaching assistant, but he did not see why they should not; she should apply.

Although all his early Yale graduate students were men, later he had many women graduate students. Among his close women friends, two were famous in the worlds of anthropology and literature: Margaret Mead and Rebecca West. Others were colleagues at Yale, in psychology, literature, art, and other fields. He also kept in touch with several Cambridge University women friends, including Anna Bidder, Penelope Jenkin, and Cecilia Payne (later Payne-Gapochkin), all of whom had notable scientific careers, the last at Harvard in astronomy. He also kept close ties with a great many of his graduate students, both male and female, and their families after they left Yale. He was a tireless letter writer, and much of his correspondence is still extant, including copies of his own letters. In Hutchinson's later years at Yale he befriended not only some of his excellent women students but also women on the faculty, particularly at his college, Saybrook, and at the Elizabethan Club.

Grace Pickford, Scientist Companion and First Wife

During her lifetime Grace Pickford became a very well-known scientist. She became the first female full professor of biology at Yale. Pickford had many collaborators, many undergraduate and graduate students, and many loving friends; I have interviewed some of them. I knew about Grace Pickford's research before I went to Yale to work on this book, but I did not know until then that she had been Hutchinson's wife and early scientific collaborator. Grace Pickford always used her own name from the time she came to Yale. She and Hutchinson first appeared in print as a couple, in fact, in my chapter in the book *Creative Couples in the Sciences*.[2] By that time both had died.

In his later years, Hutchinson wrote *The Kindly Fruits of the Earth*, an autobiography of his life through his early years at Yale.[3] Yet nowhere in this book does he mention Pickford, who was his fellow zoology student at Cambridge University, who married him in South Africa, where they did joint research, and who came to Yale with him in 1928. They even published several joint papers after their arrival at Yale. They were constant correspondents during his North India Expedition in 1932; in his letters he often asked her to critique his research.

Grace Pickford was born in Bournemouth, England, in 1902. Her father, William Pickford, was a journalist, editor of the *Bournemouth Guardian,* and owner of a printing business. He encouraged his daughter's natural history interests. Grace and her younger brother Ralph collected fossils with their father on the cliffs at Lyme Regis.[4] Grace's mother took over the printing business when her husband died young. She believed women should be educated and have careers, and she encouraged Grace's interest in a university education. In preparation for university, Grace attended Bournemouth Municipal College. One of her teachers there, Joseph Omer-Cooper, was later a fellow zoology student at Cambridge and a close friend of hers and Hutchinson's.

From 1921 to 1925 Grace was a student at Newnham College, one of the two separate women's colleges at Cambridge. By 1920, as noted earlier, women could attend Cambridge University classes, including the biology laboratories. Before 1914, the Newnham and Girton science students, all women, did their practical studies at the Balfour Laboratory in Cambridge.[5] The Zoology Department seems to have been remarkably egalitarian by Hutchinson's and Pickford's student years. Because of her earlier training at Bournemouth, Pickford was able to take her natural science tripos (examinations), part 1 after only two years, in 1923, and read advanced zoology for part 2. But women could not receive Cambridge degrees at that time; not, in fact, until 1948. Half a century after that date, in 1998, Cambridge University held a graduation ceremony for those earlier Newnham and Girton women. Hutchinson's younger sister, Dorothea, and his and Grace's friend and classmate, Anna Bidder, then ninety-five, both attended. For Anna it was seventy-three years after she and Grace should have received their Cambridge degrees.

At the time she completed her Cambridge University studies, Grace was engaged to Evelyn Hutchinson. This may not have been common knowledge, but, as already noted, E. A. Butler wrote to Evelyn before Evelyn left Cambridge for Naples to congratulate him on his engagement.[6] Grace was awarded a Newnham College Traveling Fellowship, which she held 1925–27, and with this funding she went to South Africa where she worked on earthworms and freshwater oligochaetes, relatives of earthworms. She and Hutchinson, who received an instructorship at the University of Witwatersrand in Johannesburg, got married in Cape Town. After Hutchinson was fired from his teaching position at Witswatersrand, he was invited to do research by Professor Lancelot Hogben in Cape Town (see chapter 4). Hogben sent Evelyn and Grace out to do limnological research together on the pans or shallow lakes in that region and elsewhere in South Africa. They subsequently published several joint papers in this field, limnology, which became a large part of Hutchinson's life work. Grace did some further limnological work in Virginia after coming to Yale, and she published one more paper with Hutchinson, but thereafter her work was in totally different areas of biology.

In 1928 Grace Pickford came to Yale with Hutchinson. She had no position, but at Hutchinson's request, she had been assured of research space. She continued her work on earthworms and their relatives for a Ph.D. under the direction of Alexander Petrunkevitch. He was a Russian immigrant who had studied in Germany with August Weismann and had subsequently become a spider expert and a professor at Yale. Grace and "Pete," as Petrunkevitch was affectionately called at Yale, remained lifelong friends, and later in their lives became housemates.[7] Pickford received her Ph.D. in 1931. Her thesis was finally published in 1937 as a 621-page monograph.

Hutchinson, meanwhile, was busy teaching a variety of undergraduate and graduate courses; he never earned any further degrees. But he did support Pickford in her academic career, writing to her from North India in 1932 to urge her to finish and publish her earthworm monograph. She had discovered evolutionary aspects that he thought well worth publishing. Grace was at this time considering a medical career. Evelyn wrote: "I heartily approve of your going into medicine. It will make you happy. You are made that way just as I am made to mess about with water and such things that are in it." He offered her monetary assistance "before I get sacked of course."[8] Pickford did not go to medical school; she did publish the monograph.

Hutchinson's letter was one of many that he wrote to Grace from the 1932 Yale North India Expedition, for which he was the chief biologist. The letters, all addressed to "Dear old thing," are quite affectionate.[9] He was making a special study of the high-altitude lakes in Ladakh near the Chinese border (see chapter 6) and sent her frequent letters about what he was doing, including his interesting data. He often apologized for the data he was not able to collect; Grace was clearly his research colleague and severest critic; he needed her approval. Sadly, her replies are gone. Grace destroyed most of her personal papers.

Their divorce, however, was imminent. Hutchinson wrote from North India (Ladakh): "Modern children's stories should end 'And then they got divorced and lived happily ever after.'" He also wrote that he was planning to marry a woman named Joy and "do some reproductive collaboration in a few years." The divorce was regretted but not contested on Grace's part. She told one of her Cambridge friends that they either never should have married or never have divorced. She told a friend from Yale, perhaps more tellingly, that it should have been an annulment, not a divorce. Since Grace agreed not to contest the divorce, Evelyn went to Nevada, where a six-week residence was sufficient to obtain a divorce. During that period he did an important piece of research on arid lakes, related to the work that he and Grace had done in South Africa (see chapter 8).

Grace was a person who had warm, close friendships both at Cambridge and at Yale. Some of these friends were unhappy about the divorce. Hutchinson's old friend Joseph Omer-Cooper, who had started the Biological Tea Club at Cambridge University with Evelyn and Grace and his own future wife, Joyce, clearly did not approve. He wrote to Evelyn saying that he and Joyce had four children, which was extremely satisfying, "better than an F.R.S. [fellow of the Royal Society]. Some time ago I heard that you had acquired a second wife. I hope that she proves, pleasant, docile and fruitful. It is a very great pity that polygamy is not permitted. . . . A sequence is far less satisfactory in every way."[10]

One man, however, John Nicholas of the Yale Zoology Department, heartily approved of the divorce and remarriage and of what he viewed as the results. He wrote about Hutchinson to the future Yale president Charles Seymour:

> He married . . . a fellow [Cambridge University] student in biology who is brilliant in her own right. During their early years here she showed a brilliance surpassing his and a rare aptitude for [research]. Her time was free for research while his was taken with students and teaching—her publication list grew while his lagged. She became more and more eccentric. . . . The result was divorce. Hutchinson remarried almost as soon as the divorce was final which raised the eyebrows of our department conservatives, but I think his career since has justified his judgment. . . . He demonstrated a drive and punch he never had before.[11]

Nicholas was actually recommending Hutchinson to be master of a Yale college. Ironically, it was Nicholas himself who received the master's post. Nicholas's judgment and facts were not always reliable. Hutchinson had actually published twenty-five papers between 1928 and his divorce from Grace Pickford. The "raised eyebrows" of department members were probably a reliable observation; divorce was not common nor widely approved of in 1933.

Grace Pickford's Scientific Career

Eccentric or not, Pickford stayed at Yale as a research scientist for the next fifty years. In the thirties, forties, and fifties she was a fellow, a research assistant, and later a research associate at the Bingham Oceanographic Laboratory but not a faculty member. There were neither undergraduate women nor women zoology faculty members at Yale in that era. Pickford did teach at Albertus Magnus College in New Haven from 1934 until 1948, as an assistant professor. She eventually became a lecturer at Yale in 1957 and an associate professor in 1959. Nevertheless, she always had research space at Yale and for many years her own research laboratory. She also was able to apply for her own grants even though she was not actually a faculty member.

In the 1960s Clem Markert became chairman of the Biology Department, after the zoology and botany departments had merged. During Markert's chairmanship several women, women who had long been research associates and had their own grant money, were no longer allowed to apply for research grants. Lab space was to be almost entirely limited to faculty. By this time, however, Pickford was a Yale faculty member with access to labs and research grants; she was very successful at obtaining government research grants. In fact, her research career is quite amazing. She did research on three continents and at sea, including the deep sea. She became a leader in several seemingly unrelated fields, although there were several partly overlapping stages.

In South Africa she worked on both the taxonomy and the evolution of oligochaetes, the group to which earthworms belong. Although her dissertation was on South African worms, she continued her studies on the North American oligochaete fauna and published a major work on them.

Limnology was also an early research subject of hers in South Africa, when she was working with Hutchinson. Colleagues of Pickford's at Yale suggested to this author that the amount of credit due her for Hutchinson's early ideas in limnology might be hard to ferret out, but that many of the ideas were probably hers. I have no evidence that this is true, but Hutchinson and Pickford bicycled to Wicken Fen while they were Cambridge undergraduates and explored the aquatic life there, and all of the South African pan work was done jointly. As one who has done much joint research in the field, I am certain that they must have discussed their ideas and methods. Indeed, they were still discussing them by mail in 1932.

Pickford left limnology and after 1937 published little on earthworms. She worked as a research assistant for Petrunkevitch on his spiders. Pickford worked out microanalytical methods for studying the digestive enzymes of spiders. This work required special equipment; she was clever at inventing it. This experience at the microlevel helped her with her later research on hormones.

Next she moved to cephalopods, the most intelligent class of molluscs and the one to which the octopus and squid belong. She studied the many specimens at the Peabody Museum at Yale. Pickford became a world expert on the octopus group and produced twenty publications while teaching at Albertus Magnus. Marcella Bovary was at the college then, as was Margaret Wright, a close friend of Pickford's. Wright was the first woman

26. Grace E. Pickford with two sailors on the Galathea deep-sea expedition in 1951.
From Patricia Brown (1994).

to have a graduate teaching assistantship at Yale, albeit in wartime. She later became a professor at Vassar College. Wright related that during the early forties, long after Grace's divorce from Hutchinson, the three of them—Wright, Pickford, and Hutchinson—went out to the Connecticut state forest on Sundays and collected spiders and dipped in ponds for water bugs and beetles.[12]

In 1951 Pickford was invited to go on the Danish deep-sea Galathea expedition to study octopods of the Indo-Malayan region. She was one of the first women scientists to go on such an oceanography expedition. At that time women were still not allowed on board U.S. Navy ships at sea. Pickford had studied and written about a creature with the exotic name *Vampyroteuthis*. It was a deep-sea cephalopod, and she had never seen it alive. She had the thrill of doing so on this expedition—a wonderful event in her life.[13]

Pickford entered the field for which she is best known, fish endocrinology, serendipitously. During World War II, Dan Merriman of the Bingham Laboratory, later its director, led a project of trying to use "trash fish" from Long Island Sound as a food source. One of these was the killifish or

Fundulus.[14] Pickford was involved in this project and became interested in the use of the growth rings on killifish scales as indicators of seasonal growth changes. She started to study the physiology of these changes. Shortly thereafter A. E. Wilhelmi at Yale Medical School was purifying beef growth hormone. Pickford obtained some of it and together with E. F. Thomson studied its effects on killifish, that is, the endocrine physiology of a vertebrate. It is interesting to note here that Hutchinson, the recipient of a Rockefeller Fellowship in 1925 specifically to study endocrine physiology, in his case of an invertebrate, the octopus, was never successful in this field.

Thus began Pickford's thirty years of fish and comparative endocrinology—all supported by the National Science Foundation after its establishment. The problem was one of microanalytical techniques again, as with her early spider work. Fish do not have much blood. She used methods developed by pediatricians just after World War II for blood work on premature babies. Her spider enzyme skills helped. Several people have described her lab with its ingenious microadaptations of apparatus. She had a number of graduate students. The first was James Barrow, with whom I have talked about Grace Pickford. He is the man who provided a "distinguished scientist in residence" position for Pickford at Hiram College in Ohio when she retired from Yale. He did his Yale Ph.D. research with Pickford on a trypanosome parasite of newts. "Did Grace Pickford know about parasites?" I asked. "She knew everything," he answered.[15] According to endocrinologist J. N. Ball, all comparative endocrinologists visiting the United States came to see Pickford and her lab. Some stayed as collaborators; she had particular skills in collaborative research.

Pickford described much of this work herself in a 1973 talk, "Introductory Remarks," the title perhaps a take-off on a famous earlier paper of Hutchinson's named "Concluding Remarks."[16] This talk, given at a symposium on fish endocrinology, was later published in *American Zoologist*. She described herself as an "old zoologist" (she was then over seventy) who got into comparative endocrinology when "exploratory research was still of the essence." "Now," she said, "we must be endocrine biochemists and molecular biologists." Medical research in endocrinology had greatly advanced so that "students of hormones of the lower vertebrates must . . . keep abreast of the mammalian literature." But she had long been a champion of comparative embryology.

Grace claimed to have remained a killifish researcher—a "one fish man [*sic*]." She did occasionally work on other species of fish and on sharks, including stingrays, as well as on the "living fossil" fish, the coelocanth obtained frozen for the Yale Peabody Museum.[17]

Fish use hormones, particularly pituitary hormones, on which Pickford and James Atz wrote the bible, a huge 1957 publication discussing many aspects of the lives of fish. For example, pituitary hormones are involved in the spawning reflex, thyroid control, and pigmentation changes. As regards pigmentation changes, one pituitary hormone stimulates melanin syntheses; another disperses the melanophores (cells with black pigment). Other hormones control osmoregulation of water and salts and the response to stress.[18]

In her 1973 talk Pickford told about some of her unexpected findings: for example, killifish, accustomed to salt water, died in fresh water but could be kept alive with injections of killifish pituitary glands. There was some special hormone that enabled them to retain salt. What was it? She tried a lot of mammal hormones on killifish with their own pituitary gland removed. Prolactin worked! At first this was disbelieved, but it led to research in many laboratories.

Another Pickford experiment showed that killifish engaged in crazy behavior with a particular pituitary preparation. Dan Merriman recognized what was going on; the hormone had induced spawning behavior. It had kicked off a response in a "spawning reflex" center. The technique for studying the peptides that caused this effect was difficult, and nothing was done on this problem for fifteen years after its discovery. By this time Pickford had retired from Yale and was doing research at Hiram College in Ohio and mentoring undergraduate students. Barrow later reported that when he first invited her to Hiram she said, "I wouldn't be caught dead in that little community." But she came, set up a laboratory, and stayed seventeen years. She had many visits from Yale scientist friends and other scientific colleagues. One Hiram undergraduate in that lab, Michele Macy, became interested in the crazy spawning behavior and was sent to the University of Alberta for the summer to learn the needed techniques to work on what Pickford called "one of the unresolved problems I brought with me to Hiram."

Another problem Grace Pickford wanted to solve had to do with stress and cold-shock syndrome. Epinephrine (a catacholamine) release stimulated the entire sequence of events in intact killifish. She spent many months working out a technique to measure the cortisol circulating level in this small fish with little blood. Pickford was extremely good at working with her hands on technical problems in her research.

By looking at the papers cited in Pickford's "Introductory Remarks," I discovered that Pickford had collaborated on the fish hormone work with thirteen different biologists from many institutions and countries, from the mid-fifties to the mid-seventies. Pickford was the first author of almost all of the papers. When she moved to Hiram College at the age of sixty-eight, she started a whole new field: the control by the pituitary of the hypothalamus. She inspired many undergraduate students there. One of these, Dennis Taylor, a biologist and administrator at Hiram, remembered, nearly thirty years later, her devotion to both her research and to her students.[19] She published papers until 1984, two years before her death. That amounted to a research career of nearly sixty years, nearly as long as Hutchinson's.

Pickford's concluding remarks to younger workers at the symposium cited above were as follows: "I am sure that when you reach the retirement age you will also be bristling with unresolved problems. . . . I hope you will be as fortunate as I have been in finding a slot where research can be continued."[20]

Pickford received awards late in life. Biology Department chair Clement Markert convinced her, kicking and screaming according to two reports, to teach the freshman biology course. She finally agreed and was a great success. Donna Haraway, then a Hutchinson graduate student, was Pickford's teaching assistant. She told the author that she loved working with her. Subsequently, Markert pushed for her promotion to full professor, making Pickford the first woman biologist to reach that rank at Yale, but that was only a year before she retired. Yale also awarded her a Wilbercross medal and held a memorial service for her—at which both Hutchinson and Barrow spoke. They, together with Willard Hartman, another former Hutchinson student, later Yale zoology professor, and Grace's longtime friend, scattered her ashes, as she had requested, in the Yale bird preserve, where she had taken many of her students.

A new award, the Pickford Medal, is now given every year to younger scientists at the International Symposium on Comparative Endocrinology. Pickford was able to go to Hong Kong in 1981 to present the first two medals when she was nearly eighty. Hiram College has a Pickford

scholarship for an outstanding woman student. The community of Cephalopod scientists recognized her much earlier in her career, naming an
octopod genus after her in 1953: *Pickfordiateuthis.*

What sort of person was Grace Pickford? A scientist always. She was
an extremely hard worker with "an unflagging drive to perfection," wrote
J. N. Ball in her obituary.[21] One of her friends, Margaret Wright, told
me how she stayed at Yale weekends to care for the killifish. Although
Pickford's immediate family are all gone, I have interviewed many of her
friends: Hutchinson's sister, Dorothea; her Cambridge comrades, Anna
Bidder and Penelope Jenkin; her fellow teacher at Albertus Magnus, Margaret Wright; her first graduate student at Yale, James Barrow; Denny
Taylor, who worked with Pickford as an undergraduate student at Hiram;
and several close Yale colleagues, including Willard Hartman and Charles
Remington. Kindness and willingness to make time for friends in need, in
spite of her stringent research schedule, were cited by all. Several called
her shy. One informant said she could be gruff at times to others in the
lab, perhaps those who did not live up to the high standards she set for
herself.

James Ball wrote of Grace's many nonscientific interests: history, art,
architecture, and music, especially Bach, a joy she shared with Evelyn.
She kept in touch with family and friends, and was always responsive to
their needs. This is clear from the few surviving letters of hers that others
have sent this author. She felt affection for her students at three different institutions, and they for her. Ball concluded that she was not only an
outstanding scientist. "She was an outstanding human being . . . one of
those rare people whose lives make the world a better place."[22]

What about Grace's appearance? She was small and sometimes wore
shorts or slacks at Cambridge in the early 1920s, according to Anna Bidder. At Newnham College, however, Grace had to wear long skirts. In the
early twenties she and Evelyn would bicycle together. Grace would go to
his house, and he would toss out a pair of short pants for her.[23]

At Yale, Grace usually dressed in men's or boys' clothing and had short-
cropped hair. She was once mistaken for a man and asked to leave the
women's bathroom at the Bingham Laboratory. I had never seen a picture
of Grace, but one day I found, in Penelope Jenkin's file in the Hutchinson
archive at Yale, a photo taken at an international meeting of a very large
woman and what appeared to be a short boy. I recognized Grace Pickford

immediately. Margaret Wright said that at Yale Grace loved to dress like a boy. When asked why, Wright answered, "I suspect she wanted to be a boy." This may not be the whole story of her dress, but both Bidder and Ball suggested that she took pleasure in defying convention. Convention, said Ball, had discriminated against her early in her career simply because she was a woman. Convention, if we can call it that, discriminated against her at Yale as well. The pattern of her career was very similar to that of other women scientists at Yale, research associates on soft money who continued to publish over spans of several decades. This was true for Grace Pickford even after she was internationally known for her work in endocrinology and when foreign scientists flocked to her lab.

Few Pickford letters are extant, but one is especially revealing of her kindness. When her former brother-in law, Evelyn's brother, Leslie Hutchinson, died in 1974, Grace wrote to his widow, Hannah:

> This brings you my deepest sympathy, but I am happy to know that you have children and, I believe, grandchildren. . . . I vividly remember the visits that I made to your home on the Isle of Wight . . . how kind you all were, and how understanding. I also remember our student days when Leslie, although not a zoologist, used to come with us on field trips, as to Wicken Fen. Because of his ulcer he always had a thermos flask of special warm soup or porridge. . . . Nevertheless, he was the most lively and cheerful member of the group and I have always thought it was a privilege to know a young man who could face his problems so unconcernedly. Yet he must have suffered so much . . . With affectionate memories, Grace.[24]

At least two large questions remain. Why did Evelyn and Grace separate and divorce, and why did Hutchinson leave her out of his autobiography although she was with him on three continents and throughout the period of which he wrote? Grace was always Hutchinson's intellectual companion and, at least in the early days, his field comrade and research collaborator. She was clearly an affectionate person, and the only one of his three wives for whom most of Hutchinson's family had truly positive feelings. He was affectionate toward her as well and concerned about her welfare. But something was missing from the marriage—the children that Evelyn wanted certainly, but probably something more. Margaret Wright said: "I can remember very few occasions when she actually referred to the marriage. But this one day when we were walking along looking at a wonderful view, Grace said, 'I don't know, either we should never have been married or never have been separated.'"[25] According to Wright,

Grace was "much more of an intellectual companion. She was his intellectual equal, but not really a wife."

But Hartman, Remington, and other Pickford and Hutchinson colleagues agree that none of this should have been cause for her omission from his autobiography. Grace had been kind to Margaret, Evelyn's second wife, even allowing her to stay at her house while Evelyn was in Nevada getting their uncontested divorce. Nevertheless, it is well known that Margaret was terribly jealous of Grace. For example, when Hutchinson's sister, Dorothea, visited Evelyn and Margaret at Yale, Margaret told her that Grace's name was never to be mentioned in their house. (Dorothea, however, was Grace's old friend. She telephoned Grace and went to see her on that visit, though she never told either Margaret or Evelyn.)

Grace and Evelyn had remained friends up to the time of the publication of *Kindly Fruits of the Earth* in 1979. Grace, who by then was at Hiram College, was very hurt by her complete omission from the book, even in his account of the Biological Tea Club of which the two of them had been cofounders. Grace's Cambridge friend Anna Bidder told this author that Grace never wanted to speak to Evelyn again after the autobiography was published.[26]

Hutchinson himself corroborated the above view in a letter to his sister Dorothea. Dorothea and Evelyn became frequent correspondents and confidants after Margaret's death. In answer to why he had left Grace out of the book, Evelyn wrote to his sister that he had either to offend Margaret by putting Grace into the book or offend Grace by leaving her out. He chose the latter. Margaret was his wife, the wife he adored, the woman he lived with for fifty years. Evelyn wrote to Dorothea about Grace and the autobiography: "I always regarded the statement in the preface, 'Everyone who has helped me during three quarters of a century is included in the dedication' as covering her great contribution in South Africa even though the marriage as such did not exist."[27]

Margaret Seal Hutchinson, the Wife of Fifty Years

Margaret Hutchinson is a rather mysterious character. She and Evelyn met on the ocean liner when he was returning to New Haven via England after the North India expedition in 1932. One version goes like this: Margaret was singing on the boat and Evelyn fell in love with her. On arriving in New Haven, Evelyn announced this to Grace, and he went off to Reno to get a divorce.

Some parts of this scenario are apocryphal. No one I interviewed knew what actually happened on the boat, and Evelyn himself did not wish to talk about either Grace or the early days with Margaret, although he did discuss her later illness with me. It is known, from Evelyn's 1932 letters to Grace from Ladakh, that Grace and Evelyn were planning to divorce before he met Margaret. It is true that Margaret was a singer; she continued to sing and to give performances in their home for many years. Margaret and Evelyn were married in 1933. She died shortly after their fiftieth wedding anniversary in 1983.

Although they did not have science in common, Evelyn and Margaret shared other interests: art, music, and literature, and probably religion. They went to the High Episcopal church in New Haven, where there was much ritual and incense. David Boulton, the minister of that church, remarked on Margaret's interest in music in his eulogy for her: "I first knew Margaret as a musician. Many years ago, when I was a student, I heard her sing Elizabethan and Jacobean songs and the wonder of it was that she was not performing, she was delighting in the music and asking us to share the delight." Boulton listed other things that delighted Margaret: "flowers, their color and form and scent; the intricacy of gold work and the light . . . at the center of gems, the touch and shimmer of fabrics or a cat's fur." She clearly delighted in beautiful things, a delight she undoubtedly shared with Evelyn.[28] Several people used the word "refined" to describe her.

Margaret apparently grew up in the English countryside. There are very few extant letters, but she corresponded with one close friend, Gwenllian Knight, wife of the vicar of Kewstoke. Kewstoke may have been Margaret's childhood home, or at least she and Gwen, from the one extant letter to Margaret from Gwen, seem to have been childhood friends.

Margaret and Evelyn visited the Knights in 1953 in Kewstoke, Weston-super-Mare, Somerset. Margaret was baptized at that time by the Reverend Knight in his church. She had planned another visit for 1970, but Mrs. Knight died that year. The Reverend Knight had predeceased her. The new vicar wrote to Margaret about her friend's last illness.[29]

At the time Evelyn and Margaret were married, his parents, Arthur and Evaline Hutchinson, were both still alive. Margaret and Evelyn were

27. Margaret Hutchinson, Evelyn Hutchinson's second wife.
Manuscripts and Archives, Yale University Library.

visiting them in Cambridge for Christmas, and Dorothea, Evelyn's sister, though living away, came home to visit. As Dorothea recounted, she wanted to get to know Margaret. Something came up and Dorothea said to Margaret, "Oh, did you used to do that sort of thing when you were a child?" Margaret failed to answer. Afterward Evelyn took Dorothea aside and told her not to talk to Margaret about her childhood; it made her very unhappy. Margaret had two older sisters, but other details are lacking.[30]

Margaret was loved and cherished by Evelyn. They were viewed by others as a very close couple. According to Willard Hartman, a Hutchinson graduate student and later a zoology professor at Yale, "She was ex-

ceedingly feminine. They were utterly in love, beginning to end, I would
have to say. . . . He was so cherishing of her, much more than kind. Extremely protective; his world seemed to revolve around her."[31]

Evelyn, as we know from his earlier letters to Grace, was hoping to become a father. Margaret had several miscarriages according to one member of Evelyn's family. Dorothea remembered a letter from Evelyn from
the early days of their marriage, saying that he was sorry he had not written. They had had great hopes that Margaret was pregnant but "then it all
went off." He was obviously very disappointed, Dorothea recounted.[32]

Evelyn may have been the only person to whom Margaret revealed
herself. Few people other than Evelyn seemed to have cared for or appreciated Margaret. She does not seem to have had close women friends
in New Haven. Most of Evelyn's family, including Dorothea, actively disliked her or at least found her to be cold. Dorothea wrote, in a letter to this
author: "I must admit I found it surprising that someone who apparently
was as cold as she [Margaret] appeared to be, could have evoked such a
very loving response from him. I may be biased because I was never able
to feel at all close to her myself. She made it quite clear that she disliked
me and always seemed to expect me to behave badly."[33] This seems odd
since Dorothea herself had studied at Cambridge University, had an exciting and well-respected career as a psychiatric social worker in London, and had been brought up in a fairly upper-class Cambridge family—
perhaps not wealthy but certainly well behaved!

Margaret was older than Evelyn, born December 4, 1896, more than
six years earlier than he. Photographs of her as a young woman reveal her
to be an attractive and, unlike Grace, particularly feminine woman who
cared about her appearance. She dressed well and had much valuable
jewelry, some of which Evelyn ordered or had made for her.

Evelyn and Margaret usually spent June or July in England, staying in
Cambridge at a hotel, and sometimes in Southampton, also in a hotel, to
visit family there and on the Isle of Wight. The family, especially Evelyn's
nephews and niece and their spouses, felt that Margaret kept Evelyn at
a distance from them. Not until after her death did he actually stay with
the family and become close to them. Several graduate students also commented that Margaret "protected" Evelyn from them and that she took no
part in student-faculty activities at the university. They were, however, all
invited to hear her sing, formal social events at the Hutchinsons' house
that some of them would have liked to avoid. They were unnerving to self-

conscious graduate students. Other graduate and postdoctoral students found Margaret a gracious hostess. Her voice was apparently not as good in her later years as when she was young, though Evelyn was reported to listen to her with rapt attention.

Dorothea did concede that Margaret communicated better in letters than in person, and this is borne out in the letters that do survive. Several of these are to Dorothea and other members of the family. In June 1961 Margaret and Evelyn went to Princeton, where Evelyn received an honorary doctorate. Margaret wrote to Dorothea, "We cabled Leslie [Evelyn's brother] as soon as we returned from Princeton and this morning, have a loving and heart-warming reply. . . . I think the citation deeply moving and E. is very happy about it. It does give science a much needed lift in people's minds." Margaret wrote that the citation called Evelyn Hutchinson "a scientist and a humanist in the best and proudest traditions of both camps." She continued, "People in Princeton were very good and kind to us, and they were happy that your uncle, Sir Arthur Shipley [see chapter 2], had received a similar degree at the beginning of the century. Evelyn, of course, was very moved by this."

Margaret added that they were "longing for all the news we can get of Rosemary's [Evelyn's niece in England] wedding. . . . I feel hungry for everyone's views and gossip about what happens." They were not able to attend the wedding because, although they had just been in Europe, Evelyn was implored by his department head to be in New Haven before the beginning of the term, "so we just had to go ahead [earlier] and have the first holiday I have had with Evelyn for 28 years!!!" That would actually have been since their marriage in 1933. They went to Florence where Evelyn had promised not to work and kept his word, though he worked "very hard indeed" the week they spent in Rome.

There are other letters from Margaret to Dorothea, including a much later one from summer 1977, in which she sent a photograph of the dedication to Evelyn of a zoology laboratory in Johannesburg (at Witwatersrand, the same university from which he was fired as a lecturer!). She described the scene in detail to Dorothea, including the display of writings, including one by Dame Rebecca West on Hutchinson's retirement. Margaret was clearly proud of him.[34]

There are also many letters from Evelyn to Margaret, all addressed to "Dearest love." The letters are from his travels. Although they traveled to Rome and Florence together in 1977, as Margaret wrote to Dorothea,

Evelyn also made several trips without her, to both Italy and Poland. His letters to her from this period survive; hers to him do not.

When Hutchinson went to the International Congress of Limnology in Poland, he left Margaret in England. He flew from London to Warsaw and then on to Krakow. He wrote her a seven-page letter from Krakow the first night he was there. The tone of the letter indicates Evelyn's feelings for Margaret: "Half an hour on the bed, largely thinking about you and how you may be faring," he wrote. His first impressions of Poland were "totally unexpected." Both in Warsaw and in Krakow all the people looked well dressed in bright summer clothes. "The most striking thing is the gaiety and animation with which they talk among themselves and the extreme freedom of expression." This was, of course, during the cold war. "I will write of my adventures . . . as they may provide some entertainment and transfer something of this puzzling place to you."

He had been to the art museum and had seen two "superlative pictures, the Leonardo and a Rembrandt landscape. . . . The Leonardo is wonderfully beautiful. . . . She seems to exist solidly in a magical space." He continued with a whole page about the paintings and other art objects in the Krakow museum, comparing them to what they had seen elsewhere together. Several pages about his visits to some of the ninety functioning Krakow churches follow, including the sight of gypsy women "in full array" and a trumpeter blowing for a wedding. There is also quite a bit about another interest of Margaret's: women's clothes. "The very large number of attractive summer dresses must be mainly home made," since, he reported, there was not much to be bought in the shops. "Tape doesn't seem to exist, nor paper clips, rubber bands, or anything of the sort—only immense good will. I do hope that one day they will deliver me a letter [from you], and that I can get a stamp to post this one. Love, my dearest, Evelyn."[35]

There were two hundred limnologists in Krakow. Later Evelyn described the bus trip from Krakow to Warsaw, where the actual congress was held, a trip that included the sight of storks and ruined castles. Hutchinson was at that time president of the international organization and was in charge of the opening session.

In this letter, Evelyn described the music, another of their common interests, in some detail. For example: "a concert of old Polish music, two of the four works were charming but of the international Italian 17th and 18th century styles. It also appears that Telemann was a Pole."

Then came "the great surprise," a sixteenth-century anonymous work called the "Dumba, which I am told means 'meditation triste.' It was played by a string quartet, substituting I suppose for 4 viols, and was a very quiet contrapuntal fantasy of great clarity and beauty." He wrote much more about the music.

Hutchinson's presidential address was about science and the arts. "I thanked the people who arranged the concert for the music and then suggested that the clarity and elegance of the Dumba be taken as a model for all scientific discussion throughout the week." Hutchinson also wrote to Margaret about the state of Poland, "the cardinal virtues: fortitude, piety, vanity—and complete muddle."

He was happy to be able to report to Margaret that the Lago di Monterosi in Italy on which he had done research was on the list of lakes of the world worth saving for their scientific merit. It had recently been saved from being ruined by a building development, "which pleases me very much," Hutchinson wrote.[36]

Another set of later letters was sent from Italy, when Evelyn was attending a meeting in Florence while Margaret was at home. On the plane crossing the Atlantic he wrote to her, knowing her love of colors: "We have been experiencing a fantastic sunset. . . . The whole of the Northwestern sky is red about the clouds, then orange, yellow, green-blue and violet into the darkening sky. . . . I think this must be due to the extra load of volcanic dust in the air. It was staggeringly beautiful, the green a wonderful band of transparent emerald." About his anticipation of letters from Margaret, he wrote: "It will be wonderful to get a letter. I have read so many of yours to other people, but the excitement of there being one to me is very real. . . . Love my darling, Evelyn."[37]

His next letter was from Pisa. "It would have seemed so natural to come back to tell you in spoken words what I have been doing, so much that the words kept forming all the afternoon. I will try to put them on paper rather than in your ear." He had sent a cable on arrival and "hoped it reached you quite early this morning."[38]

In a later letter he described several churches and the Camposanto cloisters, where "many of the frescos, more or less restored, are back in the cloisters, along with the drawings found underneath them when they started to fall off after the fire. Some of the drawings have more life than the finished paintings. . . . I made a great effort to get one very beautiful one photographed. . . . I discovered a beautiful (c. 1620) Rhinoceros, on the west door of the cathedral, apparently a Renaissance symbol of security."[39]

He had been driven from the station to the hotel in a Victoria. "You seem just around the corner . . . which is as it should be since I have gone away. It will be wonderfully lovely to be back—I have wondered so much what has been happening to you." The next morning, after seeing "everything exciting" in Pisa, he was going to Lucca for the day. He described the Lucca cathedral:

> There was an Episcopal High Mass, . . . the Archbishop tall, thin and very devout-looking. As a celebrant he runs our rector a close second, the whole thing was done beautifully . . . the mass chanted, the music . . . mostly Bach. You would have loved it, but probably not the bus trip out there. . . .
>
> Tomorrow evening a preliminary encounter with work [in Florence] I suppose. Tuesday maybe a letter from you. How I want it. Nothing is quite real without you. . . . All my love darling Margaret.[40]

These letters detail Evelyn's and Margaret's interest in Italy's churches, art, and music, but Hutchinson was in Italy primarily for a conference in Florence, to which a number of well-known people in both science and the arts had been invited. He was staying in the Villa Medici, "a very plush establishment, but as someone else is paying for it I will enjoy a little extra luxury." The conference members worked in a large room, "traditionally Machiavelli's Study in the Palazzo Vecchio. We were welcomed by the Mayor of Florence. . . . He wants to make Florence the centre of world cooperation and fundamentally that is why we are here."

He went on to describe the participants, who included Harrison Brown, John Kendrew, "a crystallographic molecular biologist from Cambridge who comes from [my] father's group through Bernal . . . , Dubos from Rockefeller, an admirable medical man with strong ecological leanings." Also in their working group was a "Jugoslav physicist of eminence . . . with a daughter with . . . pre-Raphaelite features, a long blond bob of hair and an air of showing off how elegant a Tito Communist girl can be, and myself." His was the scientific group. Other groups included Clifton Fadiman, the writer and radio talk show host, and Robert Hutchins, "who strikes me as a fairly able animated figurehead." Hutchinson could make remarks to Margaret that he would never have made publicly.[41]

There were wonderful concerts in Florence, too, held especially for them, one an evening of choral compositions from the "XIV century to

Scarlatti, which have remained till recently in Manuscript in the Cathedral archives. This will be the first time any living audience has heard them." Another concert was held for them at the Roman theater at Fiesole. He wrote from Florence:

> I got a walk along the Arno last night before dinner and looked in at Ogni Santi to admire Botticelli's Saint Augustin—a good picture to try to live up to at a Congress devoted to hard intellectual work. Florence is as lovely as ever. [They had been there earlier together.] I will leave this letter open in case a letter comes for me this afternoon. The whole thing here is clearly very worthwhile. In spite of this and everything else delightful it will be wonderful and glorious to put my arms round you.

Later he added, "[Your] letter just came. It has given me enormous pleasure. I am so glad things are going well. The lilies sound lovely. Harrison Brown much touched by your greetings. . . . Love again."

He also wrote from Florence that he had managed to get to the Uffizi at lunchtime and during a rest period to see a special show of seventeenth-century Genoese paintings and all the paintings they knew best, including Botticellis, Filippo Lippis and the Leonardo *Annunciation*. The afternoon session was exciting: "It is extremely interesting to see the enormous difference in approach that the philosophical background to national cultures make. . . . There was virtually a head-on collision [British and German] about free will. I have reread your letter and find it a delight, as on previous readings." She must have written about their cats; they always had their beloved cats. Evelyn wrote, "I saw at a window a very old man, a Siamese cat, and two of the latter's natural children, clearly by a black cat, dark with some of Homo's [a Hutchinson cat] stance. They were all sitting looking down the street."[42]

This sampling of letters reveals much about their relationship. It was soon to be sorely tested by Margaret's illness.

As caring for his wife took more of his time, Hutchinson had less time for his research. In April 1981 Hutchinson wrote to a friend in the Geological Survey: "When I retired I thought I should have no difficulty in completing *The Atmosphere* as well as my *Treatise on Limnology*. Unhappily my wife has been seriously ill for the past few years and the opportunity to do the necessary library work has been enormously curtailed. It is now obvious that I shall not be able to finish what I set out to do."[43]

Margaret's illness was not diagnosed until 1981; the diagnosis was Alzheimer's disease. A Swedish colleague wrote to Hutchinson in 1982:

I was horrified to hear about Margaret's diagnosis. My sister's father-in-law suffered the same disease. Only after many sad years did the gentleman's wife finally turn over his care to an institution. . . . She has returned to the living after those nightmares you must understand. . . . She realized, as perhaps you must also, her husband really couldn't tell much difference, whereas her life's strength was ebbing. . . . Dear Evelyn, please consider caring for Margaret with the help of an institution designed for that purpose. It is perhaps not my business but I love you—so I have made this statement.[44]

Evelyn did know those "nightmares"—once Margaret managed to get out of the house and wandered down the street, to be rescued by the neighbors—but he did not institutionalize Margaret. Evelyn stayed home with her at least half of each day. Devoted students and others would come for part of the day so that he could go to his Yale office and do some writing. He once told me (this author) that he had driven her to a concert in someone's home. She seemed to greatly enjoy the music, he related, but afterward had no memory of the place or the music. But he felt she knew him, knew his touch. She stayed at home with him until the end.

He wrote to Yemaiel, his and Margaret's adopted daughter during the years of World War II, in December 1982: "Margaret's condition is definitely worse. She has almost stopped reading and is apathetic for a large part of the time. . . . I think I can manage fairly satisfactorily for the next few months, with luck perhaps more. I am still able to do a little writing, the doctors encourage this as it helps me to do all the other things that I am involved with as nurse, cook and ladies maid."[45]

He managed for more than just a few months. Margaret died on August 24, 1983. Evelyn wrote to Sybil Marcuse, his old friend and a former Yale musicologist, thanking her for her letters: "The earlier one . . . arrived just before Margaret began to get really seriously ill, and she as well as I could enjoy it. . . . At the end of July she began to fail badly, but could have a tiny party for our Golden Wedding, with a 2-year-old neighbor who provided great delight. She lost all power of speech early in August and died as quietly as anyone could, pulling at my hair gently with traces of a smile before her breathing became irregular and stopped."[46] This letter found its way to the author via Evelyn's sister Dorothea, who commented, "It gives a very touching description of her last hours. It shows his very tender and loving relationship with her." After discussing her own difficulties with Margaret, real, not imagined, Dorothea added, "I think the most important thing is that Evelyn had such a rich capacity for love and tenderness in all his relationships."[47]

Evelyn saved many condolence letters. One is from his former graduate student and eminent ecologist, Fred Smith, and his wife, Peggy: "We have thought of you much more often than we have communicated. And now we are grieving. You have always been a special couple." His Yale colleague Luigi Provasoli and his wife, Rose, who had retired to Italy, wrote: "Dear Evelyn, in your time of deep sorrow we are with you with all our hearts, affection and feelings. Dearest Margaret will always be alive in our memory as the gentlest, most kind and adorable being we have had the privilege to know, admire and love."[48]

The letter from Edwine Martz, Margaret's longtime accompanist, is telling: "I want to express my sincere sorrow over Margaret's death. I always think of her in her prime . . . with her lyrical performance of Elizabethan songs. I admired her greatly for her trained musicianship. . . . I appreciated her indulgence in working with me on the recitals we gave together.[49]

Patrick Finnerty, a former student who had worked on animal population cycles, did not hear of Margaret's death until he read a dedication to her in one of Evelyn's papers. He wrote:

> This was the first I heard of your dear wife's death and the news sent me off into memories of daylilies and Burmese cats, lovely operatic arias in your home and our discussion of Piaget over champagne. . . . I was and remain deeply touched by the mutual love and sympathy you two shared for half a century. How fortunate you two were to find each other and to be able to construct that rare life of mutual support based on shared respect and love. . . . I know that while Margaret's absence leaves an unfillable void, she could not have left a greater legacy than your life together.[50]

The most helpful letter may have been the one from an old friend from England: "The sad news of Margaret had not reached me [earlier]. You have my deep sympathy. . . . The desolation of watching the spirit of a loved one day by day taking leave of the body is one of the very bad experiences. I am deeply sorry you have had to go thro' it, and hope that the balm of time does something to heal the wounds."[51]

After Margaret's death, Evelyn was able to get on with his life. He wrote to Yemaiel in London. "It is now two weeks since Margaret's death. I still feel paralyzed in daily life, but as the loneliness becomes chronic rather than acute, I think I shall be able to do what I hope to do." He hoped to come to England to visit and to do some work. "It will be wonderful to see you. My deepest love to all of you."[52]

At an earlier time, Margaret had taken care of Evelyn. One funny story was related to the author. Evelyn at one point had stomach ulcers. "Margaret was very proud of the fact that she had managed to get rid of them by giving him a proper diet . . . nothing but bland. . . . I ran into them again in London . . . and had lunch with them. The lunch has been prearranged [in a restaurant]; Margaret and I had a perfectly normal, very nice, lunch. [Evelyn] had porridge with a poached egg on top."[53]

In fact, Margaret never cooked at all except early in their marriage, according to some reports. As he had at Cambridge, Evelyn made boiled eggs for breakfast—for the few visitors, including Margaret Mead and Ruth Patrick, who stayed with them overnight.

Yemaiel, the Temporary Daughter

During World War I, a young woman who was working as a technician in pathology with a military hospital came to live with the Hutchinson family in Cambridge. Her name was Gwendoline Rendle, contracted to "Gwendle," and she became an older sister to Evelyn, in his later description. His actual younger sister, Dorothea, said that during the "Great War" people had to take in war workers, and the Hutchinsons were asked to have her. "Gwendle was a perfect dear. We loved her very much and kept up with her." Gwendle's daughter, Yemaiel, told the author that she thought Gwendle was in love with Evelyn at that time. Dorothea was doubtful. "She was about eighteen when she came to us. Evelyn was only fourteen; if it had been twenty-four and twenty-eight it wouldn't have mattered, but fourteen and eighteen?" It would seem more likely that Evelyn was in love with her at the time. At any rate they remained friends throughout their lives.[54]

Gwendle eventually married, or at least spent the rest of her life with Moishe Oved, an older, well-known Jewish poet and philosopher. She was known as "Sah Oved" (but not to the Hutchinsons, who continued to call her "Gwendle"). She became a professional jeweler and silversmith.[55] Among the couple's friends were a number of writers and artists, including Jacob Epstein, who did drawings of their daughter Yemaiel—a wonderful sketch of her as a baby with her father; a drawing of "Yemaiel's journey"; and on her return from America, a "portrait bust of a little girl which expresses in an exquisite manner the gaiety and joy of extreme youth," according to a London news clipping. For this bust Yemaiel sat for Epstein seven times in December of 1944 and January 1945.

In 1940 there was widespread fear of Hitler invading Britain, and many English children were sent to live with families in the United States. Gwendle's half-Jewish child, Yemaiel, was one of these. Evelyn had written to her mother, offering to have her stay with him and Margaret.

Sah Oved (Gwendle) kept a book, mostly of letters about Yemaiel, from June 1940 until her return to England in 1944. This book reveals much about the Hutchinsons, particularly Margaret, as well as about Yemaiel, whose mother wrote in Yemaiel's book in June 1940: "War disastrous. Decide we must accept Evelyn Hutchinson's offer of hospitality for you in America. . . . It will be better for you to be safe there in such a cultured and loving home, but my heart breaks."[56]

There is a photograph of Yemaiel's father, who was called "Abba," with Yemaiel when she embarked on October 24, 1940. She left during the "Battle of the Atlantic" just after a boatload of children had been torpedoed. Her mother wrote that it was only "afterward [that] I heard the boat had been attacked by U-boat torpedoes and bombed."

Yemaiel arrived safely in Boston on November 15 and is shown in newspaper clippings in the arms of Evelyn Hutchinson. The photographs have captions such as "Yale Professor adopts Refugee" and "Yemaiel Oved, two-year-old son [sic] of an English poet, is shown in the arms of Professor George E. Hutchinson, of Yale, who has adopted him for the duration of the war." Another clipping reads, "The youngest unaccompanied refugee was 2-year-old Yemaiel Oved, daughter of a London poet." Hutchinson looks very happy, Yemaiel much less so. Her mother wrote she had had three weeks' in passage on the *Baltrover*, which managed to evade Nazi airplanes and submarines.

Yemaiel told me in London more than fifty years later that she had been sent because her parents were very worried. She was their only child after many miscarriages, and her father was Jewish. In Yemaiel's words: "I lived with Margaret and Evelyn for three-and-a-half years. Evelyn was father and mother to me. He did all the looking after; he bathed me and fed me, everything. . . . I have memories of Margaret having tea on the sofa. I mustn't disturb her; she's having a rest."[57]

Evelyn and Margaret tried very hard from the beginning to impress upon their small temporary daughter that they were not her real parents. The real parents back in London sent letters and presents and were in turn sent photos and even a movie of Yemaiel. Yet the book of letters from America gives a rather different picture of Margaret from the one Yemaiel remembers.

It is clear from Margaret's letters, which Yemaiel's mother saved and put into the scrapbook, that regardless of whether Yemaiel and Margaret were emotionally close, Margaret did care a lot about Yemaiel and about her emotional and other development. For example, Margaret wrote: "Dec. 19 [1940]: Yemaiel is now emerging as the sweet and vastly interesting little girl you know, and the exhausted and frightened child is receding rapidly into the background." And "Yemaiel is rapidly learning to do a great number of things for herself." They gave her a doll's bed for Christmas for which Margaret made sheets and knitted a blanket. She later made doll's clothes. On New Year's Day 1941, Margaret wrote, "These days Yemaiel goes in for winter sports in a big way, sleighing . . . in cold bright weather and sparkling with joy." Margaret did cook meals as well as treats for Yemaiel. "I made junket today for the first time for Yemaiel. . . . She said 'Have that 'nother time.' What more could a chef ask?"[58]

That same January Margaret wrote to Gwendle, "You will be interested to know that the tempers—now much less frequent—are something she shares with other English refugee children," apparently including the queen's "small nephew." In this letter she has encouraged Yemaiel to call her "Margaret" instead of "Mummy" so that she will be clear who her own mother is. That may have been hard for both of them.

In February Margaret wrote, "Last week she took a fancy to a red cap Evelyn wears on the lake in the cold weather, so I bought her a red velvet hood . . . and she looks lovely in it." In this letter Margaret also discussed Yemaiel's language development, her discovery that she could make rhymes, her creative use of building blocks, and her emotional responses: "Her emotional life is much more harmonious." That same month: "Your small daughter's musical education seems to have taken quite a stride. Yesterday she hummed quite accurately a phrase from 'The Little Nut Tree' and she had recognized a Bach minuet that she had heard earlier." In March 1941, Margaret wrote to Gwendle: "Yemaiel does miss you very much indeed, although it is almost entirely unconscious. I feel she is much puzzled that you do not at least come to visit. . . . This last week

28a. Two-year-old Yemaiel Oved, the youngest unaccompanied refugee, after arrival on the British steamer *Baltrover*, in the arms of Professor G. E. Hutchinson of Yale, who adopted her for the duration of World War II. Manuscripts and Archives, Yale University Library.

28b. Yemaiel as a child in the home of G. Evelyn and Margaret Hutchinson. Courtesy of Yemaiel Aris.

she has talked an unusual amount about Abba, and actually stood at the French window one day and told me which path he would come along—and that did make my heart turn over."[59]

There seems to be affection between them. "One of my great pleasures," Margaret wrote, "is when she comes into my bedroom in the morning . . . and pats me on the face or cuddles her face against mine." By April Yemaiel was going to nursery school. In June Margaret wrote down in notes, in a letter to Gwendle, a tune that Yemaiel had made up, and they also dug up a small patch of garden for Yemaiel. By September of that year the doctor "thought her very different from the screaming frightened child of last November" and in "splendid condition."

But in October of that year, Margaret related, "Your postcard to Yemaiel with a picture of Emmanuel College [Cambridge] brought forth the query, 'Is that where they make shredded wheat?' She was abashed when we laughed and rushed to Evelyn to be comforted." Perhaps he really was her emotional anchor.

In early 1943 Margaret asked Yemaiel whether it was "muddling to have a mummy and daddy on both sides of the sea?" She answered, 'Yes, nice muddling.'" Margaret was indignant that Yemaiel's mother, on seeing a movie of Yemaiel they had sent, called her "plain." "She is lovely to look at with her colouring and lovely eyes, indeed all of her. . . . I think you must be in horror of motherly pride if you think your little girl plain." Margaret herself sounds guilty of "motherly pride" in reporting two "firsts" for Yemaiel: seeing her first play and reading her first word. "[Yemaiel] has never been so moved about anything as the play." Margaret wrote further: "Evelyn and I are teaching her to read in small doses, and she is extremely thrilled. Her first discovery was 'the.'" By May 1943, Yemaiel, at five, was reading on her own, much to Margaret's delight. No chocolate eggs were available for Easter that war year, but Grace Pickford sent Yemaiel "a beautifully painted egg."[60]

Evelyn wrote fewer letters, or fewer were saved, but they also indicate his delight in Yemaiel. In March 1941 he wrote to Gwendle: "Lots of snow and bright sunlight, so I took Yemaiel for another ride about the estate. The last few days have been by far the mot rewarding since she came here. The burden of the journey seems to have been lifted from her and she is now the child you used to describe before she came over." Later in 1941, when she was three, he wrote, "Yemaiel is extremely observant, and is always finding spiders, caterpillars, etc., and asking questions about them that I would not dare set in a Ph.D. exam. I am slowly beginning to teach her letters. . . . I think she has some glimmering of the idea of written words." A year later he reported that Yemaiel said, "The other children ought to know how to swing high on the swing and how to tie; I show them how to, but they don't learn." Evelyn's comment: "I know the feeling only too well professionally."[61]

In April 1943 he wrote, "I think we can say that Yemaiel has been less trouble and more rewarding than the great majority of children of her age. . . . She is certainly not plain." That term from her real mother seems to have rankled both him and Margaret. "In fact, she is much the prettiest of the small children in her school." There is a prophetic letter from Evelyn written on May 8, 1943:

> We have just returned from taking Yemaiel to a matinee of the Ballet Russe de Monte Carlo. . . . To Yemaiel it was all fairyland. . . . She was particularly entranced by the Christmas part in Nutcracker. She was for some time unconvinced that the dancers were real people and asked how they worked. She went out to have an ice in the interval, but although ordinarily ice cream takes precedence over everything, she was worried lest we should not be back in time for the curtain. She sat on my knee all through the performance, being too low in her own seat, and never fidgeted once. I think she was the smallest member of the audience.

Yemaiel had told me (the author) about this performance of *The Nutcracker* before I found Hutchinson's letter. She said, "Evelyn took me to see *Nutcracker*, the ballet. That I can still remember. I thought they were all dolls that were wound up. I thought they were wonderful. Evelyn said no, they were grown-up people. From that moment, that is what I meant to do."[62] And that is indeed what Yemaiel did; she became a professional ballet dancer.

Yemaiel, aged five-and-a-half, went home before the war was over on the Spanish ship *Megellaines* "with full lights and her neutral nationality clearly marked." The ship brought home about two hundred British children and women from Canada and the United States. The trip began on February 5, 1944. The *Megellaines* sailed from New York to New Orleans, remained off Trinidad for two and a half days while a spy on board was arrested, then continued to Lisbon, where it docked. The passengers then traveled by air to Ireland and Croyden, London. That was on March 5 — a whole month later! A news story, not entirely accurate, reported that "50 unaccompanied British children arrived in Lisbon March 1 from the US and Canada." The newspaper article, datelined Lisbon, reads, "Professor Hutchinson of Yale University . . . broke down completely when he had to say good-bye to the children [sic] he had adopted."[63]

Whether Yemaiel had ever seen this newspaper clipping or not she commented, "I was very anti-coming back. I got imprinted on Margaret and Evelyn. . . . It must have been terrible for Evelyn; he was broken-hearted at losing this little girl at five."[64] Was Margaret brokenhearted as well? It is hard to imagine that she would not have greatly missed the charming child to whom she had given more than three and a half years of her care.

Both Evelyn and Margaret tried to stay out of Yemaiel's life after her return to London so that she could reunite with her real parents. That was very hard for Yemaiel, especially with her father. "I was just under six when I came back. I wasn't at all keen on my parents. That transition was very difficult. I got quite fond of my mother eventually." Yemaiel said she always liked Evelyn better than her father. But when asked whether he wrote to her after she returned to England, she said, "Yes, but I got rather rebellious about Evelyn. That was fairly natural, but my mother got very cross about it. I wasn't very keen on writing to him."[65]

Later, when Yemaiel did go to New Haven to visit, she was not invited to stay with them; she said that Margaret never wanted anyone in the house, that nobody was welcome to stay with them. There were a few exceptions to this rule, notably Margaret Mead, but even Evelyn's sister Dorothea was not invited to stay in their home, nor was their young nephew Andrew when he came to New Haven to visit. He was sent to the YMCA!

Yemaiel did not come to really know Evelyn until his later years, when he came to visit after Margaret's death and stayed with her and her husband, Ben. When asked if they had interests in common, Yemaiel said that they went to art galleries together as adults. She also remembered that "as a small child he used to take me to ponds to look at beetles. As a little girl I was very interested in natural history; that was largely due to Evelyn." And later, "He was a fantastic parent and he wanted to interest me in science." About her mother's jewelry making Yemaiel said, "That was her passion, her life." She made jewelry for Margaret "as a sort of repayment for their having me. . . . My father had a jewelry shop on Museum Street and had a lot of American dealers, so there were people coming backwards and forwards. He would give it to somebody to take it for the Hutchinsons."[66]

In London Yemaiel proceeded to become a ballet dancer. She started to study ballet at age eight, which was considered a sensible age to start. As an adult she danced with the Edinburgh International Ballet, Sadler's Wells Opera Ballet, William Gore's London Ballet, and the Amsterdam Ballet. Just before she married, she danced in Canada for three years with the Royal Winnipeg Ballet. She did not come back to the United States until 1966, when the Royal Winnipeg Ballet performed at Jacob's Pillow in Massachusetts. Evelyn came to see her there, but not Margaret. She went to New Haven afterward for a day to visit them.

After three years in Canada, Yemaiel married an actor, Ben Aris, in 1966. Both Evelyn and Margaret came to her wedding. When I interviewed Ben much later, he pointed out that Yemaiel as a small child was first torn from her mother and father and later from Evelyn. "She had her emotional legs cut off twice." She did not trust grown-ups, he said, none of the four—her birth parents or Evelyn and Margaret. "She became a very reserved, private person."[67] Yemaiel said that when she was a child Evelyn was "loving but gave me space." Her real mother, she felt, did not give her any emotional room on her return to England.

Ben and Evelyn became good friends. Ben told me that Evelyn made "ordinary people [as he referred to himself] feel special." This was true, he said, right from their wedding in 1966. Ben's comment on Margaret was: "music and Siamese cats instead of sex." Ben and Yemaiel told me some family folklore: Evelyn's father on meeting Margaret remarked, "If he wanted a family, I don't think he has done any better with Margaret."

These comments, however, may have been mistaken. Margaret was much more of a private person than Yemaiel became. As noted earlier, she came across as cold, at least in contrast to Evelyn, but Evelyn may have been the only adult she was close to, felt warmly about, and revealed herself to; he certainly felt warmly about her. Judging from her letters, Margaret did feel warmly about Yemaiel as a small child, but at the same time she wanted to make clear to Yemaiel that her real mother was back in London waiting for her.

Evelyn's earlier emotional attachment to Yemaiel revived in the 1980s. His letters to her are addressed to "My darling Yemaiel." He wrote to tell her of Margaret's diagnosis of "Alzheimer's syndrome." He also wrote to her about the death of his close friend, writer Rebecca West, in early 1983 (see chapter 13) and about the solace of "having many young friends with many talents, one an authority on Tibetan art."

In a letter written after Margaret's death he told Yemaiel about the painting Christina Speisel, a new Haven artist, was doing of him for Saybrook College. It had been commissioned by the anthropologist and former Yale student Myrdene Anderson, who studied reindeer in Lapland. Evelyn was considering having a sculpted head of himself, which Christina had made, cast in bronze for Yemaiel. "You are the only person whom I would want to have this, but with a couple of Epsteins embellishing your living room" he wrote, "it might lack ultimate distinction." He did refer to her time as his adoptive daughter.[68]

He later wrote, "I have just had my annual Christmas treat when Phoebe [Ellsworth] and Sam [Gross] bring the now 3-year-old Sacha to see me. . . . You have given [me] full practical experience. I managed to get her a toy owl which was a huge success."

He also wrote about a delightful trip to Florida he had taken that December to the seventieth birthday celebration of his second graduate student, Edward Deevey. This included a daylong scientific meeting as well as "a canoe trip down a charming semitropical river with gallery forest on each side, a large heron (the American great blue heron about half as large again as the European) flew ahead of us." This passage is of interest because Evelyn, on his visits to Yemaiel and her actor husband Ben in London, talked to Ben at length about birds, which became a new special interest for Ben. Subsequently Ben and Yemaiel went to both Madagascar and India, primarily to watch birds.

In January 1985 Evelyn wrote in some excitement to Yemaiel and Ben that he had been elected an honorary fellow of the Zoological Society of London. The elected fellows comprised "twenty-five people who are supposed to have been the best zoologists of their time, still around." There was to be a ceremony on May 8, and he was coming to London for that occasion. Significantly, Evelyn wrote to Yemaiel: "I assume daughters are welcome."[69] Yemaiel would play a crucial role in caring for Hutchinson in London during his final days (see chapter 17).

A Change of Life: Anne Twitty Goldsby Hutchinson

In February of the same year, 1985, nearly two years after Margaret's death, Evelyn wrote to Yemaiel, "I am most anxious to see you, but for both happy and unhappy reasons I am having to change my plans." He continued:

Ten days ago I went to . . . Philadelphia to see Ruth Patrick [one of his long-time limnology colleagues and friends] and to . . . lecture at the Philadelphia Academy and its adjunct, the Stroud Laboratory about thirty miles away in the country, a very good hydrobiological research institute run by a couple of very rich people with a quite distinguished staff. Mrs. Stroud is a very discriminating art collector and they wanted a talk on birds and insects in illuminated manuscripts [about which Hutchinson had written over a long period of time]. The first third of the lecture went very well. Then I suddenly found myself feeling faint and collapsing onto the floor. I sat there for three or four minutes, the faintness passed off, the audience was wafted back to their seats, and I finished the lecture satisfactorily. . . . It was by that time snowing and the idea of going home by train was not appealing. Fortunately the director had to drive to New York and deliver some papers so he and a graduate student drove me about an hour and a half out of their way to New Haven, which was most kind of them.

He was to have supper that same February night with a friend of many years, Anne Goldsby, who lived around the corner. The description of the life-changing event follows:

When I got back a friend, Anne Goldsby, who lives a few houses away, looked after me most beautifully. I had known her for quite a long time very casually, but recently we had seen quite a lot of each other. When the weather ameliorated and I could get to the doctor and have an X-ray, etc., it appeared that nothing was broken. Some of my ribs were badly bruised by my hitting the corner of a table [when I fell]. . . . Six days later I am much better but by no means fully recovered. However all this gave Anne and me a chance to fall in love completely. I know I am in some ways far too old for this sort of thing but we want to share the rest of our lives together and hope to get married in April.

He added, "May we spend a belated honeymoon with you later in the year? . . . I will write again when you have caught your breath."[70] Thus far he had not told any of the Hutchinsons.

Thereafter he wrote to his sister, Dorothea, in Cambridge. The first part of this letter is similar to the one he wrote to Yemaiel. Then he added: "She looked after a rather battered man and we soon decided that we should spend what is left of our mutual lives (she is much younger than my 82) together. We shall get married on March 23. . . . She is 'black,' which in her case means a lovely coppery gold skin, very intelligent — and immensely charming and universally loved." Evelyn wrote about their plans for a delayed honeymoon in England in June or July. He added:

I know that this will at first startle you. I have been however enormously rehabilitated by what has happened and know that Margaret must approve— She had known and liked Anne years ago. I am obviously learning a great deal about race relations, particularly as Anne as a student was one of the first people involved in student demonstrations in the Deep South, which started the whole movement. Everyone I know has been enthusiastic in their congratulations to us. . . . Our only difficulty is how to have a very quiet and simple wedding at home with such an array of friends anxious to attend.[71]

It must have been quite a shock for Dorothea to read that they were to be married the following month and that Anne was black and much younger than he. She was actually in her forties. Evelyn had earlier described another black woman friend to Dorothea as "café au lait." In this letter he reported that Anne had "a lovely coppery gold skin." Anne, he wrote Dorothea, was herself a biologist, a Stanford University graduate who had run the large elementary laboratory in the Biology Department at Yale and subsequently taught high school biology in New Haven. She had been twice married previously; her second marriage was to Richard Goldsby, also a biologist and master of a Yale college. Hutchinson ended his letter by writing, "Please rejoice with me about Anne and do not worry about my fall."

He asked Dorothea to let the family in England know about Anne. He wrote that he was not fully recovered from his fall and "could not write to all of them at once," but it is clear that he knew that some of them would disapprove. He needed Dorothea's support. Dorothea did prove supportive. Some other members of the family were indeed dismayed.

He married Anne on March 25, 1985, with many of their approving friends attending. Anne and Evelyn each had a chance to make a statement. Anne chose to read the passage from the Book of Ruth, "Whither thou goest there also will I go." Evelyn spoke about racial diversity as a "source of richness and delight." It was supposed to be a small wedding but Evelyn reported receiving phone calls from the family in England while "70 guests were eating a stand-up champagne lunch."[72]

Why did Evelyn marry Anne? From a somewhat cynical point of view, he surely needed someone to take care of him, and Anne showed herself able and willing to do so. His doctor, as he related to both Yemaiel and Dorothea, had told him not to travel alone. On the other hand, from the testimony of several friends and colleagues, Evelyn and Anne were very happy together during the first period of their marriage. They appeared, even to neighbors who saw them walking down Canner Street in New Haven together, to be very much in love.

Why did Anne marry Evelyn? That question is harder to answer. She was nearly forty years younger than Evelyn. There is no question that she revitalized Evelyn. She certainly liked to have a good time and enabled him to do so. Security may have been important, as well as the status of being married to a famous scientist. She shared Evelyn's interests in biology, something that Margaret had never done. Several members of Evelyn's family, however, were convinced that she married him for his money. Ironically, he did not have a great deal of money at the time of their marriage. He would have much more soon after, as the recipient of the Kyoto Prize, a major monetary award, which he shared with both Anne and with his family, including his four nephews and one niece. Hutchinson's new wife was also "very intelligent and immensely charming and universally loved." Chapter 17 continues the saga of Evelyn and Anne.

Evelyn received a number of letters congratulating him on his marriage. Perhaps the most interesting one is from former Yale graduate student Dan Livingstone, later a Duke professor. Livingstone was doing research on the fauna of the crater lakes of Cameroon. He wrote that it was wonderful to receive the news, sent his hearty congratulations and wishes for "every happiness," and added, "Man was not meant to live alone. What is true for an ordinary man is even truer for a research biologist. One can maintain a kind of suspended animation without a woman's company, but in such a zombie state one should expect neither health, happiness, a high output of research papers, nor new ideas."[73]

CHAPTER TWELVE

Good Friends:
Margaret Mead and Gregory Bateson

E velyn Hutchinson had many close friends throughout his life; he had a special talent for friendship. Foremost among his well-known friends and longtime correspondents were American anthropologist Margaret Mead; Gregory Bateson, who was once her husband; and the English writer Rebecca West. These friendships were both personal and professional. They stayed at each other's homes in New Haven, New York, and England. Hutchinson helped to edit Margaret Mead's books; they had much correspondence on a variety of topics. He exchanged hundreds of letters with Rebecca West over a long period of time.

A surprising number of Hutchinson's fellow Cambridge University students, at least three women and a larger number of men, remained lifelong friends in spite of the fact that he left England after he graduated from Cambridge University, returning only at the very end of his life.[1] One of those fellow students and good friends at Cambridge was Gregory Bateson, whose father, the well-known biologist William Bateson, was an old friend of Evelyn's father, Arthur Hutchinson. The Hutchinson and Bateson families were part of the Cambridge intellectual aristocracy. Their circle of friends included the Darwins, Huxleys, Mitchisons, and Haldanes, families that produced generations of scholars with wide-ranging interests and considerable education. These families were aristocrats not by wealth or titles but in culture and education.

Evelyn wrote about the carnage of World War I: "Any day one might learn that someone close had been killed." The two brothers of Christina Innes, his earliest schoolmate, were both killed in the war as was John Bateson, Gregory's older brother. Martin, the middle Bateson son, committed suicide as a young man, on John's birthday. Both John and Martin overlapped with Hutchinson at his elementary school, Saint Faith's, and taught him to collect beetles. During World War I Gregory and Evelyn were young teenagers, but Evelyn wrote that every schoolboy felt that if the war went on indefinitely "his turn would one day come."[2]

While Gregory Bateson and Evelyn were classmates at Cambridge University, they, together with Grace Pickford and others who became distinguished biologists, founded the Biological Tea Club in 1922 (see chapter 3). It is recorded in the Tea Club minutes that Hutchinson and Bateson had a contest in which they were timed while eating cream buns; Bateson won. There was of course serious biology done as well. Among the guest speakers in Hutchinson's and Bateson's time were Joseph Needham, embryologist and historian, and Paul Kammerer, of the infamous midwife toad.[3]

Later, on the American side of the Atlantic, Margaret Mead and Hutchinson served together on the staff of the American Museum of Natural History, Mead as a curator of anthropology throughout her career from the 1920s on, Hutchinson as a consultant in biogeochemistry in the 1940s and 1950s (see chapter 10).

The Macy Conferences and the ISS

Mead, Bateson, and Hutchinson all participated in the Josiah Macy Conferences, particularly the "Feedback Conferences" in the late forties and early fifties. A 1948 letter from Margaret Mead to Hutchinson reads: "Dear Evelyn, Are you planning to stay with us over the Feedback Meeting? It would be two nights this time. . . . You know we always love to have you and I am not sure the Feedback meetings would have a proper quality if you weren't a simultaneous house guest."[4]

Hutchinson later wrote that he had thought a great deal about the Macy Feedback Conferences. He thought the important thing was to see how the central mathematical group, including Norbert Wiener and Claude Shannon (of the Shannon-Weaver equation in ecology), fit in with the social sciences group, which included Gregory Bateson and Margaret

Mead, and connected to neurologists like Gerhardt von Bonin (a special guest, not a regular member) and to the psychologists and psychiatrists, including Lawrence S. Kubie. John von Neumann, who later wrote *The Computer and the Brain,* seemed to Hutchinson to function as a mathematical critic. As Hutchinson saw it, his role was always "that of a student trying to learn quietly and unobtrusively from this extraordinary group of genii. G.B. and M.M. [Gregory Bateson and Margaret Mead] were of course close personal friends. Wiener liked to hear about any branch of science. I once met him getting on a train in Washington and he asked me all the way to New York about modern classifications of invertebrates. Kubie the psychoanalyst I got to know later. . . . The paper I gave in the interim summing up at the N.Y. Academy was unexpectedly well liked and may have given the impression that ecological biology was more important in the meetings than it actually was."[5]

Although Hutchinson may have thought of himself as a learner in this group, he was very much a contributor. When the transcript for the conference on cybernetics at which he presented his paper was published, it included both the presentations and the "interruptions by questions and answers" by participants, inserted where they occurred. To eliminate these remarks would destroy "the give and take discussion which is the heart of our conferences."[6] In spite of Hutchinson's claim of trying to learn quietly and unobtrusively, seventeen of his remarks were recorded for this one conference!

He wrote to his former graduate student Tommy Edmondson just after the 1946 Macy conference, which was on circular causal systems in the biological and social sciences. This conference included discussion of population cycles, which were of great interest to ecologists. The circular causal system material was to be mimeographed for private distribution, and Hutchinson wrote that he would loan his copy to Edmondson.[7] In 1948, Hutchinson published his own paper on the subject, "Circular Causal Systems in Ecology," a revision of his New York Academy of Sciences talk.[8]

During the late forties Hutchinson and Mead were also both active members of the Institute for Intercultural Studies (IIS). Mead, Bateson, and Hutchinson were on the board of directors, as was anthropologist Ruth Benedict, Mead's early mentor, and Lawrence K. Frank, convener of the Macy Conferences. Margaret Mead had founded the IIS in 1946.[9] There was a tradition, probably started by Ruth Benedict, of helping to finance work by young anthropologists. Benedict gave Mead a $300 fellowship out of her own pocket when Mead graduated from Barnard. The corporation Mead started aimed to continue this tradition, financing research using a cultural approach and applying it to contemporary international problems.

Often the IIS funds were from private benefactors, including Mead, who donated earnings from speeches and her book royalties, and other anthropologists. One IIS task was to interview émigrés from seven foreign countries, including the Soviet Union, after World War II. Even the Office of Naval Research gave a large grant, for the study of "Contemporary Cultures."[10]

Philleo Nash, a subsequent vice president of the IIS, said that the early meetings were exciting. "We weren't a conspiracy. . . . We were just having big thoughts." But it was the McCarthy era, and "big thoughts" were sometimes misjudged. Moreover, another function of the IIS, and one that interested Hutchinson, was to keep in touch with Russian academics, including scientists. Most of the affairs of the IIS were conducted from Mead's office at the American Museum of Natural History. The institute was still dispensing funds to the young and holding annual meetings attended by Mead in 1978.

Family Relationships

Mead often stayed with Evelyn and his wife Margaret when she came to New Haven to give lectures or attend meetings. Hutchinson also knew and liked Margaret's and Gregory's young daughter, Cathy (Mary Catherine Bateson), to whom he sent a Christmas present in 1947. Margaret wrote to Evelyn: "Cathy enjoyed the magnifying glass so much. It has just the combination of science and delicacy which appeals to her imagination. We spent a quiet and uneventful Christmas as a rather large family with which Gregory, however, bore very well."[11]

Hutchinson remained the good friend of both Mead and Bateson, and of their daughter, after the Mead-Bateson divorce in 1950. At about the time of the breakup of their marriage in the late 1940s, Margaret and Gregory had together visited the Hutchinsons' home in Connecticut. Hutchinson remembered political arguments during their visit: they had political differences about the part that each had played in the war. Both had used anthropology in war psychology particularly against the Japanese, Bateson in the Office of Strategic Services (OSS), a U.S. intelligence agency during World War II. Mead's wartime work was in relation to the United States armed forces. Mead and Bateson were away from each other and from their young daughter for long periods during World War II. In 1942 Mead went to Washington, returning to New York for weekends, and then in 1943 to England. Bateson went to Burma to take part in psychological warfare; he did not return until the Pacific war ended, when Cathy was five. In the summer of 1946, Margaret, Gregory, and Cathy were together in New Hampshire, where Gregory taught Cathy natural history with a microscope and telescope.

Margaret had three husbands, of whom Gregory was the last; Gregory had three wives, of whom Margaret was the first. Hutchinson, of course, did not conform in this respect either. He also had three wives, although his second marriage lasted fifty years.

Margaret and Gregory, although both lived unconventional lives, were very different. Their daughter recounted an incident that occurred twenty-five years after their divorce. The two of them were to speak at the same symposium, and Margaret tried to talk Gregory into putting on socks. Margaret cared about how she was perceived, whereas Gregory was content to flout conventions. Margaret felt that "proper behavior was important because it allowed choices and opened doors." Evelyn was more like Margaret, unconventional in terms of ideas but more conventional in terms of manners and dress.[12]

Evelyn Hutchinson and Margaret Mead also shared a love of the ritual, if not necessarily the theology, of the High Church Episcopalians. "The Christian tradition was passed on to me [by my mother]," wrote Mary Catherine Bateson in her book, "as a great rich mixture, a bouillabaisse of human imagination and wonder brewed from the richness of individual lives. . . . She taught me to follow the intricacies prescribed by tradition." Hutchinson, to the surprise of many of his students and friends, went regularly to a High Church Episcopal service in New Haven, one pervaded by ritual and heavy with incense. Margaret Mead took Cathy on Good Friday to the "highest of high-church" Episcopal parishes in New York, popularly called "Smoky Mary's."[13] It is of interest that Margaret Mead's great-grandfather had been vice chancellor of Cambridge University when that university required Anglican (Episcopal) orders. Gregory and his parents, in contrast, were atheists. William Bateson essentially declared science his religion.

Gregory Bateson and Hutchinson

Some of Gregory's ways of thinking were closer to Hutchinson's than to Mead's perhaps because of their similar scientific training at Cambridge. Gregory, however, left zoology for anthropology early in his career, though he later did research on dolphins. Bateson's views were close to some of Hutchinson's ideas and hypotheses, although Hutchinson usually collected more concrete data to support his ideas. Both were concerned with patterns in nature, surely one main aspect of ecology. Gregory's view of pattern was one that connected "all living beings in formal similarities of growth and adaptation, the dolphin and the crab and the flower, and by which they are united in ultimate interdependence in the biosphere." Hutchinson was also much concerned with the biosphere and in fact wrote an article about it as early as 1948, "On Living in the Biosphere."[14]

Two theoretical areas that interested Bateson were cybernetics and information theory. Some of the questions he asked were very similar to those posed by Hutchinson in terms of the stability and regulation of living systems. Feedback is involved here, whether in terms of predator and prey numbers in populations, or homeostasis in terms of blood sugar levels. These ideas were transferred by Bateson into the cultural realm. In the same sense that natural communities with many interconnections may be more stable than those with fewer, Bateson thought that only a fully elaborated cultural system could be stable.

Bateson had gone from biology to anthropology and from Cambridge to New Guinea, a way of decisively leaving home and family that was similar to Evelyn's leaving Cambridge for South Africa. Hutchinson turned there from taxonomic zoology to limnology, the study of many factors, both physical and biological, involved in lake ecology. When Bateson moved to anthropology from biology he "found a field with very little in the way of analytic models. . . . At Cambridge, anthropology was still very much a branch of natural history . . . with emphasis on description." For forty years, Gregory struggled with the question of how to provide anthropology with the clear, taut framework of ideas that would make it truly a science.[15] One can make exactly the same statement for Hutchinson and theoretical ecology, and for a similar number of years.

Mead and Hutchinson

The friendship between Evelyn Hutchinson and Margaret Mead was also special, both professionally and personally. Hutchinson was important to Margaret as an advisor and as an editor of what she was writing. In spring 1948 she sent him a manuscript to edit; it was the manuscript for the first part of her book *Male and Female*. This book was published in 1949 and went through three printings by 1950.[16]

He reviewed the manuscript and sent both praise and criticism. He wrote: "I have read it several times with great interest and first would like to tell you how very much I admire it. There is a quality which is to be regarded as literary rather than scientific. . . . The nearest I can get is to speak of it as an affectionate approach. . . . Rebecca [West] gets the same effect in some of her reporting, but I doubt that any man could. It is curiously persuasive and in view of the significance of the material, extremely important." There were substantive comments as well. He wanted her to amplify her discussion of the difference between isolated and peripheral communities in chapter 2. His letter continued: "It is the best general statement of a phenomenon which has always struck me, since I was in South Africa, and which in its appallingness has great ethical implications." He also wrote that some of her speculations "though guarded, will no doubt infuriate some reviewers."

When she sent him the rest of the manuscript, she asked him to do copy editing. He wrote her of some need for rewriting: "The greatest difficulty lies in too long sentences—too many dashes—and a tendency for the tenses to get out of control." There follow several pages of comments. For example, he felt that chapter 12, "Our Complex American Culture," raised a question of "the greatest importance." He found an underlying ambiguity also implicit, he wrote, in Rebecca West's *The Meaning of Treason*. The question he refers to is, why one accepts one's own culture. Hutchinson wrote to Mead:

> You accept your culture because
> (1) It is yours, it is where you have to start from, whatever you might feel.
> (2) It has manifest potentialities for doing the kind of things you want to do (and I want to do, etc.).

But, Hutchinson insisted, these two are not the same, the first being inevitable in any culture, but the second might not be true. In Nazi Ger-

29. Margaret Mead. Beinecke Rare Book and
Manuscript Library, Yale University.

many the "alternatives . . . were (a) Not wanting to do what you think is
good. (b) Becoming one of the idealists of a false ideal. or (c) Concentration camp." He went on to discuss this at some length, suggesting that she
at least modify the stress of the chapter by changing the order of the ideas.

Evelyn did line-by-line editing for each chapter, with substantive cor-
rections and additions, particularly for chapter 12. For example:

P.5, L. 13–14. Not yet. The best that can be done is to transplant a new cornea into a sightless eye. (It should read) "The installation of the cornea of

the recently deceased in the sightless eye of a living man." The whole eye works so far *only* in salamanders [his emphasis].

Or, a comment on chapter 13:

> What you are talking about is a second-order similarity. It would tighten the argument to generalise it this way. Remember that men may not take babies in books too seriously, but some men might take the ontogenesis of culturally significant behavior with all the seriousness that a baby rightly demands.[17]

Mead answered this letter, writing, "Always and again many thanks." She said that additional people would read the whole manuscript "after which the cotton wool sentences will be polished up, and I hope that the places where female psychology has been unfair to males will be all tidied up also. I am unendingly grateful to you for the trouble you have taken. I will try to write out the democratic—or perhaps we should say the open-ended ethic—a little more clearly."[18]

It is difficult to judge the substantive changes that were made in the manuscript of *Male and Female,* but the writing in the finished book does seem "polished up," lacking both overlong and "cotton wool" sentences. The literary quality and "affectionate approach" are evident. For example, in the "Complex American Culture" chapter, Mead wrote: "Besides the traditional foods in the rural child's lunch-basket, there must be a piece of store bread or that child will be ashamed and not take its lunch to school. Subtly, insistently, continuously, the standard American culture is present, to rich and to poor, to newcomer and even to aborigines whose ancestors roamed the Plains before the Spaniards brought the horse to the New World."[19]

Mead did acknowledge Hutchinson's help. In her "Notes to Chapters" she wrote:

> My method of work has been to rely heavily on discussion with individuals in each field whose approach I knew and trusted, reading such individual studies as they recommended to me as especially pertinent. . . . I tried to ask selected scientists questions in such a way that I would be led to materials which contradicted or illuminated the problems on which I was working. I am personally indebted to Lawrence K. Frank, . . . Earl T. Engle, William Greulich, Gregory Bateson and Evelyn Hutchinson.[20]

In 1948 Margaret wrote to Evelyn that she was thinking of submitting a new manuscript on Balinese children to Yale University Press, since it had particular relevance to the Gesell Clinic publications. The Gesell Clinic was at Yale and had cooperated on the project with Mead. Evelyn was doubtful. He wrote that he had been told that the press shied away from plates for pictures, and that since Harpers did Gesell's other publications they might be more interested. He might, however, have more information when he came to New York City on March 30 and would see her then. He was coming in the same week to see Rebecca (West) and would call Margaret if he had learned anything significant.[21]

In February 1951 Margaret Mead was still working on the book about Balinese babies. She wrote to Evelyn, "I have been thinking over what you said the other night, I am sorry that it has to be done in a hurry." She had decided that if she couldn't do it right, she would postpone the book, even though that would disturb her collaborator, Frances Macgregor, who was interpreting the photographs and writing the captions. This book was to be called *Growth and Culture: A Photographic Study of Balinese Childhood, Using Gesell-Ilg Categories of Analysis* and was based on photographs by Gregory Bateson. Margaret and Gregory worked together in the field in Bali. As she and her second husband, the New Zealander Reo Fortune, had done earlier, Mead and Bateson attempted to make ethnography more scientific, documenting the fieldwork with film.

For the book, Mead was to write: the introduction, a chapter useful to both parents and teachers on the "extent to which growth is one dimension of personality," and a chapter on the Balinese background of the children presented. Some of the plates show children "standing with support, standing alone, . . . squatting," but many of the plates show the "ways in which Balinese children differ conspicuously from ours." There would be an appendix, of particular use to the student. Here she wished to make her point about awareness—"that the public who use results in the human sciences *must* understand how those results are obtained." She was sending Evelyn a very rough draft (since it was a holiday and "there are neither secretaries nor heat in the Museum,") of the first five pages of this methodological appendix. "Would you glance through it for feel? If this feels right to you—as seen against the plan of the book as outlined on the first page of this letter, then I think I am set." She asked him to comment as soon as possible and seemed to be counting on Hutchinson as her principal advisor. "I can't hope for any help from Gregory." Bateson was not coming to the ongoing "feedback" meetings, and "he won't do anything with a manuscript sent from here if he isn't going to see me." He had

had the first Balinese film since the previous September without writing the script. "But," she wrote, "I have a good deal of faith that methodological points you would make will be congruent with what he would make so that the work won't get too terribly out of step to confuse students, or—if he ever pays attention again—to make future collaboration impossible."[22]

Hutchinson answered that he wanted to explain his remark about hurry. "I hope you don't mind what is coming, written as it is with all the affection and admiration that you must know I have for you." He then wrote:

> Last summer I reread, when at the beach, *Coming of Age in Samoa*. It is extraordinarily good. [This is the only mention of the beach in his sixty years of letters from Yale.] Since you wrote it, a lot of things have happened and you have learnt from them enormously. Your current work has developed beyond the Samoan work, but I felt and still feel that the full reward of experience has not been reaped. I think this is because a number of circumstances have forced too much on paper too fast. The Mountain Arapesh, moreover, is better than anything you have written on Bali. Yet both, theoretically because their culture is so rich in extraordinary things, and practically because they are so different from us and therefore so illuminating, the Balinese must provide the most important material you have. It therefore seemed to me extremely important that this new work be done with as much quietness of spirit as possible.[23]

He went on to make some minor criticisms of the appendix and a criticism of one of her major premises, the latter from a scientific point of view. Mead had written in the previous letter, describing the plates and their captions, "The basic argument seems clear enough, that from an initial lack of firmness in handling the infant ('low tonus'), there is some loosening . . . of the customary antagonism between flexors and extensors, and a corresponding freedom for parts of the body . . . to be in opposed positions, one flexed and one extended with a variety of results for the development of the Balinese baby."

Hutchinson's next comment is truly negative; he seems to be telling Mead, "You can't say that!": "The thing that strikes me about the whole story most is its improbability. . . . Extensors have antagonized flexors in the vertebrate trunk since vertebrates began and in the limb since the pentadactyl [five-parted] limb was invented in the Carboniferous. To build a whole point of view on a denial of this . . . would seem *a priori* fantastically unlikely."

He does go on to ask if there are "hints of how widespread it may be," for example, in India or Java, and whether it "has survived Mohammed-anization." But he sounds clearly doubtful about the whole viewpoint. He ends the letter by telling her that her lecture at Yale was an "enormous success" and had a very positive and valuable impact on the conservation students.

Margaret Mead and Frances Cooke Macgregor had difficulty find-ing a publisher for the Balinese babies book. It was published in 1951 by Putnam's, the publisher of an earlier book by Macgregor, under the title *Growth and Culture: A Photographic Study of Balinese Childhood,* and it is mainly an analysis of Bateson's photographs of eight young children. It is hard to determine to what extent Hutchinson's objections were ad-dressed. It was not one of Mead's more successful books.

Margaret Mead was one of the few women who were welcome to stay at the Hutchinsons' house. Dorothea, Hutchinson's sister, was not one of those few, as we have seen. Mead once wrote to Hutchinson, "I have to go up to New Haven for a meeting on Saturday the 24th. If there is any chance of seeing you and Margaret I can stay over until late evening." She probably stayed for the night.[24] Another time she wrote that the Applied Anthropology Executive Board meeting was to be at Fred Richardson's where he had picked her up the previous year in the middle of the big snow. Mead stayed with the Hutchinsons when she was in New Haven, and Evelyn stayed with her in New York City. She was still sending and he critiquing manuscripts in 1962.[25]

The Hutchinson-Mead letters discuss a variety of subjects, mostly concerning her work rather than his. Hutchinson, however, had a long-standing interest in homosexuality, partly from the viewpoint of natural selection. A paper that he wrote about uniovular (identical) twins reared apart who were both homosexual elicited a long and interesting letter from Mead.[26] Although she agreed, in part, with a genetic explanation, she suggested that different hereditary traits operate differently in differ-ent cultures:

> I would hypothesize that although the chances would be high that uni-ovular twins would be homosexual if reared apart but in the same cul-ture, this would not necessarily be true if they were reared apart in differ-ent cultures. . . . One may be tagged as potentially homosexual for being either too much like a member of the opposite sex or at the extreme range of one's own—the man's man who can't bear women in any shape; the

"feminine woman" who is passionately preoccupied with feminine ana-
tomical problems. . . . One may also be tagged because of a temperamen-
tal quality which in that particular culture is only appropriate to members of the opposite sex. I realize that I am not saying anything here that
I haven't said before, but perhaps placing it against the context of your
paper, it may be more meaningful.[27]

Margaret Mead's Samoa fieldwork and her ground-breaking book
Coming of Age in Samoa was criticized in the 1980s, a half century after
it was completed. Derek Freeman wrote a book, *The Making and Unmaking of an Anthropological Myth,* citing shortcomings of her research
in Samoa.[28] Freeman's book was, however, in turn negatively critiqued by
anthropologist George E. Marcus in the *New York Times.* Marcus's review
began, "This [Freeman's book] is a work of great mischief." Ironically, both
Margaret Mead and Derek Freeman were disciples of Frank Boas's views
of anthropology.

Hutchinson and Gregory Bateson each discussed Mead's other famous book, *Male and Female* (1949). Bateson, her recently divorced husband, claimed there were too many value judgments. Hutchinson had
a more positive view. He thought that the background of Western culture might be illuminated by anthropology of the right sort. Mead attributed traits of a variety of people to socialization and cultural influence, as
against extreme hereditarianism, thus continuing the views of the Boas
school, elucidating, as Hutchinson noted, that "what seemed to be biology turned out to be learning."[29]

Gregory Bateson rebelled against the English culture of his childhood, even though he thought stability might be good in principle. Late
in his life he became a critic of modern American culture and almost a
cult counterculture figure for the young of California, where he lived and
taught. He died in 1980.

Margaret Mead, unlike Bateson, accepted her American cultural
roots, which enabled her to reach more people with her cross-cultural
books, even though her views were hardly conventional. "Margaret always
gave the culture in which she had grown up the benefit of the doubt,
Gregory did not," their daughter wrote. "Much of Margaret's popularity
with ordinary people has been based on the fact that she affirmed and re-
spected their ways of doing things . . . even when she did not herself conform."[30]

Mary Catherine Bateson: With a Daughter's Eye

Hutchinson knew Mead's and Bateson's daughter from an early age and continued that friendship throughout his life. In early 1950, Hutchinson was asked to write a recommendation for Cathy, then ten years old, for admission to the Brearley School in Manhattan. He replied "with great pleasure":

> I have known Cathy Mead Bateson since she was three months old, her father for twenty-five years and . . . her grandfather, the great [geneticist] William Bateson was an old friend of my father.
>
> Cathy is a most remarkable child. She is as unusually intelligent as one would expect from her ancestry, but is not in the least tiresomely precocious. She is indeed very good fun both within her own age group and in any other. . . . She has an originality of outlook in anything that takes her fancy. . . . Her mother has given her a great sense of social responsibility and this she expresses in unusual and interesting ways. I am extremely fond of her and regard her as a close friend.[31]

In some ways, Cathy's childhood was similar to Evelyn's in Cambridge, England—at least during the relatively short period when she had two resident parents. Amateur theatricals, a feature of the Hutchinson household, were part of it. Gregory bought Cathy professional theatrical makeup, grease paint, nose putty, and even materials for making false beards for one of her childhood birthdays; Margaret provided gaudy artificial jewelry. Cathy wrote that her mother "went out of her way to create opportunities for me to be in contact with different kinds of people, and often when I developed a new interest would send me to someone who could present it to me more personally."[32] Evelyn's father did the same for him; someone with whatever new interest he had could be found among the family friends in Cambridge.

Mary Catherine Bateson became a noted anthropologist and writer. Her book on her parents, *With a Daughter's Eye,* quoted above, added new insights and was well reviewed.[33] It is clear from this book as well as from Jane Howard's biography of Mead that Hutchinson, though a personal friend and an important reader and consultant on some of her work, was only one of Margaret Mead's many good friends. Ruth Benedict, Marie Eichelberger, and Larry Frank were important throughout longer periods of her life.[34]

Cambridge Students, Yale Students, and Other Good Friends

Hutchinson had other good friends as well, some of them well-known biologists, such as David Lack, Ernst Mayr, and Stephen Jay Gould. His fellow students at Cambridge, including Anna Bidder, Penelope Jenkin, Cecilia Payne, Joseph Omer-Cooper, as well as Gregory Bateson remained lifelong friends. Many of his students, some of whom became leading ecologists of the next generation, remained his good friends and frequent correspondents. These include Robert MacArthur, Larry Slobodkin, Ed Deevey, Tommy Edmondson, Thomas Lovejoy, Alison Jolly, and Karen Glaus Porter. There were a great many more scientists who corresponded with and visited him, including many Europeans who wrote to him in several different languages. He also had good friends at Yale in a great many fields: art, music, psychology, literature, and history of science. He left a complete set of his papers to two friends who were not his students but whom he knew well while they were at Yale: Phoebe Ellsworth, later a professor of psychology at the University of Michigan, and Marjorie Garber, a Shakespearian scholar at Harvard.

Hutchinson's most famous and very close woman friend, however, was the English journalist and novelist, Dame Rebecca West, the subject of the next chapter.

Fond Correspondents:
Rebecca West and
Evelyn Hutchinson

R ebecca's death is a shock—mostly a dozen times a day when I realise I can't write to her." So wrote Evelyn Hutchinson to his World War II foster child Yemaiel when the English writer Rebecca West died in March of 1983. She had been Hutchinson's close friend and constant correspondent for more than thirty-five years.

"She clearly died of overwork, at 90, on the Ides of March, which I am sure would have seemed wryly amusing and most appropriate to her," Evelyn continued.[1] Hutchinson arranged with the staff of the Yale University Library to put on an exhibition in her memory, with various annotated and other special books that he had helped them collect. Later that week he learned that he had been named one of her literary executors.

A Meeting of Extraordinary Minds

Rebecca West was born in England in 1892 but grew up in Edinburgh. Early in life she was a well-known radical feminist reporter in London. She became one of the twentieth century's most celebrated and prolific writers of novels, reviews, and nonfiction, as well as an international reporter. In 1959 she was made a Dame of the British Empire (DBE), henceforth Dame Rebecca West.

Rebecca West had an array of important people in her life. One was Harold Ross, editor of the *New Yorker,* for which she wrote, and which

30. The young Rebecca West. Beinecke Rare Book and
Manuscript Library, Yale University.

often brought her to New York. Most important in her early life was H. G.
Wells, celebrated writer of the late nineteenth and first half of the twentieth century, who fathered her son, Anthony. Later there were Lord (Max)
Beaverbrook, newspaper magnate, and other equally well-known lovers.
Finally, Henry Andrews became her husband of many years. But, at least
in the second half of her life, Evelyn Hutchinson was one of her very
closest friends.

Rebecca and Evelyn exchanged hundreds of letters, from the 1940s
until Rebecca's death in 1983. It seems at first an unlikely friendship, the
scientist and the feminist and celebrated writer, on opposite sides of the

Atlantic. She was ten years older than he. But it was a true meeting of extraordinary minds, and it came about from his writing, not hers. Hutchinson's first book, *The Clear Mirror: A Pattern of Life in Goa and in Indian Tibet*, published in 1936, was beautifully written and something of a literary success. He subsequently wrote semi-popular reviews and essays for the *American Scientist* under the general title of "Marginalia" from 1943 to 1955, and some in later years as well. Scientists in all fields knew of Hutchinson through reading these *American Scientist* essays. Many of these essays were collected into books. Rebecca West was of course a much better-known writer than he. After she read one of Hutchinson's articles in *American Scientist*, in January 1947, she initiated the correspondence, writing:

> I am going to America . . . on February 13, and I would so much like to meet you. I have nothing to mitigate the impertinence of this request except that I am sure I should like to hear you talk. I am going to be in New York until the end of the month, then go to the Middle West and . . . to Bangor [Maine] and then come down from Boston about March 19, stay in New York for a few days, and go down to Washington from March 24 until April 7. Is there any point in that itinerary when I could meet you? This is a bold letter, but I stick to it that I would like to hear you talk.
>
> I am, very gratefully, Rebecca West[2]

On February 13 Rebecca telegraphed Hutchinson: "REGRET INSTEAD OF QUEEN ELIZABETH HAVE SEPTIC FOOT BUT HOPE TO SAIL MARCH 15." Evelyn had invited her to spend a weekend with him and his wife Margaret at their home in Woodbridge, Connecticut. On April 1 Rebecca sent another telegram: "SORRY CANNOT COME DOWN THIS WEEKEND HAVE HAD BAD FLU AFTER PROPOSING MYSELF FEEL THIS IS UNFORGIVABLE."[3]

On April 17 she wrote again: "Am I dreaming or did we talk on telephone about this weekend? If it was dream all right by me. Will ask myself later but could come this Saturday." She did. The meeting led to a close and long-lasting friendship.[4]

According to Victoria Glendinning, author of a perceptive Rebecca West biography, Rebecca's weekend with Evelyn and his wife came at the end of her 1947 trip to the United States. Both Evelyn and Margaret had read Rebecca's recent book *The Strange Necessity*, in which a major tenet is "the indivisibility of the aspiration that produces both art and science." Evelyn must have agreed wholeheartedly with this view.[5]

When Rebecca returned to England, she wrote, "I will write to you about the Grandfather [*Letter to a Grandfather*]. . . . It was a great grief of mine that Virginia Woolf could not understand it. . . . But if you like it, then nothing matters."⁶

In July of that year Harold Ross, the *New Yorker* editor, had sent her to Greenville, North Carolina, to cover a lynching. Rebecca wrote, "You will have seen the New Yorker and realised what hit me during the last month of my stay. I can't tell you what an experience it was. . . . They crawled over it for punctuation, ambiguities and the like, then a lawyer hunted over it for the legal aspects of the case. . . . I had to reframe sentence after sentence, because the lynchers were acquitted."

She had worked very hard and there were threats from the lynchers. Her letter went on and on and it reads like a novel, including misadventures on her return home:

> I had to get to London to sit on a Committee that concerns itself with refugees. We returned home to find black smoke pouring out of the kitchen. . . . The refrigerator had gone on fire and set the kitchen in a blaze, which was being dealt with most heroically by an SS man who is one of our German prisoners. He said the most pathetic thing. When we were thanking him, and putting in our few words of Roumanian because he is really a Roumanian who was kidnapped as a boy by the Hitler forces, he said, "And *now* may I stay here for ever?"

He couldn't, of course, but there were always such characters inhabiting Ibstone House, Rebecca's and Henry Andrews's farm in Buckinghamshire. Many of her letters to Hutchinson read like her novels. In the same letter another character appears, a Lithuanian displaced person after World War II whose husband had been taken away by the Russians five years before and shot. She was about to remarry, but she had just heard that her husband was alive in Siberia. The Russians wouldn't give him up, saying that he committed a crime and was to serve a long sentence. "Just how does a young and pretty woman take that, however much she was in love with her husband? . . . I can't tell you what you meant to me in America. I am pouring out what has happened to me."⁷ Rebecca continued to do so until her death.

West: Radical Feminist Reporter

Rebecca West was actually a pseudonym. She was born Cicely Isabel Fairfield in Richmond-on-Thames, Southwest London, in 1892. Her family, well versed in literature, music, and art, was poor, but her mother struggled to give the three daughters a good education. When her father left the family, Cicely, the youngest daughter, was ten. Her mother moved with her daughters back to her native Edinburgh. There Cicely entered George Watson's Ladies College as a scholarship student. While there she joined the Women's Social and Political Union, wore its badge to school, and had several run-ins with the headmistress. She left the school at the age of sixteen, though not for this reason; she was diagnosed with tuberculosis. She recovered by working outdoors in a market garden outside of Edinburgh. She had plenty of time to read, but tuberculosis marked the end of her formal schooling.[8]

She then auditioned for and, surprisingly, won a place at the Royal Academy of Dramatic Art in London, and thus at seventeen she became an actress. She was not a great success, but her short acting career included Ibsen's *Rosmersholm*. Her pseudonym, Rebecca West, comes from a strong woman character, an advocate for emancipation, in that play.

Why did she need a pseudonym at this early age? Cicily (her spelling) Fairfield had been a women's rights activist from the age of fourteen, and her first publication, at sixteen, was a letter to the *Scotsman* advocating women's suffrage. Two years later, in 1911, she began to write for the *Freewoman*, a radical new feminist paper. It was started by Dora Madson and others to widen the feminists' scope to issues beyond the suffragist movement. The *Freewoman* was attacked by censors and the press and banned from many newsstands. It was also banned from the Fairfield household; Rebecca's mother objected to its candor about sex. Rebecca published her first article under her own name, as Cicily Fairfield, a month before she turned nineteen. When her article was to be featured on a poster, she chose to have it appear under the name Rebecca West.[9]

According to Rebecca, the greatest service this paper did was through its "unblushingness." The *Freewoman* mentioned sex loudly and clearly and in the "worst possible taste" and did an immense service by shattering the romantic view of women. Christabel Pankhurst, according to West, did a similar service with her articles on venereal disease, at which "England fainted with shock," but then experts came forward to try to do something to prevent these diseases.[10] Pankhurst had to edit the *Suffragette* from exile in Paris, and the *Freewoman* became the *New Freewoman* in 1913; Rebecca wrote for that, too. It had become the *Egoist* by 1914. Male writers like Ezra Pound and Ford Maddox Ford contributed to both papers. Rebecca West was for a time literary editor of the *Egoist*. She was by then twenty-one.

West wrote serious and unsettling articles for these papers, including many book reviews. Her views were socialist as well as feminist. In 1911 she spoke out about women in the labor market. The Victorian era "brought women the most humiliating and the most hungry period of oppression they had ever endured," she wrote. Women must be able to dictate the conditions of their own labor. Socialist writers were there before her, of course, including George Bernard Shaw and H. G. Wells. Shaw wrote of her journalism, "Rebecca can handle a pen as brilliantly as ever I could, and much more savagely." She herself praised Shaw's work and use of language, but in a 1913 review of his play *Androcles and the Lion* in the *New Freewoman*, she criticized some of his arguments. She referred to Chekhov's *The Cherry Orchard* in this review as well as to the Christian martyrs. West had garnered an astonishing literary education by an early age, mostly through her individual reading. One wonders if her readers were as erudite.[11]

West wrote about H. G. Wells's work as early as 1912 in the *Freewoman*. She was critically reviewing a book, not by Wells, but by I. M. Kennedy titled *English Literature, 1880–1905*. West praised Kennedy at the beginning of the review as a scholar but not as a literary critic and was indeed quite "savage." Not only did Kennedy prefer George Gissing to Oscar Wilde, Shaw, or Wells, she sneered, but he wrote "in the solemn yet hiccupy style peculiar to bishops, with a 'however,' or 'indeed,' or 'of course' interrupting every sentence." She quoted Kennedy's attack on H. G. Wells and his assumption that anyone who had a mind interested in mechanics is just the opposite of an artist. "Therefore," West wrote, "because Mr. Wells has a scientific imagination, he cannot have an artistic imagination." Rebecca pointed out her favorite entry in Kennedy's index: "Sex, The unimportance of," and went on to denigrate Kennedy's view of females generally.[12]

Shortly thereafter H. G. Wells got his own comeuppance from Rebecca. In a review of his novel *Marriage*, she wrote, "He is the old maid among novelists; even the sex obsession that lay clotted on Ann Veronica and *The New Machiavelli* like cold white sauce was merely old maids' mania, the reaction toward the flesh of a mind too long absorbed in airships and colloids." There is also Wells's habit of "sputtering at his enemies." The review is long and interesting, however, and full of feminist questioning; for example, "I wonder about the women who never come across any man who is worth loving (and the next time Mr. Wells travels in the Tube he might look round and consider how hopelessly unlovable most of his male fellow-passengers are)." She suggests that women should go to work rather than endure the "domestic drudgery that men have thrust on the wife and mother." Through work, the stronger women would develop qualities of decency, courage, and ferocity, though Wells's heroine might be "sucked down to prostitution and death."[13]

Private Life with H. G. Wells and Son Anthony

Rebecca's writing attracted Wells's attention. He himself sometimes wrote for the *Freewoman*. It was, in part, a literary courtship. She first met him in 1912 when she was nineteen; he was in his forties. They became lovers the following year. It was a long relationship but one Wells never publicly acknowledged. After the birth of their son Anthony they often lived together, although he remained married to someone else. Rebecca and Anthony were hidden away, usually far from London. It was not an easy life for her.

But even after they became lovers, she continued to publish her critiques of Wells's writing. In 1912 she started writing for the *Clarion*, a socialist paper and the most popular radical journal of the time. She also continued with her feminist activities, in print and in person. She strongly supported education for working women, including a labor college for women, Bebel House. She attacked both eugenics and Labor Party members who were still "ape men" on the subject of women.

It was a time when "suffering, humiliation and death" were necessary to win such a simple right as the vote. Violence against activist women was quite common. She was herself struck in the throat by a policeman and pelted with herrings when she spoke to striking dock workers. Rebecca saw Margaret MacMillan, an activist working for nursery schools and infant welfare, hurled down the steps of the House of Commons. In prison suffragists were force-fed. Emily Davison died under the feet of a horse, Rebecca West reported, rather than at the hands of the prison doctors who had forcibly fed her after she had thrown herself down five flights of stairs and was badly injured. At twenty, Rebecca West had had enough experience of the world for "any ordinary lifetime." There was already a police file on her in the Home Office.

She kept her pen sharp, however. When a popular dancer was denounced by the bishop of Kensington for her chorus-girl costume, Rebecca denounced the bishop and told him he would be better occupied providing clothing for Mary Brown and her four children. It didn't matter whether she danced in a "crinoline or a cobweb, compared to the real problem of underfed and under-clothed working people in London. Poverty is the most obscene immorality."[14]

Starting in 1914 Rebecca also wrote reviews and articles for the *Daily News*, as did Bernard Shaw, H. G. Wells, and Arnold Bennett, whom she called her literary uncles. Some of West's male fellow journalists and editors on the *Daily News*, as well as George Lansbury of the *Daily Herald*, were themselves feminists. The *Daily News* had been started in 1846, with Charles Dickens briefly its first editor. In 1909 it had a circulation of nearly 400,000. Rebecca now had a much larger audience for her journalism than the earlier papers had provided her. Her reviews continued to be outspoken; she approved of Theodore Dreiser's subject matter but called him a sloppy writer. She even criticized Tolstoy. She also wrote positive reviews; Ford Maddox Ford's *The Good Soldier* got high praise. Rebecca herself wrote a book of literary criticism on Henry James.

She continued to attack anti-Semitism and other societal injustices, but the major issues in her essays were women and work. Important changes were needed; women themselves needed to effect them. Rebecca wrote for a number of other papers as well, but not all her editors were like-minded. At *Everyman*, she noted that the unsympathetic editor had cut down and emasculated her review of H. G. Wells's novel *Marriage* because he considered it "incompatible with revealed religion, authority and ethics."[15] Most of Rebecca's radical feminist journalism was written during a relatively short period of time, largely between 1911 and 1917. Rebecca herself preserved the words a critic of that period had written about her: "She has moved with confidence from fiction and literary criticism, where women before her have worked with genius, to paths where few female feet have left their mark—moral philosophy, psychobiography, political history. Boldness is Rebecca West's strength. . . . She polished the weapons of invective and denunciation into the tools of a fine art."[16]

One wonders what Hutchinson made of the young Rebecca's life and writings. In 1911, when she was already a published journalist, he was only eight years old. His mother had been a feminist on a bicycle, who made speeches and tried to make changes in some of the areas of Rebecca's anger, especially in the welfare of children. But Evaline Hutchinson was no radical, nor had she Rebecca's gift to "polish invective and denunciation" into a fine art.

Evelyn Hutchinson wrote, in a private notebook, about Rebecca's journalism. Her early works "struck her contemporaries as revolutionary, not only in their feminism but in their socialism. Later works may appear more conservative. This is partly due to the welfare state program in Britain having achieved many of the social aims of her early life. . . . Mainly on feminist issues did she feel that inadequate progress had been made."[17]

Rebecca West's private life over the next two decades was rarely happy. She and Wells had their good times and their terrible ones. He was "Jaguar" to her "Panther," and there were letters with his able feline cartoons. Wells also wrote and illustrated a wonderful book for his small son, Anthony, one of my (this author's) favorites as a child, about a small boy who wanted a live elephant as a pet. "Pride goeth before a fall" was its prime motif. He was less kind to Rebecca. He never left his wife, Jane; he, nevertheless, had other affairs. Rebecca's "two greatest loves were physically unprepossessing men with brilliant minds and stony hearts: H. G. Wells and Lord Beaverbrook," wrote one reviewer of Victoria Glendinning's biography of Rebecca West.[18]

Rebecca raised Anthony, but under difficult circumstances. He did not find out that Wells was his father until he was sent away to school. He resented his mother and later wrote hurtful autobiographical novels, a sad interaction never fully resolved.

Political Reporter and Novelist

Rebecca West married Henry Andrews in 1930, which was also the year of her first visit to the United States. It was "hardly a marriage of passion," but it lasted until the end of his days. "Rather like a dull giraffe, sweet, kind, and loving," Rebecca said of her first impressions of Henry.[19] Much later she said that Henry "had a very good, amusing, well-stored mind on art and architecture, and from a detail could tell you not only the rest of a painting but the museum it hung in." Henry had been interned in Ruhleben, a notorious concentration camp during World War I. It is clear from her letters to Hutchinson that, at least in their first years, her husband provided stability in her life but left her free to pursue it and to travel, with or without him.

West continued to be a political reporter, often of important events such as the Nuremberg trials. She spent much time in Yugoslavia, and the book she produced, *Black Lamb and Grey Falcon,* had a major revival in more recent years when Yugoslavia, as one country, fell apart.[20] Her book, which, in my view, combines history, politics, vivid characters, and personal impressions, helps to explain more recent events there.

As in her youth, Rebecca continued to live on her writing. In addition to her essays and reviews, she also wrote much-read novels. The *New Yorker* was good to her and she to it. Harold Ross sent her to cover the trial of the British "Lord Haw-Haw," who broadcast Nazi propaganda during World War II. She wrote an important book, *The Meaning of Treason,* that included him and other British traitors. This book, though controversial in England, put her on the cover of *Time* in 1948. The cover labeled her the "World's Number 1 Woman Writer."[21] The book was a best seller in the United States. Harold Ross told her that she had invented a way of covering trials that made them as gripping as a film or a novel. Truman Capote later wrote that West was one of the inventors of the nonfiction novel in her reports on these treason trials and on the Nuremberg trials.[22]

She received many honors, including membership in the French Légion d'honneur. The most celebrated of these honors made her Dame Rebecca West, a somewhat ironic honor for the passionate supporter of the working woman, the Rebecca of her younger years. Her "sheer life force" never left her, however. When I read a late book of hers, *1900,* with its wonderful photos and her personal commentary on the arts and politics that revolved around that particular year, I felt that the voice of the young Rebecca was still there. Written when she was eighty-eight but about the year when she was eight years old, the book and its writing style showed both the humor and erudition of her earlier writings, interspersed with many caustic comments on the position of women and the foibles of men.[23]

Faithful Correspondents: Best Friends?

What was the attraction between West and Hutchinson? They both had remarkable minds. They both seemed to have a need to communicate in writing. Their letters discuss ideas, but they are also personal. And they met many times in person on both sides of the Atlantic over a period of years.

After *Rebecca West: A Celebration,* a collection of her writings, was published in 1977, West discussed past and present essayists with Alex Hamilton, a reporter for the *Arts Guardian.* When asked what essayists she valued, she answered: "There are people whose minds you like to live with. As the best instance, Evelyn Hutchinson . . . professor of biology at Yale, great linguist, classical and modern, who in *The Enchanted Voyage* will describe . . . a special church for criminals, or more recently has done a book on bird ecology and population, too mathematical for comfort, but in the middle of all this will suddenly tell you how the first person to discover the principle of territoriality in birds was Oliver Goldsmith."[24]

Within a year after their first visit in 1947 Rebecca and Evelyn had became close friends. In January 1948 she asked him to write an introduction to her "Omnibus," a collection of her works, which, however, was not published until almost thirty years later. He saw her in New York in April 1948. She had been in the United States reporting on Henry Wallace's presidential nomination at the Progressive Party Convention.[25]

Returning to England on the *Queen Mary,* she wrote to Evelyn and Margaret that she was in distress. Her husband, Henry Andrews, was impossible to live with, disorganized, and incompetent, making her life chaos. Much of his eccentricity was later attributed to arteriosclerosis. But after this admission, Rebecca wrote to Evelyn not only about her intellectual concerns and professional life, but also about her family problems.

Hutchinson wrote to Rebecca later that year about his application for a Guggenheim Fellowship.[26] He hoped to travel, particularly to Italy, the following year in connection with his projected book on lakes. He did indeed receive the fellowship and traveled both to England and to Italy. He also wrote that he and Margaret were very happy about Truman's presidential victory over Republican Dewey. That was the 1948 election, in which the *Chicago Tribune*'s headline prematurely proclaimed Dewey's victory.

In January 1950 Rebecca wrote a long letter covering many topics ranging from a book by Ruth Benedict to the British National Health Service. One topic on which she needed Evelyn's help concerned "cowdoctoring." The cattle on their farm had Johne's disease, which was going through their herd. A pathologist had suggested streptomycin. She wanted to know if Evelyn thought this would be good for Johne's disease, and if so what it would cost to get some sent from the United States. Her cows were worth at least a hundred and twenty pounds apiece, she wrote. Evelyn went off to consult American veterinarians.

Rebecca also wrote that she was doing an article on the British National Health Service for an American magazine. She had concluded from her investigations that Britain had no choice: "There would be no teaching hospitals in ten years time. . . . No G.P.'s could afford to wait till age thirty before earning their keep, no surgeons till forty under post-war conditions." Her question, "How is the State going to find the money?" went unanswered. "But almost nobody, except very old doctors like Lord Horde, who must be nearly eighty, is against the N.H.S." She wrote that one ingredient of the excessive cost was abuse of radiography, "which is used to such an extent that it must logically follow that before the discovery of X-rays everybody died as soon as they were born."[27]

Rebecca told Evelyn that she had a number of letters from H. G. Wells that she had offered to the British Museum but got a "churlish form letter answer telling me that I must send them in this and that form in order that the Keeper of Manuscripts could decide whether to advise the Trustees to accept them." She was not at all happy about this and asked Evelyn if he thought Yale would accept them, on condition that they were not published for fifty years after her death. Later that year, 1950, Hutchinson arranged for the deposition of her letters from H. G. Wells and other letters and personal papers in the Beinecke Library at Yale, on the condition that many of the most personal not be used until after the deaths of Rebecca, Henry Andrews, and Rebecca's son, Anthony Wells. Later she allowed some of them to be used by a biographer of H. G. Wells.

Hutchinson became her bibliographer, collecting her work, often from unusual places, for the Beinecke. After Rebecca's death, Hutchinson added all his letters from Rebecca, thirty-five years of correspondence on a tremendous range of subjects, philosophical and personal, to this collection. Hutchinson's letters to Rebecca West, however, are not at Yale; they were sold by her heirs to the University of Tulsa.

Several of West's good friends and supporters died in the 1940s and 1950s, including Alexander Wolcott in 1943 and Harold Ross in 1951. By then Hutchinson clearly had become not only friend and correspondent on a huge variety of subjects but also a concerned advisor. For example, in 1953 Rebecca wrote a series of anticommunist articles for the London Sunday *Times*, which were sold to and, without her permission, reprinted by *U.S. News and World Report*. These articles were denounced by some as pro-Joseph McCarthy and the House Committee on Un-American Activities (HUAC). *U.S. News and World Report* was known as pro-McCarthy. Arthur Schlesinger, Jr., an anti-Communist liberal, wrote her a letter, denouncing McCarthy as an opportunist who attacked other anti-Communists, such as Charles Bohlen and even George Marshall of the Marshall Plan. West wrote that she also believed McCarthy ideologically dangerous, but she did not believe he would last, but would exhaust the patience of the Republican Party chiefs. He did indeed. He attacked the army and was condemned by the U.S. Senate in 1956, but not before he ruined the careers of many innocent Americans.

Hutchinson was much distressed by the position in which Rebecca West found herself in relation to McCarthyism. He wrote that it was strange that "I should tell you, of all people, of the danger of small dicta-tors. . . . That I risk being thought impudent by someone I love and admire greatly is a measure of how serious I feel this matter to be." Nothing would please him more than "to read some day in the next week or two a letter over your signature, say in the New York *Times*, indicating that it has been brought to your attention that your recent series has been used in a way that seems, in spite of your explicit remarks thereon, to indicate that you defend the activities of Senator McCarthy, and that you wish to dissoci-ate yourself from such innuendo." He wrote that he felt this way because it was extremely "painful to your friends to find your authority implicitly involved in this way . . . but chiefly because, when it is absolutely essential that the ship remain on an even keel, it is being rocked in your name."[28]

She did, in part, take Hutchinson's advice and wrote a letter to the *New York Herald Tribune*, but the dispute with Arthur Schlesinger con-tinued. Even though she thought McCarthy "ideologically dangerous," she also thought communism a clear and present danger to the world. Eventually John Gunther got Rebecca and Schlesinger together and an "amicable friendship" resulted.[29] In any case, Hutchinson provided im-portant advice at a time it was sorely needed, although the whole incident remained painful to some of Rebecca's friends.

The Relationship of Science and Art: Hutchinson and West

In 1953 Hutchinson published *The Itinerant Ivory Tower: Scientific and Literary Essays.* It was dedicated "to Yemaiel Oved, for what she taught me long ago." On the title page is a quote from Simone Weil, expressing Hutchinson's own philosophy: "La vrai definition de la science, c'est qu'elle est l'étude de la beauté du monde" (The true definition of science is the study of the beauty of the world). All but one of the essays had appeared in various journals from 1945 onward, most of them in the *American Scientist* but others in *Ecology*, the *Scientific Monthly*, and elsewhere. One, "The Dome," was new. It was an appreciation of Rebecca West and her work, more literary perhaps than Rebecca's own critiques of other people's work. Some other essays involve philosophic aspects of science. The theme is the relationship of science and art. He wrote:

> Toward the end of *Black Lamb and Grey Falcon* when bombs are falling on London, the light begins to appear and we realize that we have had it all the time. It appears in part in a most unexpected form, as an aria from *The Marriage of Figaro*. This opera is one of the most mysterious of works; from the moment of its opening one feels something of extraordinary importance is happening. . . . The implied importance of art emerges as a theme over and over again in these books, but it is most clearly expressed in *The Strange Necessity*, a work of immense originality and importance.

He felt that critics had not given the latter book by Rebecca West its due. Here Hutchinson's feminism, clearly present and expressed throughout his years at Yale even before he had women graduate students, comes out. He wrote that most critics had been unaware of the importance of *The Strange Necessity* as an exploratory theory of art. "The reason for this is doubtless as silly as it is illuminating. It is probably because it was written by a woman. . . . Practically all considerations of aesthetics previously written were written by men."[30]

The letters beween Hutchinson and West continued throughout the 1950s, 1960s, and 1970s, often referring to the important events of those years—from McCarthyism and the cold war to the assassinations of President Kennedy and of Martin Luther King, Jr., and the Vietnam War, from the Black Panthers and student unrest to Agent Orange, and eventually to the Iranian revolution. Interspersed are reports of honors given to each, anguished discussions of Rebecca's family problems, and commentary on mutual friends and on Hutchinson's students and Yale colleagues. The variety of intellectual topics is enormous.

Margaret wrote some letters to Rebecca, too, on books, on the music she was singing, on the garden and the cats, but sometimes about Evelyn. For example, in late 1957, when Hutchinson was preparing for an international limnology conference, she wrote: "Evelyn chose to put into Italian the most brilliant lecture I have ever heard. I heard it in English of course. . . . What a lecture to give in a foreign tongue. Evelyn is quite a lad, isn't he?"[31]

From Palermo Evelyn wrote to Rebecca that he had finished the bibliography of her work, and several copies were sitting on his desk in New Haven: "It is far from perfect; only authors who don't write are easy to keep track of." He had found all sorts of unexpected writings in his search. He added, "You would find Palermo a marvelous place for unexpected sculpture."[32]

Evelyn and Margaret Hutchinson were in Rome for a month in 1958, more than thirty years after his first visit, during the time he was working at the Naples zoological station. He was working in Rome on his lake study. They then went to Florence, where Rebecca gave them an introduction to Bernard Berenson. Evelyn wrote from Florence relating that the Villa Berenson had wanted Hutchinson on the phone, causing a great rise in his stock at the Hutchinsons' pensione! He also wrote Rebecca much about excavations in Rome and the art in Florence, adding, "Florence we are loving, and have plenty of time to explore it, as I am not trying to do more than 3 hours of writing a day."[33] By July, still on his Guggenheim, they were in Oxford having lunch with Hutchinson's old friend, the English evolutionary biologist David Lack, together with Dillon Ripley and Hutchinson's former student Robert MacArthur. What a collection of ecological and evolutionary biologists! Evelyn and Margaret would see Rebecca the following week.

In November, back in New Haven, Hutchinson was writing to Rebecca about the dangers of the increase in carbon 14 and strontium 90 as a result of the testing of nuclear bombs.[34] Margaret wrote Rebecca two letters in the summer of 1959. They are quite literary; one wonders whether she thought these might be saved for posterity. They had visited Dillon Ripley at his estate in Litchfield, Connecticut: "I almost felt I was back in England on Sunday when we drove out to Litchfield. . . . Sitting out on the terrace looking down a steep enough slope I felt I was looking at a Constable. . . . Then, by the pond and looking at the house at the top of the slope, and yet another slope and woods behind it, that I was looking at an eighteenth century water colour."[35]

Margaret reported to Rebecca that she had offered the Yale art gallery a collection of her rings, "eighteenth century dating back to well BC." One was a seventeenth-century astrolabe ring. She concluded "and now I want a drink outside in the garden, Vermouth with a twist of lemon peel and no ice."[36] Margaret also included in one of her letters a review by Evelyn for the *Herald Tribune* of *The Road to Man* by Herbert Wendt. Both Evelyn and Rebecca wrote reviews for this newspaper.

Later in 1959 Evelyn wrote to Rebecca about his experience as a major speaker at a United Nations oceanographic conference. It was in the assembly hall, "a somewhat terrifying experience." Russians were there, and his speech was simultaneously translated into Russian.[37]

Personal experiences also appear in these letters. The Hutchinsons' cat, Sambo, died, and they got a new blue point Siamese kitten. Rebecca was having trouble with her now-grown son, Anthony; this intensified when he began to write semi-autobiographical novels about her and about his father, H. G. Wells. Rebecca also recounted her happy experiences to Evelyn. The two of them shared an intense interest in art. Rebecca wrote about a trip to Paris, reporting that when she was asked to speak about art in French on the radio and on television, she found to her delight that she could do so. She had been to see a collection of Van Gogh's and his friends' paintings, including "a Cézanne never before exhibited, beautiful beyond words, the supreme Cézanne, dahlias on a table." Another collection she went to see contained "many Rouaults that I would have crossed the world to see." She had written a preface to a collection of Picasso drawings and was asked to follow this up with a preface to a volume of Braque drawings. "I felt dubious, but Picasso liked my preface so much that I risked it."[38]

Rebecca was invited to give the Terry Lectures at Yale in the spring of 1956. Hutchinson had arranged this for her. She wrote to him: "Many thanks for the Terry Foundation Lecture. The news came to cheer me at a moment of the most acute depression. . . . It is something that I should be recognized as a writer whose views are worth hearing."[39]

Rebecca's son, Anthony West, did publish his semi-autobiographical novel, *Heritage,* called by Rebecca a "public degradation." She sent a telegram to Hutchinson in October 1955 reading: "PLEASE ENQUIRE IF YALE COMMITTEE WISHES CANCEL LECTURES IN CONSEQUENCE HERITAGE LOVE REBECCA"[40] The lectures were, of course, not canceled, but she later wrote to Evelyn not to expect too much from her lectures: "I have been defeated by forces that were too strong for me."[41]

She came to New Haven in April 1956 with her husband, Henry. Her lectures, later published as *The Court and the Castle,* "proved to be a triumph." Rebecca at sixty-four was as exciting and critical as she was at nineteen, although now no longer a revolutionary, only a reformer.[42]

Rebecca herself, in spite of her bitterness at Anthony's novels, wrote

31. Rebecca West at work. Beinecke Rare Book and
Manuscript Library, Yale University.

her own plainly autobiographical novels. "Rebecca's bravura performance
at Yale," as biographer Rollyson called it, occurred just as she had completed *The Fountain Overflows*, the first of a trilogy about the "Aubrey"
family, with herself as "Rose." This novel was both a popular and critical
success, a best seller, but it was hard on some members of her own family,
particularly her older sister.[43]

It was in 1958 that Rebecca was named Dame of the British Empire
(DBE). In February 1959 she visited the queen. She sent the Hutchinsons
a whole description, some of it humorous, about being "damed" and her
conversation with the queen and with Princess Margaret. Rebecca, the
former revolutionary, was henceforth to be "Dame Rebecca."[44]

Rebecca's journalism career, always of interest to Evelyn, was not yet
over. The London Sunday *Times* sent her to South Africa in 1960–61 to do
a story about the National Party. It was the era of the pass laws and the
extension of apartheid. Rebecca kept a journal and sent letters to both her
husband and to Hutchinson. She met Nadine Gordimer and attended the
treason trial that had gone on for three-and-a-half years. There had originally been 150 defendants, now reduced to 30. Eventually all were acquit-

ted but some, like Zulu chief Albert Luthiulu, were essentially held under house arrest.

Rebecca went to the "squatter slums" in Cape Town. She then stayed with writer Lulu Friedman, whose husband, Bernard, along with Helen Suzman, had helped found the Progressive Party, dedicated, Rebecca reported, to eradicating apartheid and developing a nonracial government decades before that actually happened. She interviewed the other side as well, including an eighty-seven-year-old Cape Town doctor who told her, "I've treated lots of them [blacks] and they're no more like human beings than zebras."[45]

In March 1961 the Sharpeville massacre occurred. Police fired into a large crowd of unarmed blacks who were protesting the hated passbook laws, killing 69 and injuring 180. On April 21 a state of emergency was declared and both the Pan African Congress and the African National Congress were banned. Hendrik Verwoerd, the prime minister, was shot in the head in an assassination attempt. (He recovered from this injury but was assassinated in 1966.) It was dangerous to be there; another woman journalist had been arrested. Rebecca West stayed through all of this.

Later, Rebecca gave a "smashing indictment of apartheid" on the Edward R. Murrow television news show.[46] When Rebecca returned to England in April she noted that it was fifty years since she had received her first journalism check. She was seventy; the South Africa report was her last major piece of political journalism.

Hutchinson received some materials of Rebecca's for his bibliography of her writings from her husband, Henry Andrews. He replied to Henry, writing about his own research: "I am at last getting a great deal of time to do my own work and am assembling all sorts of fascinating details into a picture of the history of Latium as reflected in what fell into Lago de Monterosa [sic], with similar things to come from Yucatan and Persia."[47]

In 1961 Hutchinson also wrote to West about his scientific interests and his colleagues at Yale. His research associate, Ursula Cowgill, had traveled to Guatemala to try to obtain some further material for their research. He added that Alison Bishop (later Alison Jolly) had finished her doctoral dissertation on lemurs. "She and two other women in their 20s went to two [animal behavior] congresses at the London Zoo and in NY and apparently made everyone else working on primate behavior look quite outclassed." He told Rebecca of Jane Goodall's "extraordinary observations of wild chimpanzees sucking up termites . . . through straws, broken from a hollow wild grass," suggesting, he thought, that our human ancestors may have used "perishable materials many million years before they started on stone." He also told her of a psychiatrist friend who had been to Nigeria and seen a "famous illiterate native 'doctor'" treating mental patients with a *Rauwolfia* decoction, "which is of course the source of reserpine, the main tranquilizer of the past few years!"[48]

By the end of 1962 another book of Evelyn's collected essays, *The Enchanted Voyage,* was published, dedicated to his Sicilian friend Laura Mangione. He had done much of his research in Palermo. At the end of the year Margaret wrote to Rebecca that Yemaiel had been doing successful solo ballet, bringing the house down with a solo in *The Nutcracker* and "having the real gratification to so young and modest an artist of having a young man stand up in the stalls and shout for joy."[49]

Rebecca made a trip to Barbados with Henry in 1962 and afterward went by herself to New York City, where she saw Hutchinson. When she got home, she found that Henry had been diagnosed with cerebral arteriosclerosis. Rebecca also wrote about her own illnesses. Early in their friendship she had written, "I have been struggling to write a book, run a farm, drop in now and then at Nuremberg [where she had covered the Nuremberg trials] and have gastro-enteritis . . . at one and the same time."[50]

Evelyn also wrote about his hospital stays—often with humor. In a 1963 letter from Grace-New Haven Hospital he wrote: "I got whisked off here last week from the doctor's consulting room without even being allowed to go home, owing to the customary interaction of my age and my sex in impairing my personal plumbing. I am assured that I am benign throughout—in the limited sense in which the word is used in a carcinoma-conscious world. . . . I am being operated on tomorrow."

His hospital stay had been a "wonderful time for miscellaneous reading." The miscellany included *Measure for Measure, Troilus and Cressida,* and a book on the archeology of Palestine. From the last he reported to Rebecca that "Mousterian flake scrapers remained in fashion for 40,000 years whereas nothing in this [hospital] room is the least like what it was in the 1890's." Meanwhile the whole thing was "perfectly wretched for Margaret," which seemed to be the case whenever Evelyn was ill.[51]

The next month he wrote that, after an admirable recovery from the operation, he had a relapse when a blood clot got caught in his lung, but now all was well. He added, "Leading the life of a gentleman of leisure has moreover been very pleasant and I have got some writing done." None of

Rebecca's illnesses were "very pleasant" and their spouses' illnesses were nightmares for each of them.[52]

Hutchinson wrote to Rebecca about S. Dillon Ripley, the director of the Peabody Museum at Yale and a close friend. They had met when Ripley was a Yale undergraduate. Ripley, who came from a wealthy family, had traveled with them to India. Very few people were allowed into Ladakh in northern India in those days. Only he, at thirteen, and his older sister went into Ladakh. Later, at Yale, he saw and recognized a mask that Hutchinson had brought back from Ladakh while on the North India expedition and came to see Hutchinson about it; they became friends. Ripley majored in history, not biology, as an undergraduate. But he decided not to go to law school but to become a biologist. It was 1936, during the Depression, and "this was not a popular subject in my family." It was not very easy to persuade his father, but he started off to Columbia Graduate School to major in zoology. Soon after, one of his sister's friends, who had studied anthropology, was about to sail with her husband on their schooner to New Guinea. They invited Ripley to come with them as zoologist. "It seemed like a bolt from the blue and I jumped at it like a starving trout."

He prepared himself for this trip by going to the American Museum of Natural History in New York and working on birds with the then-young curator Ernst Mayr. Ripley went to New Guinea and Sumatra, studying and collecting birds, not returning until 1939, when he came down with severe malaria. Hutchinson stayed in touch with Dillon Ripley, continually encouraging him in his work.

Ripley then went to Harvard instead of to Columbia for graduate work. For research with a collection of Indonesian birds he received his Ph.D. in 1943. During World War II he served in the Office of Strategic Services, in charge of American intelligence in Southeast Asia. "I worked in the areas that I knew best, from New Guinea to Sumatra." He had learned the local languages when he had studied birds there. "I worked with agents organizing the underground," he told me.[53] He later became a professor at Yale and director of the Yale Peabody Museum of Natural History. Dillon Ripley was appointed secretary of the Smithsonian Institution in summer 1963. On hearing this news, Evelyn wrote to Rebecca:

> I suppose it is the most responsible job of its kind in the world as all the museums and galleries as well as the zoo in Washington are under the aegis of the Smithsonian and its secretary. I think he is the perfect man,

having a good feeling for art and archeology as well as natural history
and all the right connections. It is an awful blow for us personally. . .
There is a very real problem generated by the approaching extinction of
the learned man of wealth who has professional standards, public spirit
and amateur enthusiasms.[54]

S. Dillon Ripley's grandfather, Sidney Ripley, had been president of the
Union Pacific Railway.

In October 1963 Hutchinson wrote to West about the honorary doctorate of science he had received from Princeton University. As noted earlier,
his revered uncle, Sir Arthur Shipley, had received an honorary doctorate
at Princeton more than fifty years before. Evelyn wrote to Rebecca, "It
also meant rather a lot here in view of the tangled attitude to science in
general and to my sort of biology particularly, that has grown up in Yale."
The "tangled attitude" to Hutchinson's science is ironic, since he was then
at the height of his profession as an ecologist. The two papers, "Concluding Remarks" and "Homage to Santa Rosalia, or Why Are There So Many
Kinds of Animals," he had written in 1957 and 1959 (see chapter 14) and
the research that followed from them had made him truly famous. But
molecular biology had taken over by 1963. Hutchinson was right; ecology
was about to be largely swept away for some time at Yale. He himself
stayed at Yale almost another thirty years, but except for Willard Hartman, all of Hutchinson's graduate students who had become Yale professors left Yale.[55]

Spring of 1970 found Hutchinson writing about events at Kent State
and in Cambodia and their repercussions at Yale, and also about his hernia operation. On the latter subject Rebecca responded that her sister,
Winifred, would have been especially sorry for Evelyn because she con-
fused "hernia" with Hermia of *Midsummer Night's Dream* and "thought
of anybody who had a hernia as suffering from unfortunate love (which
would I suppose have meant that Margaret would have left for fresh
woods and pastures new."[56]

In 1971 Hutchinson officially retired—at least from teaching at Yale.
His last year kept him very busy. He wrote to Rebecca:

I had to have my seam resewn [hernia operation] which was very hard
on Margaret. . . . As soon as I was out of hospital, term began. Everyone
wanted to learn from me before my retirement, so my main ecology course
which should have 35 has 85 students, all of whom need a lot of individual
attention, . . . most exhausting. A small group in the lab is looking at Sci-

ence and Art. I suggested they read [Rebecca West's] *The Strange Neces-sity* . . . a great and unexpected excitement to them. . . . Is Dame Rebecca ever coming to New Haven? I was offered a post-retirement job in UCLA, but I think we shall stay here. We like it.[57]

He was offered a great many post-retirement positions, but he remained at Yale, where he still had an office during his last year of life.

In response to Hutchinson's description of the spectacular retirement party for him at Yale in May of 1971, Rebecca wrote: "I have received your letter regarding the celebrations and am laughing like a horse because you do not seem to have known how much people love you. I am en-chanted by your note that the [graduate] students departed with your blessings clean-shaven as Chihuahuas and returned [looking] like Amish or Eastern Orthodox priests."[58]

Two publications were produced as part of Evelyn's retirement cele-bration. Rebecca wrote a special essay for one of them, at the request of Edward Deevey, Hutchinson's former graduate student and afterward a Yale professor. After Markert became chairman of the Biology Depart-ment, Deevey left Yale for Dalhousie University and later the University of Florida. Deevey had written to Rebecca, "I think I am not mistaken in discerning a 'Hutchinsonian influence' in your own work. . . . I mean that there is a very strong philosophical connection between your work as an artist and Evelyn's as a scientist." Deevey felt that their joint themes in-cluded the joy of creativity. "Knowing Evelyn's heart and mind as I believe you do," he continued, could she reflect on the similarities and differ-ences "in style and substance between your work and that of a truly great natural scientist."[59]

In her essay, Rebecca West did so: "There are many strands in the links that bind Evelyn Hutchinson and myself." She continued, noting magical experiences, "For Evelyn is a descendent of Merlin, a white warlock." She recounted an incident that, she claimed, happened when Evelyn came briefly to England to talk to the Royal Society on behalf of the preserva-tion of the island of Aldabra, home of giant tortoises; the island was in danger of destruction (see chapter 15). She wrote:

He had taken off in a plane to England from America for only a few hours, in order to vote at the annual general meeting of a learned society, in order to strengthen its hand in defense of an island in the Indian Ocean, where rare animals and plants were making a last stand, but were now threatened. . . . He did me the compliment of seeking me out as a companion for the hour he had before the meeting, and ran me to earth at my hairdresser's. He arrived there a quarter of an hour before I did, and by the time I got there the staff had realized that this was a visit from Merlin. . . . He got well away on a description of the singularly architectural courtship of bower-birds to the two young women, a subject which had come into his mind by an indirect route starting at some wigs in the window. . . . When he had left, the manager and the receptionist and the manicurist all breathed, "Who is he?"

According to Rebecca, Evelyn had a special gift for understanding women, including her.

> They need not call for liberation when they are in contact with him, for he never enslaves them. He thinks of them with perfect justice. He knows us and likes us so well that he can laugh at us, and has for ever preserved the rugged quaintness of the dons' and professors' wives in Cambridge who started the suffrage movement. In a sentence [of Hutchinson's] that is a comic masterpiece in brief: "When they rode through the town on bicycles it was difficult not to believe that men were responsible for all the evil in the world."

At the end of her essay she wrote, "Often I wonder what is happening to civilization," and she recalled a line that Denis Saurat wrote about a Serbian critic, poet, and partisan during World War II: "'If there are twenty people like this scattered between here and China, civilization will not perish,' and it is Evelyn who is in my mind. . . . The classical scholar, the art historian, the humorist, the wit, the liturgist, the kind friend to the perplexed, there they all are. But of course it is the scientist who is the master self."[60]

From reading Rebecca West's tribute, it is apparent that she and Hutchinson had a special understanding of each other. She had read Hutchinson's scientific essays, starting in 1947, and he had recounted his own work and that of his students and colleagues to her in his letters. West was herself no scientist, but unlike Margaret Hutchinson, she seemed to have an interest in and an appreciation of science, or at least the kind that Evelyn did.

Evelyn's younger brother, Leslie, wrote to Rebecca, thanking her for her tribute to Evelyn and saying that he and his wife, Hannah, enjoyed every word of it:

> at times roaring with laughter, at times nearly weeping with nostalgia. Growing up with Evelyn as an elder brother was a wonderful experi-

ence. . . . He always knew so much more and could "penetrate" so much
further than the rest of us that I worshiped him as Merlin from very early
days. He also had and has today to an even greater extent, the quality of
treating his dumbest companions as equals and making them believe they
had some gifts! I too loved your reference to the Ladies of Cambridge
[who included Evelyn's and Leslie's feminist mother], a sentence I have
cherished as a . . . description of a very special species.[61]

The year after his retirement, Evelyn wrote to Rebecca about an
impending sad event, the death of Robert MacArthur, his outstanding
former graduate student and later Princeton professor (see chapter 14):

We are having to get prepared for the shock of the death of the ablest
man I have ever taught, a person of enormous integrity and sweetness,
as well as being the most important investigator of the more mathematical aspects of ecology. . . . He was taken ill the week before my party last
year and found to have a badly metastasized renal carcinoma. Since then
he has written a book and several very important papers, as well as going
with his children to look at birds in a group of islands off Panama. We are
trying to get him a National Science medal.[62]

The last sentence of this letter is ironic, since Hutchinson himself was
later awarded a National Medal of Science which he turned down because of his disapproval of President Richard Nixon. In this letter he
wrote about the impending presidential election of 1972, noting that
Nixon "lacked any consistent ethical base." Robert MacArthur died on
October 31, 1972. Writing after his death, Hutchinson told Rebecca that
he had been to see MacArthur at his home in Princeton six days earlier.[63]

Evelyn and Rebecca often wrote to each other about the characters
in their lives. In Rebecca's case, these characters often appeared, though
slightly disguised, in her novels. Hutchinson's characters only appeared
in his letters and, of course, in real life.

Evelyn described his friend and neighbor Dorothy Horstmann, whose
house had been broken into. She was, he wrote, "the leading investigator
of the epidemiology of German measles from the point of view of birth
defects—a very eminent and lively woman and probably the only member
of the US National Academy of Sciences to wear bright blue eye shadow."
Hutchinson also wrote to Rebecca about Marjorie Garber, Shakespearean
scholar and close friend and later a Harvard professor: "the first woman
fellow of Saybrook [College] with a regular faculty appointment, a very
lively-minded and attractive young woman with a deep interest in poetry

and painting." Garber would soon be traveling to London, and Hutchinson was recommending her to Rebecca as a visitor.[64]

Margaret, who was soon to succumb to Alzheimer's, wrote to Rebecca that Evelyn was flourishing. "His new Ecology book is a big success and he gave a lecture on his 50 years in the lab to an enormous crowd giving him an ovation."[65]

This book, *An Introduction to Population Ecology,* went through five printings between 1978 and 1980. Hutchinson had written to Rebecca that she would like the "unrivalled collection of footnotes."[66] These sometimes take up more room than the text and are a joy to historians of science.

In 1983 Hutchinson lost both Margaret, his wife of fifty years, and Rebecca West, his close friend, confidant, and intellectual companion of thirty-six years. Margaret had been ill for many years and had needed Hutchinson's constant care. Rebecca had been ill also, but her death came as a shock to Evelyn, as he realized "mostly a dozen times a day" that they could no longer share their lives.

In her biography of Rebecca West, published in 1987, Victoria Glendinning acknowledged: "My first thanks are to Dame Rebecca West's executors . . . in particular to Professor G. E. Hutchinson." She also thanked Yemaiel and Ben Aris. Glendinning wrote, "Evelyn Hutchinson . . . became not only Rebecca's close friend and most regular of correspondents, he became the curator of her literary reputation and her bibliographer." She noted that Hutchinson had previously published an appreciation of West's work. Although some of Rebecca's letters were addressed to both Evelyn and Margaret, Glendinning wrote, "The primary friendship was between Hutchinson and Rebecca." Was it more than a close friendship? Probably not, but we cannot be sure. They met many times in New York, where Rebecca West came to confer with her magazine and newspaper editors. One cannot tell from Hutchinson's letters. But here is what Rebecca West wrote about Hutchinson fifteen years after they first met: "I love him dearly . . . and what he has done for me! I have received such generosity from people outside the sexual sphere."[67]

The letters between Evelyn Hutchinson and Rebecca West do, however, provide much insight into each of their lives and many aspects of their relationship.[68] The deaths in the same year of both his wife Margaret and of his special friend Rebecca left a void in Evelyn's life. He was eighty years old, but further publications, honors, and a third marriage were still to come.

CHAPTER FOURTEEN

From the N-dimensional Niche to Santa Rosalia

Clearly, G. E. Hutchinson spanned much of the period between the first colonization of ecology by mathematical theoreticians and the more recent one, and was on the beach to welcome the new colonists,"[1] wrote Robert McIntosh. He was not the only one to use this metaphor. It was Hutchinson's formalization of the niche that probably propelled the "pioneer existence on the intellectual frontier of ecology," cited by Platil and Rosenzweig.[2] The image of Hutchinson and his graduate students spearheading the "colonization" of ecology is striking. One could certainly argue that this process had started much earlier in terms of Hutchinson and his students, particularly with Gordon Riley and Ray Lindeman. Colonization, as McIntosh noted, implies "a relatively barren area, an influx of new talent and skills."

The new skills, of course, came from mathematics, and some of the new talent in terms of Hutchinson's students were mathematically trained ecologists, for example, Robert MacArthur (Ph.D. 1958) and Egbert Leigh (Ph.D. 1966), spanning the decade of "colonization." Larry Slobodkin, although he arrived at Yale earlier (Ph.D. 1951), was also important during this period.

New skills also involved chemistry, geology, and engineering. Lindeman and Hutchinson have been credited with founding ecosystem or systems ecology in 1940 (see chapter 10). Systems ecology later became "big science," and its engineering aspects were of less interest to Hutchinson than to his former student Howard T. Odum. Hutchinson had integrated biogeochemistry into his own research and that of a number of graduate students, including V. T. Bowen (Ph.D. 1948), H. T. Odum (Ph.D. 1951), and R. J. Benoit (Ph.D. 1957), spanning the decade prior to the "colonization of ecology."[3]

Theoretical or mathematical ecology had a long previous history, starting with Lotka and Volterra in the 1920s. It had mostly been concerned with changes in population size; with life tables of populations of single species, a subject to which Hutchinson's early student Ed Deevey (Ph.D. 1938) made considerable contributions; and with two-species interactions. Attempts were made to analyze whole communities mathematically but with little success. By the late 1950s, aspects of population and community ecology, including competition, niche diversification, and diversity, became the foremost interests of Hutchinson and many of his students. Many young ecologists, including Hutchinson's students (but many others as well), used Hutchinson's multidimensional niche concept, presented in "Concluding Remarks" (1957), as a starting point for their research. They used a great variety of organisms, from birds to barnacles to mosses. Hutchinson himself continued to draw on his own favored group, the water bugs.

Hutchinson's Multidimensional Niche Model

As historian of ecology Robert McIntosh has written, Hutchinson's "'Concluding Remarks' . . . [is] certainly one of the least explicit titles in the history of population ecology, masking . . . one of the most highly touted and disputed productions of that discipline—a formulation of niche theory."[4]

It is worthwhile to take a closer look at what Hutchinson actually said when he proposed it in "Concluding Remarks" before examining the effects of his 1957 niche model. In his remarks at the Cold Spring Harbor demography symposium, he seemed to be struggling to find some common ground among the "heterogeneous unstable population" that spoke at the symposium, researchers in the field of demography. The human demographer and the field zoologist demographer have different points of view, the former interested in the immediate future (unless human fossils are the concern), the animal demographer in problems of more general or theoretical interest. Moreover, time periods relevant for the study of one type of organism are not relevant for others. Environmental variables and catastrophes can be more devastating to insect populations than to populations of animals with longer lives and longer generation times. Laboratory workers have kept all but a few factors constant whereas field workers have had to deal with the ever-changing environment. It is in this context that Hutchinson argued that "a difference in interest may underlie some of the arguments which have enlivened, or at times, disgraced, discussion of this subject."

In introducing his formal analysis of the niche in his remarks at the end of the conference, he hoped to remove some of the irrelevant difficulties inherent in divergent points of view. His metaphor here is to a vacuum cleaner, which "when a lot of irrelevant litter has accumulated the machine must be brought out, used, and then put away." Similarly, in an empirical science one need not always keep apparent a "logicomathematical system," that is, the n-dimensional niche he is about to elaborate. Nevertheless it may be able to remove some of the "litter." The remainder of the paper is devoted to his formalization of the niche concept and its relation to the Volterra-Gause or competitive exclusion principle.[5]

It seems likely that most demographers and others present understood his presentation, which does not require calculus but only basic linear geometry. He considered a species, S_1 and two environmental variables, x' and x'', which varied linearly from x'_1 to x'_2 and from x''_1 to x''_2, forming a rectangular area. If variable x''' is then introduced we obtain a volume. The addition of further ecological variables relevant to S_1 (to x^n) produces a hypervolume, N_1 defining this species' "fundamental niche," a concept Hutchinson had earlier introduced.[6] A second species, S_2 would have a different hypervolume N_2. Hutchinson was concerned with both physical and biological variables and said the hypervolume was an "abstract formalization of what is usually meant by an ecological niche." An example of the fundamental niche of a species of squirrel is shown in figure 32 based on three variables: x' food size; x'' tolerated temperature; x''' some measure of branch density. Hutchinson then explained the "realized niche" when a similar and competitive species entered the same "biotope space"; the realized niche of each species of, in this case squirrel, would then be smaller than its fundamental niche.

As was not so clear to later interpreters, Hutchinson did point out very clearly at this point in his exposition the limitations of or restrictions on

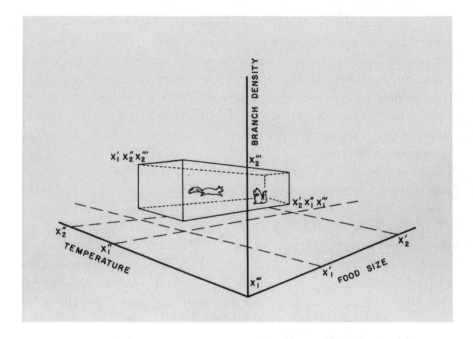

32. The three-dimensional fundamental niche of a squirrel; x′ might define food size, for example, the mean diameter of acorns; x″ temperature tolerance; x‴ some measure of branch density. Reprinted from Hutchinson (1978b), page 159, fig. 99, by permission of Yale University Press.

his niche concept. It included the following assumptions, some of them unlikely for any particular case:

1. It implied that each point in the fundamental niche of a species provided equal probability of the species' persistence, whereas some were likely to be more optimal than others.
2. It assumed all environmental variables to be linear.
3. It referred to a single instant of time.
4. Only a few species of those comprising the community could be considered at once.

Moreover, it assumed competitive relationships between these species. According to Hutchinson, these limitations were not so important in terms of his reasons for devising the model, the clarification of niche-specificity.

Why was all this important in terms of inventing modern ecology? Hutchinson's niche model and the contrasting concepts of fundamental and realized niches led to much subsequent ecological research on a great variety of organisms. It led as well to much research on the measurement of niche breath and niche overlap. In figure 33, showing the realized niches of six closely related moss species growing on Mt. Washington, New Hampshire, one can also visualize the niche breadths and overlaps for these species, at least based on the two niche factors depicted. Mathematical formulas were developed to measure these and other aspects of community ecology and of the multidimensional niches Hutchinson elucidated.

Why Are There So Many Kinds of Animals?

"Homage to Santa Rosalia or Why Are There So Many Kinds of Animals?" probably Hutchinson's other best-known paper, was published in 1959, two years after "Concluding Remarks."[7] It was originally his presidential address to the American Society of Naturalists at their annual meeting in 1958. In it he returned to the water bugs of the family Corixidae and a small pond near the church of Santa Rosalia in Palermo, Sicily. Although there were vast numbers of corixids in this pond, they all belonged to two species of the same genus. The larger species was at the end of its breeding season, the smaller one beginning to breed. The questions he asked were, why did the larger species breed first, and why were there only two, not twenty, corixid species present? And finally, why were there so many kinds of animal species in the world? At least three-quarters of these are insects, but this still leaves a great magnitude of marine, freshwater, and terrestrial animals. He assumed, as an approximate number, one million animal species, almost surely too low, according to E. O. Wilson.[8] What factors have led to the evolution of such a great number of animal species?

Hutchinson discussed this problem in the context of the evolution of sympatric species through natural selection, isolation and later range changes, and of separate realized niches.[9] The last was, in his view, determined by competition and by what he there called the Volterra-Gause principle (later the competitive exclusion principle), assuming equilibrium conditions.

David Lack had already provided evidence for these ideas in terms of birds and of essentially separate niches in regard to food and food-obtaining structures and behavior. Therefore Hutchinson used food chains as a starting point in his search for factors determining the number

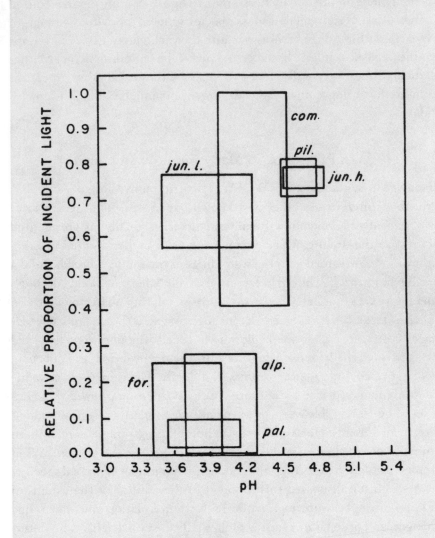

33. Realized niches of six closely related species of mosses: three species of
Polytrichum (above): *P. commune, P. piliferum,* and *P. juniperinum* at high (h) and
low (l) elevations on Mt. Washington, N.H., and three species of *Polytrichastrum:*
P. alpinum, P. formosum, and *P. pallidisetum,* in relation to illumination and soil
pH. After Maxine A. Watson, Ph.D. diss.,Yale University, 1974. Reprinted from
Hutchinson (1978b), page 163, fig. 103, by permission of Yale University Press.

of animal species. He here referred to Eugene Odum's "predator chain," from Eugene Odum's already influential 1953 textbook *Fundamentals of Ecology*.[10] He also referred to the Eltonian food chain, in which the predator at each successive level, from green plant to top predator, is larger and rarer than its prey. He calculated that if approximately 20 percent of the energy passed from one link to the next, and the predator had twice the mass of its prey, the animal that was the fifth link would have a population of 10^{-49} of that of the first link.

That, he calculated, would be the most links possible. Indeed, some such cases had been recorded. He concluded from these calculations that neither the Eltonian nor other types of food chain could provide the animal diversity actually seen. He pointed out that if natural selection increased the efficiency of energy transfer between a predator link and the one below, the food chain would be shortened, not lengthened. In the case of baleen whales feeding on copepods, there are only three links, the whales becoming primary consumers. "Mechanical considerations would have prevented the evolution of a larger rarer predator" than the whale, he wrote, until humans developed "non-Eltonian methods of hunting whales."

What determines the number of food chains in the community? That question caused Hutchinson to leave the animal world and consider the primary producers, plants. Plant diversity determines the number of herbivores, which have structural and chemical adaptations for eating particular types of plants. "The extraordinary diversity of the terrestrial fauna," he wrote, "is clearly due largely to the diversity provided by terrestrial plants." The relatively recent evolution of 200,000 species of flowering plants (including trees) and the concurrent evolution of perhaps three-quarters of a million insects, which depend in one way or another on the plants, certainly increases the animal diversity. Hutchinson continued, "The problem still remains, but in a new form: why are there so many kinds of plants? As a zoologist, I do not want to attack that question directly; I want to stick with animals." Therefore he looked for abstract properties of communities relative to plants and animals. He pointed out that predators often have alternative food sources, so that in a food web, as distinct from a simplified food chain, the predator at one level will not become extinct or exterminate a prey species at a population low.

At this point Hutchinson cited the work of two of his recent students, Larry Slobodkin and Robert MacArthur, whom he cited in the first footnote of his paper as having provided him with "their customary kinds of intellectual stimulation."[11] MacArthur had produced a mathematical proof of the increase of stability of a community with the increase of the number of food web links. Particularly if a new species enters the community to occupy an unfilled niche, stability should increase. It may also partition a niche with a preexisting species, giving the same effect. MacArthur was analyzing the process of the development of communities.[12] As the community builds up the opportunity for a new species to enter the food web become rarer. In Hutchinson's (and MacArthur's) view a more diverse community is better able to persist than one of fewer species. This is a partial answer to Hutchinson's original question, why are there so many kinds of animals (or of plants)? There must, however, be some kind of asymptote, not allowing further subdivision of niche space by additional species. What sets this?

One possibility is external control of community evolution, particularly by climate. Tropical and temperate communities have more species than those of boreal and arctic zones. The total biomass of these communities is no doubt related to such factors as growing season, which may limit the growth forms of plants, such as trees, and therefore the types and numbers of niches available for animals. Perhaps, however, the low diversity of arctic animals is a result of an earlier point in community development and/or because of more frequent catastrophes in terms of the severe climate, which could keep a community in a state of "perennial though stunted youth."

Hutchinson then asked about the size of the habitat that can be colonized, using as an example the species of voles on small islands in Britain, a question he had started to investigate while a student at Cambridge. Only two cases were known of two species of vole of the genera *Microtus* and *Clethrionomys* co-occurring on one island. No voles at all had reached some of the islands, perhaps because of the lack of favorable refuges in unfavorable seasons or because of the presence of particular competitors such as rats. Similarly, although there are four or five species of terrestrial shrews in central France, only one of the Channel Islands is home to two species; presumably for the above reasons only one species can survive on most of these islands. Lack had made similar observations for bird species on these small islands; although the species present occupied larger niches in the absence of competition found on the mainland, bird species were few. Hutchinson postulated that there was not enough space (or habitat differentiation?) on these islands for the evolution of a more species-rich and stable community.

How different do closely related species have to be to occupy different niches? The phenomenon of character displacement had recently been discussed by Brown and Wilson.[13] Characters such as beak size, they showed, tended to diverge when previously allopatric species with related niche requirements become sympatric in part of their range. Hutchinson cited examples he had collected of character divergence in metric characters related to feeding apparatus in both birds and mammals and came up with a ratio of the larger to the smaller of 1.1 to 1.4, or a mean value near 1.3. This number, though later controversial and eventually largely abandoned, gave subsequent workers something with which to compare their own quantitative results. Klopfer and MacArthur later found that smaller differences, and thus smaller numerical ratios, would allow competing species to coexist in tropical regions that were climatically stable. More species packing could result than in temperate zones.

At that point in "Homage to Santa Rosalia" Hutchinson returned to his original question about the corixids in the Italian pond. The three different species of *Corixa* found over most of Europe differ in size; the two smaller species are never found together. There are other niche differences in terms of water depth and adaptation to brackish water. In spite of some evidence of character displacement in terms of size, such differences are not sufficient for niche separation in these hemimetabolous insects with relatively long growth periods; there must also be differences in reproduction times. This is what he originally observed in the Santa Rosalia pond. The corixids story indicates quite complex adaptations necessary for niche separation, also studied by MacArthur for related *Dendroica* warblers and by Alan Kohn for Hawaiian cone shells. The question Hutchinson was concerned with here was how finely can niches in different types of communities be divided, and thus what degree of diversity can be achieved?

The mosaic nature of the environment, found nearly everywhere except perhaps in open waters, is significant in terms of animal diversity, but the diversity varies with the size of the animal under consideration. Large animals, whether herbivores or carnivores, require more different types of terrain within their home ranges than do small animals; therefore one should expect fewer large species than small ones at both trophic levels, both herbivore and carnivore. At most, two species of very large animals, such as elephants in Africa, would be likely to coexist. Hutchinson here included his environmental concerns: "I cannot refrain from pointing out the immense scientific importance of obtaining a really full insight into the ecology of the large mammals of Africa while they can still be studied under natural conditions. It is indeed quite possible that the results of studies on these wonderful animals would in long-range, though purely practical, terms pay for the establishment of greater reservations and national parks than at present exist."

As for smaller animals, he cited five or six coexisting closely related passerine birds and up to four species in one genus of small mammals. Insects, on the other hand, can have hundreds of species, especially in beetle and fly genera. Small size enables these animals to become specialized or adapted to elements of the environmental mosaic not possible for larger animals.

Hutchinson postulated, though here without specific data, that as evolution from the standpoint of community development proceeded through time, single niches would be split. With increasing diversity, new species, perhaps those with brains capable of new learned behavior, would be preadapted to filling new and unusual niches. Hutchinson stressed the stability produced by diversity, a view also held by Elton. Hutchinson cited an increase in diversity among macroscopic animals after the appearance of predators (known by their fossilized skeletons) in the early Cambrian. He then envisioned an increasing rate of diversification of animal species in the early Paleozoic, followed by a long period in which brain size increased and large-brained animals became dominant.

There may be a point at which, although each species in a biological community appears to fend for itself, the integrated aggregates increase in stability. Moreover, more complex types of behavior emerge. Is his view holistic? In part, but Hutchinson insisted that there is nothing mysterious about it; it follows from mathematical theory and can, at least in part, be confirmed empirically. "The types of holistic philosophy that import ad hoc mysteries into science whenever such a situation is met are obviously unnecessary."[14]

This paper contained much that Hutchinson himself counted as hypothesis, nothing new for Hutchinson. And as with his niche formulation, these hypotheses sent many ecologists of the sixties out to find data to prove or disprove them. This led in part to a major conference at Brookhaven National Laboratory in 1970 on diversity and stability in biological systems.[15]

Hutchinson's theoretical view of ecology was taken up in the 1960s by several of his students who had completed their doctoral studies in the 1950s and by Egbert Leigh, his student in the sixties, who worked on the ecological aspects of population genetics. In his 1961 paper, "Preliminary Ideas for a Predictive Theory of Ecology," and in his well-used 1962 book, *The Growth and Regulation of Animal Populations,* Slobodkin developed the view that a complete theory for ecology could be developed, and that as a predictive theory, it could be used to seek solutions to practical ecological problems.[16]

When MacArthur reviewed this book, he split ecologists of the early sixties into two factions, the antitheoretical and those supporting a theoretical, mathematical school. He wrote about the latter ecologists, which included himself, as those who use ecological data to test proposed theories "patching up the theories to account for as many of the data as possible."[17] Some of these theories were Hutchinson's; others were MacArthur's. Later ecologists doing "normal science," that is, not inventing new paradigms, continued using MacArthur's "broken stick model" even after MacArthur himself rejected it.

Why Are There So Many Kinds of Warblers?

In the spruce forests of Maine in June dwell many multicolored species of wood warblers (Parulidae) of the genus *Dendroica.* There are black-throated blue warblers and black-throated greens, Blackburnians with brilliant orange throats, bay-breasts and Cape Mays, yellow-rumped (formerly myrtle warblers) with several patches of bright yellow. Birders farther south delight in these birds as they forage and sing on their migration routes north. A great many avid birders have watched them, but none of us had measured their activities. Robert MacArthur did. Bird studies proved useful to test ideas about competition and niche diversification. MacArthur, who had, according to Hutchinson, helped to develop the ideas in Hutchinson's two landmark papers, analyzed the multidimensional niches of four species of co-occurring warblers of the family Parulidae. These species did not differ in some of their important niche factors, particularly the ability in terms of beaks and other physical attributes to search for and pursue their prey, insects in spruce trees.[18]

MacArthur was, however, able to observe and measure the percentage of time that each of five *Dendroica* species spent in particular microhabitats. He divided spruce trees into zones and worked out competition co-

mole of Yale on the comparative feeding ecology of tropical sympatric sea birds living and breeding together on Christmas Island.[25]

Jared Diamond studied the niche relationships of tropical island birds on the islands of New Guinea. He found species that had niches that included the niches of other species. He called thirteen species with very broad niches, i.e., highly adaptable to almost all conditions, "super tramps." Other species living in included niches were able to survive under extreme conditions of low resources, conditions under which the broad-niche species could not survive.[26] R. H. Miller found similar included niches in Colorado pocket gopher species. Sizes of the fundamental niches of these gophers were inversely related to the competitive ability of a species in its optimal environment. Thus the adaptable, broad-niched species do best in their larger fundamental niche factors but the specialized species does best in its narrower niche, included within the fundamental niches of the others. Miller also studied niches of yellow-headed blackbirds included within the broader fundamental niche of the red-winged blackbird. Bob Jaeger has also used these niche concepts in his studies of related *Plethodon* salamanders.

Hutchinson discussed many such studies in a chapter titled "What Is a Niche" in his 1980 population ecology book. Most of these studies are related to his n-dimensional niche concept although some of the terminology has changed and further mathematical models have been proposed.

Critiques of Mathematical Ecology

In the 1970s a battle raged in American ecology in terms of critiques of "theoretical bandwagons driven by charismatic mathematicians" as distinct from ecology based on "a foundation of natural history."[27]

Hutchinson had already pointed out the importance of the knowledge of the natural history of the organisms under study together with the use of theory; he wrote that MacArthur "really knew his warblers." In this same paper Hutchinson praised the mathematical ability of many of the seventies generation of ecology students but cautioned that "a wide and quite deep understanding of organisms . . . is as basic a requirement as anything else in ecological education."[28]

Each of Hutchinson's students who worked in population ecology was expected to know the particular group of organisms with which he or she was working—whether tropical birds, north woods warblers, Hawaiian cone shells, or Mt. Washington mosses. Maxine Watson, who worked on population biology of a family of mosses, Polytrichaceae, cited above, and Tom Lovejoy, who worked on the immense bird diversity in Amazonian forests, were among Hutchinson's last group of students; he officially retired in 1971.[29] Both were concerned with niche differentiation and with questions raised by Hutchinson in "Concluding Remarks" and "Homage to Santa Rosalia." By the mid-seventies the battle over theoretical ecology was in full swing. What were the views of other ecologists about the contributions of Hutchinson and his students, especially those of Robert MacArthur, to population ecology?

Even before the 1970s there was considerable disillusionment with the usefulness of the logistic equation and of Volterra's and Gause's two species competition equations. Hutchinson's associate Thomas Park, among others, and some of Hutchinson's former graduate students, including F. E. Smith, were disenchanted. Even Larry Slobodkin, earlier a supporter of mathematical ecology, had become disillusioned.

Scudo and Ziegler wrote that the golden age (their term) of mathematical ecology ended in 1940! But the logistic curve and equations of Volterra and Gause were still in use by some ecologists. One need only look at the ecological papers published in the *American Naturalist* in the sixties and seventies to see that mathematical ecology was flourishing, although it sometimes appeared that it was only one group of elite ecologists talking to each other—but at great length. A second wave of population and community ecologists appeared and wrote many papers, using mathematical equations to elucidate their results in the field and the laboratory.[30]

Robert McIntosh attributed the success of some of the second wave of "colonists" to Hutchinson's influence. He was "standing on the beach to welcome the new colonists." "Hutchinson was influential in the careers of many of the ecologists, mathematical and otherwise, who were contributors to diverse aspects of ecology." These included the next generation, for example, Michael Rosenzweig, MacArthur graduate student and noted ecologist.[31]

Robert MacArthur both contributed to and expanded many of Hutchinson's ideas; Hutchinson viewed MacArthur as the most talented of his many talented students. Hutchinson's program outlined in "Concluding Remarks" pointed to future studies of community structure, to a future evolutionary ecology. It was MacArthur who carried out much of this program, with his own innovations. Together the two of them made the new ecology intellectually exciting. Many people have expressed this view and written of the MacArthur "school." MacArthur surely applied theoretical ecology to animal community ecology in new ways. One animal ecologist wrote, "The study of communities and ecosystems was raised from the doldrums of ecological energetics to become an intellectually challenging branch of ecology."[32]

One wonders what Hutchinson thought of that comment, since he, together with Raymond Lindeman, was responsible for the origins of the study of ecological energetics, ecosystem (or systems) ecology, the very branch that was supposedly "in the "doldrums" (see chapter 9). H. T. Odum, another Hutchinson graduate student, had done exciting work in that field after leaving Yale.[33]

Sharon Kingsland, in her pursuit of the history of population ecology, has written a very perceptive analysis of Hutchinson's and MacArthur's roles. She included the arguments of some of their detractors, largely in the 1980s, after MacArthur's death. Much of the contention is based on their emphasis on competition.[34]

The diverse areas of MacArthur's contributions to both field research and mathematical models include species diversity; relative abundances of species, including his broken stick model; competition and species packing; r- and k-selection, and with E. O. Wilson, theoretical island biogeography. Detractors or no, MacArthur's work in all these fields has stimulated further research. "A theory need not be true, even better if it is false, as long as it stimulates work and is fruitful," goes a maxim of early ecologist Henry Cowles.[35] Hutchinson might have agreed. Not so MacArthur, who tried to kill off his discredited broken stick model but did not succeed; it continued to stimulate research. In fact, in a review by leading mathematical ecologist Robert May, this model of the relative abundances of species was found to have biological significance under some circumstances.[36]

"In November 1972 a brief but remarkable era in the development of ecology came to a tragic, premature close with the death of Robert MacArthur at the age of 42." This was the first sentence of Martin Cody and Jared Diamond's perceptive Preface to *Ecology and Evolution of Communities*.[37] Hutchinson was distraught about MacArthur's early death from kidney cancer. He viewed MacArthur as an intellectual son.

Hutchinson and MacArthur had corresponded and talked until nearly the end of MacArthur's life. MacArthur knew what his diagnosis was; it was a battle for survival of which he was well aware. "I have very rapidly regained my strength and am feeling fine," he wrote to Hutchinson in July 1971, but he added, "There are some new symptoms which I hope are imaginary." He was working on several manuscripts and "a moderately elementary book" on biogeography. . . . The book may be too ambitious, but I thought it might go on earning income longer than I." He wrote to Hutchinson the following year about a major operation he was to have the next week. "On the more optimistic side, I will be on some new chemical after the operation which is very experimental but appears quite promising, and if all goes well I may have more pleasant months ahead than I have been expecting."[38] Sadly, that did not happen. Hutchinson noted at the bottom of this letter that it was the last he had from him before he died.

MacArthur's wife, Betsy, kept in touch with Hutchinson, although she herself had not known him well. "I always felt close to you as Robert's teacher. . . . I feel I know you a bit through Robert. He was very proud of being your student—and humble too!"[39]

That was in the seventies, but MacArthur himself wrote about Hutchinson as teacher at a much earlier date, 1957. It was early in Mac-Arthur's career, the year his thesis work on the coexisting warblers of the northeastern spruce forests was published and before Hutchinson's "Homage to Santa Rosalia":

It is, of course, hard to characterize in a few words a great man. . . . I think Hutchinson is, above all, an intellectual. Any fact, no matter how small, and any rational idea, no matter how insignificant is treated with respect. And the people who discuss these facts and ideas are also treated with respect. . . . Perhaps [his] most significant achievements have been by using procedures of other sciences on ecological problems: He has used his tremendous memory and the methods of the historian to study the problems of the guano deposits; he used the methods of electrical engineering to elucidate the effects of time lags on animal populations; he used the methods of radioactive tracers (in a novel way) combined with a knowledge of theoretical hydrodynamics to explain motion of phosphorous in a pond.

I believe a few words about him as a teacher are in order. . . . He never made appointments. Students go into his room to discuss, not to get advice, and are treated as equals no matter how far-fetched their ideas. . . .

I believe he has influenced me the most by his genuine enthusiasm about the small, novel ideas we may have. This enthusiasm is the most sincere sort of encouragement. I feel rather foolish having said all this "in print" but I hope it may contribute to the understanding of a remarkably fine man.[40]

The two of them obviously discussed many ideas, novel but not necessarily small, some of which appeared in two coauthored theoretical papers in the *American Naturalist* in 1959.[41] Hutchinson also credited—in print—in a footnote to his "Concluding Remarks" paper the contributions of MacArthur and a number of other students to his niche theory. It was certainly a reciprocal process from Hutchinson's point of view. He praised MacArthur in a recommendation he wrote for Robert May in 1981: "I know Bob May quite well. . . . Of the people who have worked on the mathematization of ecology, he [May] is certainly one of the most gifted. Of those whom I have actually known only my exceptionally brilliant student, Robert MacArthur, was his equal."[42]

Edward O. Wilson and Hutchinson wrote of MacArthur in their National Academy memoir: "Inevitably his approach was condemned by some ecologists as oversimplification, but . . . it energized a generation of young population biologists and transformed a large part of ecology."[43]

Both MacArthur and Hutchinson played a large part in that transformation. The subsequent criticism, some of it vitriolic, was aimed more at MacArthur than at Hutchinson, but it must have hurt Hutchinson. Nevertheless, before Hutchinson's death in 1991, it became clear that MacArthur's work was still influencing ecology. "The response to Hutchinson and MacArthur continues and provides adequate testimony to their pervasive influence on modern ecology," Kingsland concluded.

> Periodic stocktakings of the role of mathematical models in ecology continue . . . to influence standards of hypothesis making and testing. Implicit in these discussions of methods and principles is the question: How many schools of ecology can peacefully coexist? But they also serve as reminders that nothing can replace an intimate knowledge of some part of the real world; that the naturalist is on equal terms with the mathematician."[44]

Neither Hutchinson nor MacArthur would have disagreed.

Was there "a road not taken" in terms of Hutchinson's research because of his intense involvement with MacArthur's work? It is possible. As discussed earlier Hutchinson never did further research in either systems ecology or radiation ecology, although he was instrumental in inventing both these fields.

MacArthur moved out of population biology into biogeography in his last years; Hutchinson continued to teach a graduate course at Yale in population biology, publishing a textbook based on that course in 1978. This book is dedicated not to Robert MacArthur but to Arthur Hutchinson, Evelyn's father. It contains Evelyn Hutchinson's and Robert MacArthur's research but also a great deal more, starting with the London Bills of Mortality and the human demography of the seventeenth century. In the final chapter Hutchinson considered some of the then-current critiques of mathematical ecology (see chapter 18 of this book).

CHAPTER FIFTEEN

Hutchinson the Environmentalist

T he most practical lasting benefit science can now offer is to teach man to avoid destruction of his own environment, and how, by understanding himself with true humility and pride, to find ways to avoid injuries that at present he inflicts on himself with such devastating energy." Hutchinson wrote that in 1943.[1] He was barely forty, but it remained his lifelong philosophy. The title of Hutchinson's partial autobiography of his early years, *The Kindly Fruits of the Earth*, embodies the same philosophy. It comes from "Preserve, O Lord, the kindly fruits of the earth that we may enjoy them in good season."[2] Few thought of Hutchinson as the activist he truly was. He chose his causes carefully and often worked behind the scenes. But he was effective, particularly on behalf of the environment and foremost for the preservation of biodiversity.

Much of Hutchinson's early research was closely related to the physical environment. He was well known in the field of biogeochemistry long before his fame as a theoretical ecologist. He and his students worked on the cycling processes of many elements in the biosphere, from carbon to strontium. He himself worked on phosphorus, nitrogen, and aluminum. He wrote semi-popular articles as well as research papers on these chemical cycles and their biological importance. His *Scientific American* article "The Biosphere" was reprinted and recycled among both college students and general readers interested in the environment.[3]

Hutchinson received a number of major environmental awards, including the Eminent Ecologist Award from the Ecological Society of America in 1962. He won both the Tyler Prize for Environmental Achievement and the Cottrell Award for Environmental Quality from the National Academy of Sciences. In his 1974 acceptance speech for the Tyler Prize, he said that the people who congratulated him about his award (as reported in the New Haven paper) were disappointed that he "had not produced a single phosphate-free soap flake."[4] This referred to the then-current "green scum" controversy—the deleterious effects of phosphates in detergents on lakes. Eventually phosphates were definitively proven by ecologists to be a major cause of the accelerated eutrophication of lakes and of the lake-surface scum caused by green algae or by cyanobacteria (earlier called blue-green algae).

Hutchinson's contributions to environmental issues were not of the "phosphate-free soap flake" type. He was an advocate and a teacher, but his major contributions were scientific, both theoretical and experimental. In his Tyler acceptance speech he said that recent newspaper articles had intimated that he was the father of ecology, but that he could not admit to that paternity. He could, however, admit to the intellectual paternity of ecologists because "an extraordinary succession of incomparable young men and women" had studied with him. Some of these graduate students later made their own important environmental contributions. Hutchinson's concern for the preservation of biodiversity, a major concern right up to his death, was also exemplified in this acceptance speech. He quoted from a book on mammals by R. F. Ewer, who wrote: "I have lived to see the cheetah and the leopard, to camp where lions roared . . . but it would be good to feel confident that my grandchildren could do the same, and not to fear that they will say: 'just think—all those lovely animals were actually alive in granny's day; it must have been wonderful.'" Hutchinson ended his Tyler Prize speech by asking his audience to "please try to keep it so."[5]

In the same year, Hutchinson also received the Cottrell Award, another prestigious environmental award from the National Academy of Sciences (NAS). He was one of five honored for significantly aiding the development of his chosen field, but the specific award Hutchinson won was the NAS Award in Environmental Quality, which included a $5,000 honorarium established in honor of Frederick Gardner Cottrell. This honor went to Hutchinson not only for his scientific contributions in ecology and limnology but especially for his continuing public advo-

35. Hutchinson receiving the Browning environmental medal at the
Smithsonian Institution, S. Dillon Ripley presiding. Manuscripts
and Archives, Yale University Library.

cacy of the desperate need to understand and to protect and preserve the
environment in which we live.

Hutchinson collected much data during the 1930s and 1940s on nutrient transport and cycling, on phosphorus and nitrogen in the biosphere,
on oxygen deficits, and on intermediate metabolism. These data came out
of his work in limnology, and particularly the research he and his students
did on Linsley Pond, the much-studied lake near New Haven. A landmark paper on phosphorus cycling in lakes, carried out with his graduate
student Vaughan Bowen, was published in 1947. Several large studies,

starting with those of Edward S. Deevey, Jr., one of his first students, but also by Hutchinson and early collaborators, including Anne Wallach and Ruth Patrick, concerned paleolimnology, especially changes over time in lake environments. Many later studies of paleolimnology, from Rome to Guatemala to Israel, were carried out by Hutchinson and his research associate Ursula Cowgill. One question pursued throughout these studies was the effect, often detrimental, of human activity on lakes, from ancient to modern times.

Hutchinson put his knowledge of Connecticut lakes and their biota to practical use in one of his early advocacy letters, written to Connecticut Governor Raymond Baldwin in 1943. He wrote against the passage of two bills that had been recently introduced in the Connecticut General Assembly. The bills, he warned, were "a matter of concern to anyone interested in the rational development of the waters of Connecticut." Their passage would greatly decrease the technical work accomplished by the Board of Fisheries and Game, by the "abolition of the pond and stream surveys." Moreover, he added, the very wide distribution of their previous survey report must have had "a desirable educational effect on the numerous fishermen and others who applied for copies, leading to intangible but nevertheless valuable growth of an appreciation of the aquatic resources with which our state is so richly endowed."[6]

Hutchinson's "Marginalia" columns in *American Scientist* were widely read and covered a great variety of topics. In the very first one, in 1943, in his statement of purpose, he warned his audience: "These notes will reflect the attitude of the philosophic naturalist, rather than that of the engineer, the point of view of the mind that delights in understanding nature rather than in attempting to reform her. . . . Certain forgotten bushels will be overturned in the hope that they cover unsuspected bright lights."

By 1947 Hutchinson was already officially involved in conservation issues; that year he gave a talk to the Conservation Section of the New York Zoological Society. Fairfield Osborn, president of the society, sent his appreciation to Hutchinson: "You gave us ideas that will be of permanent value. We are happy, too, that you have accepted our invitation to serve as a member of the Advisory Council of the proposed Conservation Foundation."[7]

Osborn enclosed a check to pay Hutchinson's travel expenses from New Haven. It was for ten dollars. The Conservation Foundation, which was to initiate research and education on all aspects of conservation from water to forests to wildlife, received its charter on March 30, 1948. Conservation of natural resources, including animals and plants, became a popular public topic after the publication of Fairfield Osborn's book *Our Plundered Planet*.

There was a growing interest in conservation. A. Whitney Griswold, later president of Yale, forwarded to Hutchinson a copy of a letter H. M. Gray of the University of Illinois had sent to Osborn, adding the note, "Hope we can keep this moving." The letter, written in 1948, discussed the need for graduate programs that would "push forward the conservation movement" and "train young men and women to work in this important field." Gray wrote that what was lacking at the University of Illinois and presumably elsewhere was a "unified, inter-disciplinary program of training and research, which might serve to integrate these various specialties [including social sciences] into some coherent body of knowledge."[8]

Nine years later, in 1957, Osborn came to Yale for a special meeting to discuss the problems of conservation programs at Yale. The meeting was attended by Griswold, now Yale's president, Hutchinson, and Hutchinson's former student Ed Deevey, by then a Yale professor, among others. At this meeting Hutchinson argued that the most urgent conservation problems were in the social and psychological aspects of conservation, and that a graduate program addressing these aspects would be valuable at Yale, "with its strong tradition in the social sciences." Osborn favored this view but thought it would be difficult to fund a graduate program in the behavioral aspects of resource utilization.[9] The Yale Forestry School was interested in gradually developing into a professional school of natural resources but did not then wish to take over the current conservation program at Yale. Hutchinson's view was a radical one at that time, but public policy aspects of conservation were later incorporated into the Forestry School's graduate program.

The Problem of Eutrophication

By the 1970s eutrophication of lakes was a major concern. Green scum was appearing on lakes due to algal or "blue-green" (cyanobacterial) blooms. Hutchinson's fellow limnologists eventually proved experimentally, to the dismay of the detergent industry, that these blooms were very largely a result of excess phosphorus, coming from fertilizers and detergents and getting into lakes and subsequently into the oceans.

As early as 1949 Hutchinson was writing about carbon dioxide.[10] He also noted that humans contribute to both the gain and the loss of phosphorus. "He quarries phosphorite, makes superphosphate of it, and spreads it on his fields. Most of the phosphorus so laboriously acquired ultimately reaches the sea."

Hutchinson wrote an article for *American Scientist* called "Eutrophication: The Scientific Background of a Contemporary Practical Problem."[11] The word "eutrophication," unknown to the general public ten years before, had become common parlance by then because of the "green scum" problem on lakes used for recreation. "Eutrophication" often describes the process in which "a beautiful lake . . . became converted into a body of water covered with decomposing . . . algae. That the word actually means 'becoming well fed,'" he wrote, tongue-in-cheek, "illustrates once again the horrid dichotomy between starvation and excess that is apt to characterize much of contemporary society." This long article detailing the history of lake studies relating to eutrophication, as well as much current research, is characteristically Hutchinsonian. His post-doc Clyde Goulden remarked about Hutchinson's "cross-referenced brain."[12]

Hutchinson was frequently asked for his advice on environmental problems, for example: what were the limiting factors for excessive algal growth in aquatic systems? "Eutrophication" became a household word, but detergent companies insisted that phosphates were not the problem. Philip Crowley, a graduate fellow in environmental science at Rice University, wrote to Hutchinson in 1970 about a graduate seminar his department was conducting. It was on the controversy surrounding limiting factors and algal blooms. He wrote, in part: "We would like to know your own views on the limiting influences of CO_2, PO_4 and other factors in algal blooms and in eutrophication generally, with references supporting your views wherever possible. Also, what do you think would be the best experimental system in which to study these limiting factors?"[13]

At the time Hutchinson was writing volume 2 of his *Treatise on Limnology*, as well as doing his own research and teaching his graduate courses. It is therefore remarkable that he took the time and care to answer such an inquiry, of which this is only one example.

Hutchinson replied in just over a week, after having consulted his "colleagues," W. T. Edmondson of Seattle and J. R. Vallentyne of Winnipeg, both his former graduate students. He wrote that they all saw two different aspects of the problem. The first was: what are the causes, in any given case of an excessive production of algae? Hutchinson wrote:

> I think it is impossible to lay down any hard and fast rule, but in general, in ordinary hard water, the inflow of additional phosphate and fixed nitrogen together is usually involved. However, since many filamentous blue-green algae [cyanobacteria] fix nitrogen, phosphorus . . . may produce the effect by itself. Where the water is initially soft, any process leading to increase the bicarbonate content may cause eutrophication. Moreover Bormann, Likens, Fisher and Pierce (Science 159:882, 1968) have shown that cations and nitrate can enter watersheds in the process of deforestation and in some cases thus may cause eutrophication. On the whole I suspect phosphorus is the most important but not the only agent.

The second question of concern was how to find out what was actually limiting growth in any particular case. Hutchinson suggested reducing essential elements until the algal populations fell, with phosphorus being the easiest and carbon dioxide the hardest to control. He suggested a paper by Edmondson in *Science* and the National Academy of Sciences volumes on eutrophication as additional references.[14] Experiments such as Hutchinson suggested, but with divided lakes given different treatments, eventually showed phosphorus to be the main culprit in eutrophication, despite the protests of the detergent companies.

That same year, 1970, I (this author) took a graduate course on eutrophication taught by Donald McNaught at the State University of New York at Albany. The graduate students would never have had the temerity to write to Hutchinson, the famous ecologist, for help, as Philip Crowley did. A year later, however, McNaught himself wrote to Hutchinson, inviting him to Albany to speak on his paleolimnological studies of Linsley Pond, Lago di Monterosi, and other lakes, "placing them in a framework of human influences which is further imposed on a baseline of natural changes." He mentioned that there were seven biological sciences faculty members and fifteen graduate students who were interested in ecology.[15]

At that time Hutchinson was turning down most invitations to speak, but he accepted this one, even though the honorarium offered was only $100. He wrote to McNaught that he wanted to "learn more of what you and your students are doing."[16]

Hutchinson came and lectured to the biology faculty and students about "the Natural History of Lakes" and also held a rap session for the fifteen graduate students and myself, then a professor but a recent graduate student. He answered our questions about many ecological and environmental topics, giving a short dissertation on each topic and citing references in true Hutchinsonian fashion. Afterward he asked us for any new research on aquatic plants, and subsequently included some of my first postdoctoral research in volume 3 of his *Treatise on Limnology*.[17]

The Problem of Overpopulation

Evelyn Hutchinson had a long-term interest in the problem of human overpopulation and its effect on the environment. In a 1969 letter to Edward L. Chapin, a professor at San Bernardino Valley College in California, he wrote that he hoped that in Chapin's study of human populations he would "take very seriously the fact that however good the contraceptive devices that may be invented, if people persist in wanting to have more than the replacement number of children . . . we shall have the population problem . . . in increasing intensity." If the desired number could be reduced to the replacement value, Hutchinson wrote, "I suspect modes of achieving it would soon be forthcoming."[18]

A number of years later, Hutchinson responded to the comments of Congressman Fred Richmond on "Food and Population," which were printed in the *Congressional Record*. In his remarks, Richmond cited a task force of major environmental organizations sponsored by the Rockefeller Brothers Fund that called for foreign aid measures that would "have the effect of reducing the number of children couples generally desire in most of the developing world." Richmond also cited a Gallo-Kettering survey of countries in Africa, Latin America, and Asia where the number of desired children was between four and six. Not only was this number much higher than the replacement number advocated by Hutchinson, but according to the survey, it would lead to a doubling of the population every fifteen to thirty years. As a result, the 1977 Environmental Fund warned that with respect to the world food crisis, "unless population growth is stabilized, the inescapable result of saving lives today will be even greater numbers of lives lost tomorrow." Only Sri Lanka was cited as a country that had a program to lower birthrates at that time. Richmond thanked Hutchinson for his comments on "Food and Population": "It is

impressive that there are Americans such as you who realize the seriousness of the population problem."[19]

Hutchinson discussed problems of population growth elsewhere, particularly in his 1978 book *An Introduction to Population Ecology,* which
went through many printings. Although his major concern was the populations of wild animals, in the chapter entitled "Why Do They Have So
Many Children?" he discussed human populations as well. For example,
he wrote:

> From a human point of view regulation by density-dependent death rates,
> removing the offspring in excess of an equilibrium number [i.e., infant
> mortality] seems wasteful and immoral. Medical science has developed
> largely as a means of eliminating this immoral waste, but without any
> compensatory regulation such elimination can cause disaster. . . . A new
> regulatory mechanism involving foresight and contraception is replacing
> density-dependent death, but in most societies this replacement is not
> being achieved fast enough, if at all.[20]

On the interesting question as to whether anything comparable to
human family limitation has evolved in other animals, he discussed, in
the same chapter, the phenomenon of territoriality, one mechanism that
limits population growth in both birds and mammals. He quoted both
Charles Darwin and Robert MacArthur on the relation of bird population
size to food availability. Darwin wrote, "A large number of eggs is of some
importance to those species that depend on a fluctuating amount of food,
for it allows them to rapidly increase in numbers." Family size (called
clutch size in birds) can also fluctuate in nature depending on environmental conditions. MacArthur wrote, a hundred years after Darwin, that
opportunistic warbler species like the Cape May and bay-breasted warblers, species that are dependent on spruce budworm outbreaks for their
food in the forests of Maine, do indeed lay more eggs than other warblers.
The bay-breast has larger clutches during budworm outbreaks. Hutchinson added that in human populations, "though fecundity is doubtless
depressed by malnutrition, and juvenile mortality greatly increased, the
relation is much less neat" than in these birds or even in some small invertebrates, such as *Daphnia,* in which reproduction is limited as food supply
decreases.[21]

As with all topics in his population ecology book Hutchinson amply
supplied the reader, whether biologist or historian, with footnotes. Many
of the early studies of the relation between clutch size and food supply

were done by Hutchinson's close friend David Lack, whose English studies he discussed and footnoted. Hutchinson also pointed out in the text that the first work on this subject was done by A. E. Moreau in Tanganyika. Indeed, one footnote on historical knowledge of territoriality goes on for more than two long pages of small print, in which we are taken back to Anglo-Saxon sources, to seventeenth-century Italian studies on the nightingale—and even to Aristotle on ravens.[22] This is typically Hutchinsonian and a boon to historians of science.

Hutchinson's ideas on overpopulation were based on his own and his students' innovations, both theoretical and applied, in population ecology of birds and other wild animals. But he worried and wrote about the negative effects of overpopulation of humans on the carrying capacity of the environment and the resulting degraded environments.

The Problem of the Loss of Biodiversity

The preservation of biodiversity and its converse, the extinction of species, were other major aspects of Hutchinson's environmentalism. Hutchinson had a very early and continuing interest in insect biodiversity, but he often wrote of larger animals, including large furry mammals. He had asked: "I wonder if you or anyone you know of has thought of looking seriously at toys [such as stuffed animals]. . . . As these are the models of what a child will grow up to be interested in, interact with and desire, there may be many keys here as to what is needed to produce a satisfactory life. What would the effect of the Giant Panda becoming extinct be on future generations? What played its part before it was discovered?"[23]

By the early 1980s Hutchinson's interest in and writings about the importance of biodiversity was being widely recognized. Ruth Patrick of the Philadelphia Academy of Natural Sciences wrote to him asking his permission to reprint his article "The Lacustrine Microcosm Reconsidered" in a collection of "Benchmark Papers in Ecology" for a volume she was editing entitled *Diversity*. She also asked him for a biographical sketch, which he wrote, citing, as of 1980, his "nine books and about 150 papers of varying seriousness and importance."[24]

In 1981 Jeremy Holloway of London wrote to Hutchinson's former graduate student, Tom Lovejoy, then a vice president of the World Wildlife Fund, about the coming Royal Society expedition on conservation and forestry in Indonesia. Lovejoy had sent him Hutchinson's reprints. Holloway reminded Tom about Evelyn and his Cambridge College, Emmanuel, which had contributed at least two important men to society. One was Evelyn Hutchinson, ecologist and conservationist; the other was John Harvard, who gave his fortune as well as his name to Harvard University.[25]

The introduction of non-native fishes into natural bodies of water and their effect on biodiversity elicited Hutchinson's concern and expertise. Kenneth McKaye of the Duke Marine Laboratory wrote to him in 1982 asking for help in regard to an unfortunate proposal to introduce clupeids, a type of non-native fish, into Lake Malawi in East Africa. He was coming to visit Yale and sought Hutchinson's advice. McKaye had written a rebuttal to the proposal and had circulated an earlier draft to the Malawi government. "I would very much like your advice on how to proceed politically and who I should contact to try to stop such an introduction." He added that Saran Twombly's data "will play a crucial role in answering some of the questions which are posed by such a suggested introduction. . . . Both the Malawi government and I are looking forward to seeing her results."[26] Saran Twombly, a professor at the University of Rhode Island, was a Yale Ph.D. student who worked closely with Hutchinson after his official retirement from Yale.

Hutchinson had earlier written to Tom Lovejoy about the biodiversity of the Lake Malawi fishes. "I think its fish fauna poses the most challenging evolutionary problem now before us." More than two hundred of the fish species described were endemic, that is, only found in that one lake, and perhaps the true number was more than four hundred. Very many species, at least one hundred, belonged to one genus, *Haplochromis*. The only comparable biodiversity of fishes is found in Russia's Lake Baikal, whose fauna is much older and whose fishes more ecologically diversified than in Lake Malawi. The incredible biodiversity in Lake Malawi was "a complete mystery."

Hutchinson was not impressed by the explanations that had been put forward for this amazing biodiversity. He did not find evidence either for specific food preferences or for the involvement of mutualism, the type of symbiosis in which both species benefit, as hypothesized by others. "Only a prolonged and very intense study will throw any light on this extraordinary situation. Moreover this is one of those problems the solution of which would illuminate evolutionary processes in general." Hutchinson was obviously interested in the evolution of biodiversity as well as in its preservation. Yet, he also wrote, these same diverse fishes provided the major protein source for the large human population around the lake. "Unwise tampering of a kind that has happened, with much good will and virtually no knowledge, in other tropical lakes must be avoided in Malawi at all costs." He cited the humane as well as the scientific reasons for preserving this amazing high diversity of fishes.[27]

Hutchinson's students had learned from him about the importance of biodiversity. In a letter to Walter Rosen of the National Academy of Sciences, written after a 1982 NAS workshop on biological diversity, Tom Lovejoy reminded Rosen of Hutchinson's NAS environmental medal and quoted Hutchinson as saying: "Living systems are always turning out to be more complicated than anybody thought. We can never hope to understand this incredible phenomenon, life, of which we are a part, if we permit a major portion of the evidence or data bank to be expunged." Lovejoy was urging the academy to set up a permanent working group on biological diversity. Late in his life, in the preface to the fourth volume of the *Treatise on Limnology*, Hutchinson wrote, "Thomas Lovejoy, in his whole professional career, has persistently worked as an agent of understanding and protection . . . of the whole biosphere."[28]

S. Dillon Ripley, director of the Yale Peabody Museum and subsequently of the Smithsonian Institution, quoted Hutchinson on biodiversity in his book *The Paradox of the Human Condition*. Hutchinson had noted that destroying species and their habitats is economically prodigal. It removed tools a scientist might need. Ripley expanded this idea:

> If endangered species and their habitats . . . can be thought of as common resources of humankind, then they . . . serve to illuminate other dimensions of the complexities of common property resources. Such species represent the conflict between personal advantage and community loss, the difference, for example, between exploitation and conservation of presently international ecosystems like the oceans or the Antarctic continent.[29]

The Ecological Societies and the Conservation Movement

Evelyn Hutchinson used his money as well as his pen to further biodiversity and his other environmental concerns. A list of his and Margaret's charitable contributions includes Greenpeace (for whale protection), the Natural Resources Defense Council, National Wildlife Federation, Nature Conservancy, Population Institute, Seal Rescue Fund, Wildlife Preservation Trust, Wilderness Society, World Wildlife Fund, U.S., and even the Friends of Madrona Marsh.[30]

In addition, Hutchinson had, from an early date, a voice in the Ecological Society and the Ecologists Union. A report titled "The Status of the Ecological Society" as of December 14, 1945, is found in his archives. Charles C. Adams, chairman of the Ecological Society, reported the "great expansion of ecology in America" during the preceding fifty years. "The early active opposition to ecology is now largely outgrown," he wrote, "through the development of important research" reported in the journals of the Ecological Society, *Ecology* and *Ecological Monographs*. "They and similar European journals had provided outlets not previously available for ecological papers."[31] Hutchinson had published in *Ecology* from 1940 on. His previous ecological papers had been published in *Science,* in *Proceedings of the National Academy of Sciences,* and in *Nature* and several other European journals.

Adams pointed out that for an emerging field to have recognition and influence, it must first build up a body of literature. Since that had already happened, "If the signs of the times are read correctly, a great expansion of ecology is to be expected in the near future, because of its integrating tendency." He continued, "Ecology's approach, with its emphasis upon direct contact with nature, and its integrative and objective methods, is gaining recognition." Endowment funds, he felt, were much needed for research grants, fellowships, and publications, and also for promotional and popular education activities. Adams understood discipline building and looked forward to the growth of ecological research foundations, laboratories, and biological stations. He thought that more rapid progress could be made if ecologists informed themselves about what other similar and successful organizations had already accomplished. He listed, among others, the American Geographical Society of New York, which had an endowment of $300,000 in 1945, and the American Philosophical Society, founded in 1743 by Benjamin Franklin, which had an annual research fund of $80,000 in 1945 and a "book value of their funds" of nearly $8 million. The Geological Society of America then had its own building and permanent staff and a $4 million endowment.

Adams concluded, "If and when the National Science Foundation becomes established and the Ecological Society wakes up, organizes and realizes its role and opportunities, *conditions would* [italics in the original] then be favorable for the society to request [government] grants for research and publication" as well as for foundation grants, but only when "we had organization, leadership, plans, and a program--none of which do we have in functional form today. We must get out of the amateur stage as soon as possible!"

Comments were solicited from the then-current executive committee, which included W. C. Allee, Ada Hayden, A. S. Pearse, and A. O. Weese. Their responses were not ascribed, but one was evidently from a plant ecologist. His flowery comment on ecology as a field read:

> Those who delved among its primary roots so long ago would now be amazed at the superstructure. There are still roots but they have migrated far from their beginnings. The shoots of Ecology are strong and manifold. Yet relatively few flowers have yet been seen or fruits tasted. . . . If mankind were to sponsor Ecology, it would doubtless rescue man from the blinding dust storm, the burning drought, the swirling flood, the disasters of erosion and enable him to live more happily with his fellowmen.

Another commenter used language like that of Frederic Clements, the guru of plant succession, who had recently died: "Here is a society which talks much of ecological succession and placidly is content to retain its leading society in a pioneer stage, with seeming indifference to working toward a mature stage! . . . We have not even reached the stage of talking about it, much less doing something—recalling Mark Twain's remark about the weather." Others quoted Ben Franklin and Thomas Jefferson.

Several respondents pointed out the need to emphasize the relations between general ecology and human ecology in order to attract donations. The seven hundred current members of the society were urged to demonstrate, as one man put it, "the fundamental importance of the ecological point of view for its bearing upon the public interests." Others emphasized technical skills and administrative leadership to attract funds from foundations such as "Mr. Rockefeller's."

As many previous historians of science have written, before the days of the National Science Foundation and other big government funders of scientific and medical research, scientists often had to look for private sources of funding, for example, the Carnegie Institution and the Rockefeller Foundation. Hutchinson had been a recipient of Rockefeller funds at Naples early in his career.

A practical executive committee member recommended adding more ecologists to this committee, including Orlando Park, Paul Sears, Victor Shelford, Carl Hubbs, "and others of like kidney." Hutchinson was not on this original list but soon became involved with the committee.[32]

The other organization with which Hutchinson was early associated was the Ecologists Union, "devoted to the preservation of natural biotic communities for scientific use." In 1946 the Ecological Society of America discontinued its committee for the preservation of natural conditions and gave up all action toward establishing natural preserves. The Ecologists Union was formed at that time to take over these activities. It was separate from the Ecological Society but had overlapping memberships, among them those of Ada Hayden and A. O. Weese, who became its president in 1948. As of its third annual meeting in 1947 it had only 191 members and $425.58 in its treasury, but by 1949 there were 300 members. This organization included such well-known ecologists as V. E. Shelford, R. F. Daubenmire, S. C. Kendeigh, and Paul Sears, as well as O. E. Jennings of the Carnegie Museum. Aldo Leopold, author of *The Sand County Almanac*, had been a member but had recently died tragically in a fire. The union was clearly environmentalist in the current meaning of that term. In 1947 its Preservation Committee recommended the following: preservation of wolves; support of McKinley National Park in Alaska; opposition to the proposed reduction in the size of Olympic National Park in Washington; purchase of the wilderness area in Superior National Forest in Minnesota for an international park; and opposition to the transfer of large areas of western grazing land from federal control.[33]

At its 1949 annual meeting, the Ecological Union held discussions about changing its name; "union" was thought to have a negative connotation in spite of the American Ornithologists Union and other scientific "unions." Some thought "ecological" too technical, although others pointed out the term was starting to be used in newspapers. Graduate students joined the organization, including John Cantlon and William Niering of Rutgers University, but there was some discussion about whether membership should be opened to nature-study people, that is, nonprofessionals.[34]

In 1950 Congressman Charles E. Bennett of Florida introduced a bill, H.R. 8513, to create a Nature Conservancy of the United States. The Ecologists Union encouraged passage of this bill, asking members to write their congressmen, but had not decided whether to unite with the Nature Conservancy if the bill were passed. The Nature Conservancy was to be a nonprofit organization with open membership. Its purpose would be to preserve small natural areas of special scientific or educational value and to preserve many kinds of natural features of the landscape; it would cooperate with state and county government agencies. The National Trust for Historic Preservation had recently been established by a similar congressional act.[35]

The Ecologists Union was a member of the American Institute of Biological Sciences (AIBS). The September 1950 *AIBS Newsletter* contained an important item: the Ecologists Union had become the Nature Conservancy.[36]

Hutchinson, who followed these events closely (as well as preserving the paperwork he received), had long been concerned about the preservation of areas of scientific interest. As early as 1943, as we have seen, he was writing to the Connecticut governor about Connecticut waterways and their preservation and his concern about the stocking of rainbow trout in Ball Pond "for which this locality is quite unstudied."[37]

In 1948 the Legislative Reference Service of the Library of Congress was studying the question of preserving North American wilderness areas. It sent out a questionnaire to members of the Limnological Society of America, of which Hutchinson was a charter member. It asked such questions as: "Does your organization advocate or approve of the preservation of wilderness areas as such? What are the determining factors which would decide whether a particular area should be preserved as a wilderness as against other uses? Does your organization engage in any activities favoring or opposing creation or extension of these areas? Please describe."[38]

If Hutchinson wrote answers to these questions, they are not extant, but both he and the Limnological Society were especially interested in the preservation of freshwater sites and in preventing damage, whether purposeful or inadvertent, to these areas and their fauna.

Hutchinson was particularly interested in "living museums" as described in a 1950 Ecologists Union publication. These were the remaining wilderness areas left from the vast resources of "primeval America," which, in the Union's view, were in need of preservation in nature reserves. They included wetlands of value: "A few landowners with rare foresight have deliberately saved parts of their properties, just to have a little untouched nature left on the place. But farmers are even now draining the last vestiges of wet prairie, and plowing them."

The Nature Conservancy, of which Hutchinson became a longtime financial supporter, eventually emerged as the major national society working to preserve natural areas, at first in the United States and more recently elsewhere in the world, especially in Central and South America. The Ecologists Union, out of which the Nature Conservancy emerged, right from its start had envisioned such preserves as samples of "wild nature," extremely important for scientific research and as refuges for rare species as well as a heritage to pass on to our descendents.

Hutchinson was involved in the education of future generations through Yale's Conservation Program. His former graduate students, particularly Tommy Edmondson, Tom Lovejoy, and Egbert Leigh, as well as Alison Jolly, furthered these causes by their own research and activism.[39]

Carbon Dioxide and Other Biosphere Concerns

In his 1948 article "On Living in the Biosphere," Hutchinson defined the biosphere as that zone of life "in which we can live and which we can explore." This zone he characterized by "temperature, which does not depart far from that at which water is a liquid, and by its closeness to regions on which solar radiation is being delivered." He wrote about the material requirements of life, including human life: "Looking at man from a strictly geochemical standpoint, his most striking character is that he demands so much—not merely thirty or forty elements for physiological activity, but nearly all the others for cultural activity. . . . We find man scurrying around about the planet looking for places where certain substances are abundant; then removing them elsewhere, often producing local artificial concentrations far greater than known in nature." All of this makes modern man an effective agent of "zoogenous erosion," affecting specific aspects of the biosphere such as arable soils, forests, and mineral deposits, which provide what "*Homo sapiens* as a mammal and as an educable social organism needs or thinks he needs."[40] Unfortunately the quantity of these needs necessarily increases as populations expand, a problem Hutchinson never stopped worrying (and writing) about.

Many biosphere processes are cyclical in nature, and some have considerable stability despite small disturbances. But there are critical limits to such disruptions. In disturbing cyclical processes we humans "usually do not know what we are doing," Hutchinson wrote. He considered several types of geochemical cycles, including those involving water and nitrogen. The cycle about which he was most concerned, however, is that of carbon, in which we have "a remarkable, perhaps a unique, case in which man, the miner, increases the cyclicity of the geochemical process."

A long discussion of the role of carbon dioxide follows in this paper, which Hutchinson originally presented at an AAAS symposium, "The World's Natural Resources." The research concerned is in part collected from previous workers. He wrote, "At the present time it appears that the combustion of coal and oil actually returns carbon to the atmosphere as CO_2 at a rate at least a hundred times greater than the rate of loss of all forms of carbon . . . to the sediments."

The "present time" was 1948, fifty years before the intense current concern with this problem in relation to global warming. The CO_2 content of the atmosphere, as had been noted by the mid-1940s, had increased at low altitudes in the Northern Hemisphere since 1890. Hutchinson thought the most reasonable explanation of the increase observed was an impairment of photosynthesis in the biosphere caused by deforestation. He concluded, "It is clear that in any intelligent long-term planning . . . an extended study of atmospheric gasses is desirable."

In a paper published the following year on the geochemistry of carbon Hutchinson pointed out that the available data on methane production by domestic animals, another concern today, had already indicated that 45 million metric tons of methane went into the atmosphere each year largely from this source.[41]

Hutchinson foresaw further environmental problems in this country as the result not only of population increase but also of changes from pioneer to settled communities. In the former, resources were originally treated as unlimited. In relation to resources in the future, which had been discussed in popular books by Fairfield Osborn and William Vogt, Hutchinson declared, "Anyone with any technical knowledge understands that the dangers described in these books are real enough, more real and more dangerous perhaps than the threat of an atomic world war." There were great difficulties, Hutchinson feared, in "forestalling such dangers in a culture that has developed under conditions of potentially unlimited abundance just around the corner." He was concerned with education, including college teaching, as a way to change attitudes and to show that "extremely difficult constructive activities are capable of giving enormous pleasure. . . . It ought to be possible to show that it would be as much fun to repair the biosphere . . . as it is to mend the radio. . . . And if we do not, then each one of us will be condemned to repeat throughout eternity:" (Hutchinson continued by quoting a poem by Edith Sitwell)

> Though cockcrow marches crying of false dawns
> Shall bury my dark voice, yet still it mourns
> Among the ruins—for it is not I
> But this old world is sick and soon must die![42]

The theme of how best to live lives of environmental quality is one that Hutchinson continually thought and wrote about. In the 1948 biosphere paper he suggested, somewhat tongue in cheek, that in a North American society in which a "considerable section of our population" was (already!) "definitely overweight," we should encourage the consumption of Mozart opera and delightful ballets, "which are geochemically very cheap," instead of hot dogs and ice cream, requiring many more geochemical resources. Hutchinson received a request in 1970 from the Economic Development Institute in Washington, D.C., for a copy of his paper "What Sort of Life Should Mankind Aim For?" delivered at an Ohio State University Conference on Quality of the Environment.[43]

The year 1974 brought Hutchinson the prestigious Tyler award, presented to him for his basic research, especially on the biogeochemistry and fauna and flora of lakes. This research was cited as very valuable for understanding lacustrine problems. His students applied his research as well as their own to environmental problems. Larry Slobodkin studied red tides in Florida. Later W. T. Edmondson and his graduate students at the University of Washington studied and reversed the pollution of Lake Washington in Seattle.[44] Hutchinson himself, especially in his later years, played an active role in solving practical environmental problems. One important example is his work on behalf of the preservation of the island of Aldabra.

Saving Aldabra

Aldabra is a ring atoll in the Indian Ocean, seventy miles in circumference, the only remaining home of the giant Indian tortoise, similar to the more famous giant tortoises of the Galapagos Islands. Many seabirds breed there as well as very unusual land birds, including a flightless rail. Ninety-two islands and atolls form an archipelago across the Indian Ocean. The most remote is Aldabra.

This coral atoll became part of the British colony of Mauritius in 1814 during the Napoleonic Wars. It was uninhabited and had been little explored. There was an early threat of a woodcutting industry on Aldabra in the 1870s, which activated the first organized protest by naturalists, including Charles Darwin. Their letter of protest expressed concern about the "imminent extinction of the Gigantic Land Tortoises." Even then Aldabra was the only locality where this tortoise was known to exist in a natural state. Their letter continued:

> The giant tortoise . . . flourished with the Dodo and the Solitaire, and while it is a matter of lasting regret that not even a few individuals of these curious birds should have had the chance of surviving the lawless and disturbed conditions of past centuries, it is confidently hoped that the present government and people . . . will find a means of saving the last examples of [their] contemporary.[45]

Woodcutting was abandoned, but the government in Mauritius did not protect the giant land tortoises. Indeed, Aldabra was leased from 1888 on to people who paid only a small rent and were free to capture and sell the marine green turtles as well as the fish around the island. One renter tried planting coconuts. Another sold thousands of green turtles, but he did protect the giant tortoises.

In the twentieth century Aldabra became part of the Seychelles, which was a Crown Colony of Great Britain. By then it was again uninhabited and had no economic value—no coconut plantations, no guano for use as fertilizer. Moreover, it was very difficult to reach, lying 700 miles from Mahé Island, site of the capital of the Seychelles. It is 260 miles off the northern tip of Madagascar and 400 miles southeast of Mombasa on the East African coast.

There had, nevertheless, been a surprising number of early scientific expeditions to Aldabra despite its isolation and the lack of regular shipping even to Mahé Island. Dr. W. L. Abbott, an American zoologist, brought flora and fauna back from Aldabra in 1892. The Percy Sladen Trust Expedition of J. Stanley in 1905 was followed by one of J. C. F. Freyer in 1908. Bristol University in England made two post–World War II expeditions to Aldabra, bringing back many specimens. Jacques Cousteau actually tried to lease the island from the British government in 1955 but was turned down. He shot a beautiful underwater film, *The Silent World*, in the waters surrounding Aldabra. The Yale Seychelles Expedition visited Aldabra in 1957.

The real threat to Aldabra came in 1965 with a secret joint British-American plan to establish an administrative region called the British Indian Ocean Territory (BIOT). The United States would pay half the cost of a military air base, to be established there. Plans were drawn up for the base and for establishing a BBC transmitter site there. By 1966 there were even more advanced plans for the air base. The BBC technical team plus the Ministry of Defense were already on West Island of Aldabra.

Tony Beamish, biologist and filmmaker, came to Aldabra in 1966 when the island was under threat and made a film that proved instrumental in the fight. Up to that time no one had filmed the birds and animals on the Seychelles islands. Beamish had hoped to do this earlier, but had been distracted by other projects. But in 1966, when the threat to the island was imminent, Beamish was most anxious to film Aldabra. Unlike most of the earlier explorers, he did not have his own boat, and he had much difficulty trying to find a berth on one from Mahé. Finally he received a telegram:

GOVERNMENT AGREES TO YOUR TRAVELLING VIA ALDABRA ON LADYESME LEAVING MOMBASA NOT BEFORE FOURTEENTH FEBRUARY PLEASE CABLE IF ACCEPT.[46]

Beamish had no idea what "ladyesme" was, but he accepted and later learned that he had been booked passage on the government boat, the *Lady Esmé*, out of Mombasa to Aldabra. He had only ten days to get to Mombasa, the East African port, and he needed maps and up-to-date information on Aldabra. Robert Gaymer and others of the Bristol Expedition provided considerable help with maps and with tortoise and rare bird localities. Gaymer also gave him a copy of his *Oryx* article, "The Case for Conserving the Coral Atoll." The current case was strong: on nearby Assumption Island all the resident seabirds as well as three rare land birds had been exterminated, and the island turned into a desert by guano hunters. Beamish was ready to make his Aldabra film.

English newspapers were reporting the fight for Aldabra: "Battle for Survival—Flightless Rail against the VC 10." The London *Times* printed an article entitled "Airfield Scheme Threat to Island." This article directly named Hutchinson: "There was a lively discussion from the other side of the Atlantic. Professor Evelyn Hutchinson, Professor of Zoology at Yale University, declared that 'the destruction of Aldabra wildlife would lead to a permanent gap in our understanding of the natural world.'"[47]

The *New York Times* of February 24, 1967, reminded readers that Charles Darwin had rallied to defend Aldabra's wildlife almost a hundred years before. The *New Yorker* wrote cynically: "The coming decimation of the atoll is itself a splendid specimen of evolution in action, illustrating, as it does the universal tendency of man's evolving machinery to destroy all forms of life that are less noisy, brutal and demented than itself."[48]

Hutchinson wrote to Rebecca West in March 1967 about the island's dire situation and his efforts on behalf of its rescue:

> You have probably seen a small article in the London Times about Aldabra Island. . . . This was why I was in London. The island is uniquely important biologically, the only reasonably virgin atoll left in the Indian Ocean and by far the most interesting of the three or four left in other parts of the world. The [British] Ministry of Defense wants to convert part into an air strip, choosing the place where the [giant] tortoises live, and to build a road all round. . . . The island has been more or less unofficially preserved, partly owing to the efforts of Charles Darwin and partly through the financial support of the late Lord Rothschild. A vast tangle of security restrictions protected the Ministry of Defense scheme and we still cannot learn if all our protesting is too late to stop contracts which may have been signed. There are several other islands long ago despoiled available in the same part of the Indian Ocean, so that no real strategic objection to abandoning the Aldabra plan can be raised.
>
>   We know that a number of the biological problems posed by the island are of enormous theoretical interest and this has led the Royal Society to take a firm gentlemanly stand, with Sir Ashley Miles going into bat, and finally beginning to realize that there is no agreement as to what game is being played. I am however by no means too discouraged. Something can be done from Washington. . . The National Academy of Sciences who sent me to the meeting at the Royal is being very active and I think rather more realistic. I have been in Washington this weekend.[49]

The outstanding British biologist Julian Huxley, Hutchinson's contemporary, wrote on Aldabra's behalf: "Aldabra is unique. . . . And it was almost lost to the world of science before the general public was reminded of its existence. A scientific tragedy of the greatest magnitude was averted only at the last minute. . . . It is in even greater need of protection today than it was when Charles Darwin and his friends first appealed for its preservation nearly a hundred years ago."[50]

The Royal Society had by then held its January 1967 conference about Aldabra and issued a press notice of "considerable concern" about the military proposals. The Royal Society announced plans for an expedition and for discussions with the Ministry of Defense. Hutchinson made his whirlwind plane trip to London specifically to attend this Royal Society meeting. (He was a foreign member.) The Royal Society did send its expedition, and subsequently Beamish's Aldabra film was shown on the BBC to seven million viewers, who could thus see Aldabra for the first time.

Hutchinson's letter to Rebecca West indicates a certain impatience with the gentlemanly behavior of the Royal Society in relation to the British government. Apparently the U.S. National Academy of Sciences had no such hesitation; Hutchinson went to the meeting in London not only as a foreign member of the Royal Society but also as a representative of the National Academy, to which he had long belonged.

Hutchinson's notes for his Royal Society meeting include a large number of scientific reasons for the necessity of preserving Aldabra. First of all, islands are natural laboratories for the study of evolutionary processes. They "must be subject to minimal disturbance over long periods of time. . . . The number of such natural laboratories is distressingly small and continually becomes smaller." The Aldabra atoll, though not untouched, was in Hutchinson's view the best remaining island for such study. "There is not the slightest chance that the interesting species would persist till natural extinction and be replaced by newly evolved forms, if a base were established. . . . All the neighboring islands have lost their endemic species through human interference."[51]

He pointed out many unique biological aspects of Aldabra, notably the giant tortoise population, which would be threatened: "The main airstrip is planned in one of the best tortoise areas." There were also declining populations of the Aldabra sacred ibis and of the flamingo, which were in need of monitoring. Although evolution is largely nonpredictive, Hutchinson noted, there were many cases of island birds losing their ability to fly, particularly true in the rail family. The Aldabra rail appeared to be in the process of loss of flight. "Of the flightless rails that were present [on oceanic islands] when European navigation began in the 16th century at least half are now extinct. No insular group appears more vulnerable, few groups would be more interesting and important to study." He also discussed the insects and the effect an air base would have on them: "Any military occupation would of course involve insecticides."

There were also many endemic plants there, plants found nowhere else in the world.

Hutchinson noted that there had been more than a century of previous preservation: "Darwin himself as well as the late Lord Rothschild did a great deal to forward the preservation of Aldabra, realizing its importance without knowing in detail what can now be done [to study its flora and fauna]. . . . We are beginning to see how to reap what they have sown. It would be a tragedy if at this time the whole system ceased to exist."

Perhaps most telling is Hutchinson's personal note of introduction to the notes that he had prepared for the British Royal Society:

> I have not expressed in the formal document my personal feelings on the matter, namely that the intended occupation of the island is a sickening and criminal attack on what I would call a natural work of art, and, bad as it is in itself, would set precedents that would impoverish the world even more completely and rapidly than is being done. I cannot believe that the people involved wish to go down in history as they well may with the simple epitaph, "They saved money."

In the end the British pound was devalued and the government decided to abandon the plan to set up a staging post on Aldabra. Beamish felt the government used the devaluation of the pound to get itself off the hook without losing face. He wrote that the advocacy of the Royal Society and the National Academy of Sciences "triumphed, however obliquely, in the end." He quoted the U.S. secretary of defense at that time, Robert McNamara, who said after the decision not to go ahead with the defense plan: "Thank God. I've had these scientist fellows on my back for months."[52]

War-Related Issues: Nuclear Wastes and Agent Orange

Hutchinson's letters to Rebecca West are full of his serious concerns, especially during the cold war. They are, however, interspersed with such comments as, "Margaret and I are still without the domestic and outside help we need. . . . We share the work and try to spare each other the job of scrubbing the kitchen floor."

In a letter of 1959 he wrote much about radioisotopes and the radiation danger from nuclear weapons. He thought that it was "rapidly becoming a political problem in the widest sense rather than [only] a scientific one." He was as much concerned about carbon 14 from nuclear

weapons as about strontium 90. He quoted AEC figures on carbon 14 to
Rebecca and added a prescient warning: "The problem of reactors going
wrong is ultimately as much of a problem for industrial atomic energy as
[for] military. This matter of industrial waste disposal terrifies me. This
is ultimately a problem of energy supply to our absurdly over-increasing
population."[53]

Hutchinson joined Yale plant scientist Arthur Galston in opposition
to the use of Agent Orange in Vietnam. Galston had earlier become chair
of the Yale Botany Department and was instrumental in the move to fuse
zoology and botany into one department in 1962. Although the two departments had not been in close contact before the merger, the two professors knew each other, but they were not close friends. As Galston said,
"Evelyn was kind to me."[54]

The two of them came together in opposition to the use of the herbi-
cide Agent Orange in Vietnam. Their objection was originally based on
destruction of the ecosystem by Agent Orange. Dioxins in it were later
found to be teratogenic, that is, causing birth defects. Agent Orange had
not originally been sufficiently tested toxicologically before being de-
ployed.

Galston wrote a letter to President Lyndon Johnson about the possible
human danger of Agent Orange. A commission on military use of Agent
Orange was set up, with Matthew Meselson as its head. Galston's graduate student Arthur Westing was sent to Vietnam to investigate. Meselson
had been a graduate student of Linus Pauling; both Pauling and Meselson
knew the president of Cal Tech, Lee Du Bridge, who was Nixon's science
advisor. According to Galston, "Nader's Raiders," a group of mostly young
lawyers, obtained a copy of a report on the teratology of Agent Orange, a
report that had not been released by Litton Industries or its predecessor.
The report got to Du Bridge who was reportedly "astounded" and con-
fronted military advisors in Washington. This led to the banning of Agent
Orange, announced in 1970. It went into effect in 1971—well before the
end of the Vietnam War. This change in policy was largely effected by
Galston and Meselson, with Hutchinson "cheering from the sidelines,"
according to Galston.

Hutchinson, meanwhile, was serving on the National Academy committee on the effects of Agent Orange; Anton Lang was the chair, but
Hutchinson was heavily involved in writing the committee's report. This
went on until the end of 1973. He wrote to Rebecca West: "I have been

so involved with the dead leaves and wood of Viet Nam and Washington that I have not been able to do anything [else]. The final report was telephoned in (on the third attempt) to a tape machine in Washington this afternoon."[55]

Hutchinson's writings about environmental issues and about biodiversity spanned nearly fifty years. His political activism on behalf of the preservation and study of biodiversity continued into his final year, 1991. In December 1990 he wrote a letter to the British Museum (Natural History) criticizing the cutting back of its research staff:

> I am utterly dismayed to hear of the impending loss of 51 of 300 curatorial and research positions, including seventeen positions in Entomology [Hutchinson's specialty]. . . . In these days of rain forest destruction and renewed interest in biodiversity, we must sample what is still there . . . before it is gone. We need experts on all the groups of organisms, but particularly on insects, since so many species are as yet undescribed. These organisms . . . are contributors to our ecology, and many may be of economic as well as intrinsic importance.
>
> Without the experts at the British Museum and similar institutions, our knowledge will slip downwards due to incompetence. Money must be found . . . so that the British Museum can maintain itself as a first class research institution as well as providing exciting and attractive exhibits that encourage the public understanding of biodiversity.[56]

The public understanding of biodiversity was one of Hutchinson's many important goals as ecologist and environmentalist. In spite of Hutchinson's awards for environmentally valuable research, many of the environmental activities described above were done behind the scenes and were not widely known, even to his graduate students.

CHAPTER SIXTEEN

The Polymath:
Art History and Many Other Fields

He knew, and everybody knew he knew, everything." One of Evelyn Hutchinson's students could well have written this line about Hutchinson; several of them said as much. It was actually intended to refer to J. D. Bernal, the English "sage of science," another polymath, expert on crystal structure, but on ancient architecture as well.[1]

Who was this man Hutchinson? Why was he so important to so many people, some of whom had never even met him? Innovative thinker, excellent intellectual synthesizer, integrator of diverse scientific fields, he was an extraordinary teacher, writer, and friend. His major scientific fields included evolutionary and population ecology, limnology, biogeochemistry, entomology, radiation ecology, and biogeography. He also had an intense interest in the history of science, publishing in this field and also mentoring an outstanding graduate student, Donna Haraway. He advised several graduate students in their behavioral ecology research, from lemurs to jackals and reindeer, a field that fascinated him, but one in which he did not do research.

Natural history and biodiversity interested him as a small child, and he was still enchanted, especially with insects and with the aquatic world, as an old man. Hutchinson wrote, at the age of eighty-six, about one of his early haunts: Wicken Fen in Cambridgeshire.

In my childhood, Wicken Fen was best reached from Cambridge by bicycle. . . . After about a ten mile ride to the north, one crossed the river Cam at Upware by hiring a rowing boat large enough to carry a few bicycles, rented from the Upware Inn, "Five Miles from Anywhere, No Hurry." . . . Once when riding my bicycle along the lode or boundary of Wicken Fen, a swallowtail approached from the opposite direction. I stumbled and dropped my net over the insect. . . . I fell off the machine and saw something fluttering. As I lifted the net, a swallowtail flew out. I had therefore been one of the few British naturalists to have captured *Papilio machaon britannicus* while riding a bicycle. Of course I was bitterly disappointed not to have retained the specimen, but now I can tell the story, adding, "without reducing the number of an endangered subspecies."[2]

But in these early years, Hutchinson had already become interested in insects less studied than butterflies, especially the aquatic water bugs of the suborder Heteroptera. On an early trip to Wicken Fen he found a rare, small but pretty, semi-aquatic water bug, a species of *Chartoscirta*. It had curious sexual dimorphism and a very specific habitat, on wet ground in sedge fen. He might have worked out its Hutchinsonian niche in his later years. But on his return to Wicken Fen in old age he could not find any specimens. He worried about extinction, even local extinction. Conservation of biodiversity, as we have seen, was one of Hutchinson's important concerns.

When one considers Hutchinson's lasting legacy, the subject of the final chapter of this book, one thinks first, of course, of theoretical ecology, limnology, and biogeochemistry, the three interrelated fields in which he made a tremendous contribution, both as innovator and as synthesizer. He himself highly valued his contribution as teacher, particularly of graduate and postdoctoral students, who themselves became many of the innovators and leaders in these fields and have since produced a third generation of eminent ecologists.

Nevertheless, Hutchinson also left an important legacy in additional fields. In some fields, such as animal behavior and the history of science, he had an important role in training graduate students, as noted above.[3] There were many other fields, however, in which he simply had a deep interest. These he researched and wrote about. They ranged from art history to prehistoric archeology to psychoanalysis to women in science.

J. D. Bernal and Arthur Hutchinson

Evelyn Hutchinson was truly a polymath, as Larry Slobodkin and I have written. We compared him to Ben Franklin, Humboldt, and Goethe.[4] It is difficult to find another twentieth-century scientist who was a comparable polymath. John Desmond Bernal was perhaps the most likely counterpart. A famous crystallographer, he was a legend to his students and associates, one of whom was Rosalind Franklin, crystallographer and codiscoverer of the structure of DNA.[5] Bernal also had a Hutchinson connection. As a young Cambridge University scholarship student, Bernal became too involved with socialism to do well on his first mathematics examination, and was discouraged about continuing. Thereafter he met Arthur Hutchinson, Evelyn's father, whom he described as a "very gentlemanly don" at Pembroke College. Arthur Hutchinson encouraged Bernal to continue with physics, chemistry, and geology in addition to mathematics, to Bernal's delight. He chose physics, and especially crystallography, a subject close to Arthur Hutchinson's field, mineralogy. That was in 1921. Two years later when Bernal had written a paper on crystal structure, Arthur Hutchinson read it and urged him to submit it for a prize at Emmanuel College. (He did win a prize and thirty pounds.) When Arthur Hutchinson later became professor of mineralogy at Cambridge, he created the first lectureship in structural crystallography to which Bernal was appointed. Bernal and Evelyn Hutchinson were both at Emmanuel College and probably overlapped as Cambridge students, but by the time Evelyn arrived, Bernal was married and no longer living in the college.

Years later in an interview for the *New Statesman*, Bernal was asked to name a field with which he was not familiar. He came up with eighth-century Romanian churches. The novelist C. P. Snow, who knew Bernal at Cambridge, is said to have used him as the model for the character Constantine in the novel *The Search*. Constantine is the brilliant but eccentric friend and colleague of the scientist narrator, who said of him, "I did not know a more remarkable mind." Constantine (Bernal), for example, amazed the narrator by his seemingly casual conversation initiated by lavatories they passed in the London underground: "Every civilization until the middle ages had sanitation of a very reasonable order. The Cretans did it very well indeed. . . . And also there was sanitation elsewhere: Sumeria: I don't mean the Sumerian Second Period, of course. But the third was as good as anything in Crete before what we might call the Cretan Renaissance."[6]

The narrator continued, "He knew, and everybody knew he knew, everything. He was equally at home in French and *Finnegan's Wake.*" As I have reported, Hutchinson gave a lecture in Italian when he was president of SIL, the international limnology society, when it met in Italy. He is also reported to have responded to questions at international meetings in several languages. Hutchinson, like Bernal, was certainly able to bridge the two cultures that were the main subject of C. P. Snow's fiction: the scientific and the humanistic. Dan Livingstone related to me that when he was at Yale as a graduate student, each fall Hutchinson was delving into some new and seemingly esoteric subject quite unrelated to their scientific pursuits. One year it was Chinese. Both Hutchinson and Bernal were leaders and innovators in their respective scientific fields, but each had wide-ranging interests and knowledge outside of the sciences.

"Marginalia"

For decades Hutchinson wrote his column "Marginalia" for the *American Scientist.* He sometimes wrote reviews of books, sometimes freestanding essays. They encompassed a very wide range of topics, from medieval bird art to prehistoric stone circles to homosexuality to D'Arcy Thompson's view of morphology. He sometimes wrote about the physical sciences as well. Stephen Jay Gould, a great admirer of Hutchinson, later started a column in *Natural History,* also influential but largely for the layperson. Hutchinson's audience was primarily a scientific one, but it included all kinds of scientists, not only biologists. I have talked to physicists who knew Hutchinson from these *American Scientist* articles, and they recall both his intriguing ideas and his literate exposition. Some of these "Marginalia" articles were collected into books, *The Itinerant Ivory Tower, The Enchanted Voyage,* and the one that excited my own interest in evolutionary ecology, *The Ecological Theater and the Evolutionary Play.*[7] Through these columns, Hutchinson's writings reached a wider and appreciative audience. Even his very first book, *The Clear Mirror: A Pattern of Life in Goa and Indian Tibet,* partly limnology and ecology but also including information on art, rituals, and other cultural aspects of the little-known places the North India Expedition visited, was later reprinted for a wider audience.

Art History and Hutchinson's Specialty:
Medieval Animal Art

One of Hutchinson's special fields was art history. Here he was able to combine, in professional publications, his knowledge of art, zoology, and the history of science. He was passionate and knowledgeable about art, especially Italian painting and sculpture, as is clear from the letters, quoted earlier, that he wrote to his wife Margaret from Florence and in his letters to Rebecca West. He published on art history in his own specialty, medieval animal art, particularly on birds and insects in illuminated manuscripts of the European Middle Ages.

Hutchinson's interest may have started early; the insects he saw at Wicken Fen varied in their color patterns in terms of sex and subspecies but also seasonally. These variations, especially of the large decorative butterflies, were depicted in medieval paintings he saw in Cambridge's Fitzwilliam Museum when he was a Cambridge student. He was especially intrigued by the illuminated medieval Psalters there.[8]

Hutchinson's 1974 article in the history of science journal *Isis*, "Attitudes toward Nature in Medieval England: The Alphonso and Bird Psalters," exemplifies his interest and knowledge of medieval animal art and its relation to the history of biology. To quote Hutchinson's introduction: "This study attempts to throw some additional light on the understanding and appreciation of nature during the Middle Ages by a scrutiny of certain [thirteenth- and early fourteenth-century English] illuminated manuscripts. . . . It has long been realized that during the thirteenth century the growth of a naturalistic tradition reflected changes in the whole outlook of medieval man."[9]

He pointed out that there had been few previous attempts to look carefully at the surviving illustrations of animals painted in the High Middle Ages from the point of view of natural history. For his discussion, he chose two bird Psalters, which contained the Book of Psalms or a selection from it. One was the Alphonso Psalter of about 1284, the other a lesser-known Bird Psalter from 1309; the latter is in the Fitzwilliam Museum.

The first Psalter was associated with Alphonso, the son of Edward I of England. It contains, according to Hutchinson, recognizable portraits of many English birds, including a wood pigeon, a crane, a green woodpecker, and a kingfisher. There are also action scenes of other birds, such as a falcon seizing a mallard duck, and of various mammals and a "fine

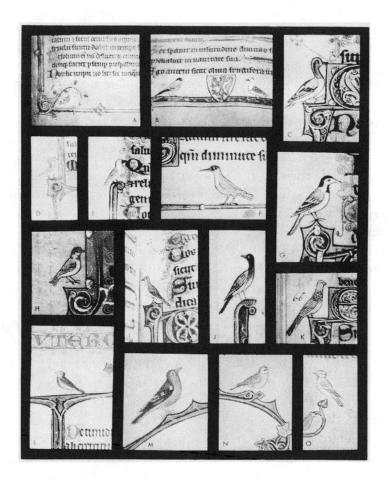

36. The Cambridge Psalter of 1309 depicted a linnet, bullfinch,
goldfinch, magpie, Chough, and many other birds, all possibly known
wild in England in 1300. Reprinted from Hutchinson (1974), plate 1,
from Fitzwilliam Museum Manuscript MS2–1954, with permission,
Chicago University Press.

green lizard," *Lacerta viridis*, never recorded from the British Isles but
present in France and the Channel Islands, indicating a well-traveled
artist.

The Cambridge Bird Psalter from 1309 also contains depictions of
birds, including a remarkably accurate painting of a woodcock and many
other identifiable species, almost all wild and presumably native in 1300.
In this Psalter, forty-four of the fifty-two bird and animal paintings can be
definitely identified. They include a butterfly, a lamprey, a pike-like fish,

and a lifelike spotted dog. Several other contemporary manuscripts that Hutchinson examined also contained illustrations of recognizable birds.

Hutchinson compiled a taxonomic table from the Psalters and other manuscripts. Thirty-seven birds are included, many occurring in several of the thirteenth- and early fourteenth-century works. Some of these birds are represented in mythological or humorous contexts, but even these are realistic portraits, indicating that the birds were well known to their portrayers. In some cases the paintings show such detailed knowledge of the birds that they were probably dead game bird specimens "on their way to the kitchens," reminding us that more recent bird artists, including Audubon, shot their birds in order to study them and paint them accurately.

Other bird portraits, like that of the smew, which visits Britain only in winter, indicate that the artists knew something about the biogeography of English birds, including summer and winter migrants. Hutchinson also discussed in considerable detail the medieval religious, mythological, and literary associations of various illustrated birds, from the crane to the goldfinch. The latter, for example, is depicted in fourteenth-century French sculptures and Italian paintings of the Madonna and Child from 1270 onward, according to Hutchinson and his many sources. (As with many of Hutchinson's writings, the most interesting details and anecdotes are often in the footnotes.)

Hutchinson also wrote about naturalistic medieval representation of both plants and animals in other forms of medieval art — in drawings and paintings and in church woodcarvings and stone sculpture. One conclusion that Hutchinson reached from his wide study was that there was not only an interest in nature and its accurate portrayal in this medieval period, but that a good deal was already known about the natural history and behavior of birds. For instance, in an East Anglican Book of Hours, an owl is depicted in an oak tree being mobbed by magpies, goldfinches, and other birds.

Hutchinson also discussed various historical figures such as Edward I, father of Alphonso, and Richard, duke of Cornwall. The latter was a friend of Emperor Frederick II of Sicily, who was well known as a student of birds. Hutchinson noted a strong Dominican influence on the Alphonso Psalter, as well as on Albertus Magnus, author of *De animalibus*. The "value of direct observation of nature rather than exclusive reliance on literary sources" was already recognized, at least in some medieval intellectual circles, Hutchinson wrote.

Hutchinson the polymath used all kinds of sources in Latin and other languages, but he made his point about direct observation well in a translation from Albertus Magnus's work on the goldfinch: "a little bird that frequents thistles. . . . What is said of this bird feeding on the points of spines and thorns, however, we have found by observation to be false. It feeds on the seed of thistles and burdock and teasel and the like. It eats also those of poppy, of rue and of hemp and whatever it eats, it removes with its beak the clean kernel from the husk and then consumes it."[10]

Hutchinson could not be sure that any copy of *De animalibus* actually reached England in the thirteenth century, but he thought this attitude toward the study of natural history on Albertus's part could well have influenced the Dominicans and the unknown artists in their charge who produced the bird Psalters. Hutchinson concluded that the naturalistic attitudes of Albertus and Frederick II, coming from Germany, France, and Sicily, influenced the English Psalter artists. But that when these continental traditions crossed the channel, "the result was something essentially English" and continued into the fourteenth-century Psalters that he studied.

Most artists were anonymous, but one artist of about 1400 is actually known to have been a Dominican monk. He was the principal illuminator of the Sherborne Missal and introduced different species of birds in a systematic fashion, as though he were attempting a serious ornithological presentation. Probably for the first time, vernacular English bird names were included. The fact that this artist was a Dominican monk was of great interest to Hutchinson as it tied this work to the naturalistic bird paintings of the earlier Psalters.

Recently, while following in Hutchinson's footsteps in Cambridge, England, I went to his favorite museum there, the Fitzwilliam, where he had spent many hours as a boy and a young man. The Fitzwilliam owns the Macclesfield Psalter, "the most remarkable English illustrated manuscript to be discovered in living memory." It was completely unknown and was discovered in 2004 at Shirburn Castle, where it had apparently been on the library shelves for centuries.[11] It is a "window into the world of late medieval England," from East Anglia, probably dating to the 1320s. It is beautifully illustrated, including many marginalia, some depicting mythical beasts, others apes and other animals not found in England, and also many native animals. Some of these, birds, rabbits, a red fox, a snail, a giant skate, Hutchinson would have recognized and delighted in. Some are not in their natural environments and interact with humans in humorous ways. The pages of this Macclesfield Psalter are "peppered with satire and scandal throughout" as Stella Panayotova, keeper of manuscripts and printed books at the Fitzwilliam, wrote. She asked:

> What was the role of, and the justification for such images in the Book of Psalms? No doubt they beautified the book. . . . But their function was hardly limited to the effect of slapstick humor. Nor was it "marginal," despite their position on the page. Laughter was not forbidden in the Middle Ages. It was part of every-day life. This holistic and healthy attitude to life, accommodating the saints and the sinners and embracing the world in all its shapes and colours, springs from the pages of the Macclesfield Psalter without prejudice or false modesty.[12]

She added that these illustrations, though perfectly acceptable to the medieval patron of the Psalter, would have offended more puritanical later owners in post-medieval times. Hutchinson died too soon to examine this Psalter, but had he known it, I am sure he would have laughed at the more ribald marginalia and have had much to share about the animals depicted.

During the fifteenth century, the naturalistic tradition in England included insects as well as birds. Hutchinson was delighted by insect paintings, and he was scientifically interested in the color variations and subspecies of illustrated butterflies and moths. He delved into these with much attention to their artistic detail and to their relation to the history of science.

A later article by Hutchinson in *American Scientist*, "Zoological Iconography in the West after A.D. 1200," includes insects. It originated as a lecture he gave to support the World Wildlife Fund and is an excellent example of Hutchinson's popular writing. He pointed out that any animal with two eyes, a nose, and a mouth—and even more primitive organisms that have an anterior end with some sort of sensory apparatus—can "provide an image into which human character can be projected." Some of the illustrations for his article are fascinating in that respect. In an illustration from a French manuscript of 1280 depicting a "successful manhunt," two rabbits are in a procession, one blowing a horn, the other carrying a man on its back. Naturalistic and recognizable birds and butterflies appear in later eras, for example, "extravagantly decorative" illustrations from a fifteenth-century book by Christine de Pisan, *Le Livre des Trois Vertus*, was noted by Hutchinson as one of the first feminist books.[13]

Butterfly paintings when they first appeared are already recognizable as to species, as is a 1325 damselfly. Hutchinson especially noted a rare yellow and black mutant of a burnet moth, together with its ordinary red and black form. The paintings were probably made from sketches in the field; the butterflies look natural, but some are more accurate than others. Modern specimens are shown for comparison with paintings from 1435.[14] He showed that anatomy had become part of animal illustration by the seventeenth century, but that ecology lagged behind. "The idea of an animal having an environment was still curiously undeveloped." He noted a winter ermine in Buffon's *Histoire naturelle*, in which the animal's coat has whitened but not the landscape!

The eighteenth century saw the discovery of many new animals, some of which soon became the subjects of hand-colored lithographs. In the nineteenth century, for example, the Australian birds discovered by English ornithologist John Gould were beautifully illustrated by his wife Elizabeth.[15]

While pursuing obscure sources, Hutchinson was able to make his scholarship fascinating, instructive, and even humorous for a general audience. He was not, however, one of those scientists who was famous in one field and dabbled and wrote popular books in quite another. Hutchinson was recognized as an expert in the field of early animal illustration. For example, he was asked to write a critique of a book about Greek zoological illuminations in Byzantine manuscripts.[16] The study of illuminated manuscripts is just one example of Hutchinson's wide-ranging interests that led to research and publications in the arts.

Sharon Kingsland discussed Rebecca West's book *The Strange Necessity*, in which Rebecca wrote on the importance of art to herself and to society. This book influenced Hutchinson. Kingsland wrote: "Clearly West's views about the purpose of art struck a deep chord with Hutchinson, and expressed very much his own views that science was important for the same reasons as art . . . that fundamentally there was no distinction between them, only the language was different." West concluded that "this strange necessity," art, was as essential as life. Hutchinson felt that both science and art were essential to life.[17]

From the Stone Circles of England to the Green Scum

Prehistoric British archeology was another of Hutchinson's interests. Some of his writing about it was aimed at mathematicians. "Long Meg Reconsidered" explores the geometry of British stone circles. The subtitle reads, "How Nearly Had Megalithic Builders Become Pure Scientists?" "Long Meg" is in Cumberland; her "daughters" form the stone circle.[18]

Hutchinson, in his articles, considered many stone circles, evaluating the previous work of Alexander Thom, an engineering professor at Oxford, as well as previous astronomical work. This article fascinated those who were interested in stone circles and at least somewhat conversant with geometry. Hutchinson, although well trained in mathematics, was not himself a mathematician. Nevertheless, he could write articles like these with authority.

Occasionally Hutchinson wrote more applied "Marginalia" articles like the 1973 article on eutrophication (see chapter 15). Hutchinson's environmentalist comments in this article on eutrophication may also not have been obvious. Fossilized organisms, such as midge larvae and diatoms, had been studied in lake cores obtained by Hutchinson and his students and associates. Some of these lakes had long ago become eutrophic or nutrient-rich and had remained in a steady-state condition for millennia. This was true of Linsley Pond, his favorite study site near Yale, for more than 7,000 years until the arrival of European colonists to New England in the seventeenth century.

What contemporary newspapers mean by "eutrophication," he wrote, is something much more extreme than any eutrophic condition in Linsley Pond before European arrival. "Eutrophy in a New England lake before 1650 probably meant good fishing; now it is apt to mean a thick scum of blue-green algae."[19]

Unlike his writings on medieval animal iconography or the mathematics of stone circles, which Hutchinson obviously researched, he could have produced the eutrophication article almost off the top (or the inside) of his head, including the fifty footnotes.

One is reminded of the comment made by one of Hutchinson's postdoctoral students, Clyde Goulden, about his mentor's cross-referenced brain. In Goulden's words, Hutchinson "had a photographic memory that was cross-indexed. He was able to relate everything in science he had ever learned. He could recall it because it was always interrelated with something else."[20] This was likely true of Hutchinson's serious interests outside of science as well, whether the marginalia of medieval manuscripts or prehistoric stone circles. In science, as exemplified in the eutrophication article, he was clearly thoroughly familiar with the early European studies in many countries and languages. Much of the research referred to in this article is that of his graduate students, from Ed Deevey's in the 1940s to Karen Porter's, then in press. The work of many other students and research associates—Gordon Riley, Vaughan Bowen, Ursula Cowgill, Ruth Patrick, Dan Livingstone, Joe Shapiro, and W. T. Edmondson—is included, as well as Hutchinson's own research. Here we see Hutchinson's teaching legacy in limnology, although that might not be obvious to the casual reader.

The Philosophical Naturalist: "What Is Science For?"

One of Hutchinson's most-quoted *American Scientist* articles was entitled "What Is Science For?" Written in 1986, it was dedicated to his close friend Dame Rebecca West and to his wife Margaret, both of whom had recently died. He also dedicated it to "Alexandra Ellsworth Gross." She was the two-year-old daughter, usually called Sasha, of his Yale colleague and friend Phoebe Ellsworth. Hutchinson hoped that the text would have become "partly self-evident and partly irrelevant" by the time Sasha was old enough to read it.[21]

In this article Hutchinson took up the serious theme of science as an activity of great social significance the results of which can be very useful but also very dangerous. This had become a serious theme among scientists, particularly physicists, because of their contribution to the atomic and hydrogen bombs. The results of science may often be unexpected. Citing the testimony of one expert, Cyril Smith, that metallurgy was largely invented to make ornaments, not weapons, Hutchinson wrote: "It is nearly always easier to break than to make something. This . . . kitchen statement of the second law of thermodynamics means ethically that we must always be careful in any application of science because incidental unintended effects are more likely to be harmful than beneficial."

An important problem, Hutchinson felt, was the public's lack of sufficient scientific knowledge to make decisions on the use of scientific discoveries. "Short of everyone becoming his own physicist or biologist . . . the role of the best science writers . . . is of the utmost importance." It is also the duty of scientists, insisted Hutchinson, to discuss their findings "clearly, interestingly and impartially." He concluded that political decisions were often made without such knowledge. "Anyone studying the political careers of dioxin or yellow rain [during the Vietnam War] in the newspapers is likely to be impressed that a political position was taken first and that afterward politicians welcome such facts, if there are any, that are in its favor."

The most important current problem, in his view, was the avoidance of nuclear war from the prospect of nuclear, biochemical, and bacteriological war, and "at the present time an absurd number of completely useless wars . . . being fought on our planet."

In this article Hutchinson turned next to the "usually delightful subject of sex." This, however, included the "curious problem of abortion." He was very worried about the failure to keep the human population under control, even though the scientific facts of overpopulation were well known. He postulated an AD 3000 population of five trillion. "It could hardly be denied that crowding would be excessive. Enormous authoritarian control would become necessary to deal with allocation of resources. It has been suggested that the hardy pioneers could go to the moon or to artificial satellites. The water bills for such people would be unthinkable."

Hutchinson's essay also included his thoughts about the evolution of behavior, and in a subsection called "The Balance of Violence and Love" he wrote: "In general there is a tendency for the more aggressive and territorial behavior to be shown by the male, the more nurturing behavior by the female. There is enormous variability in this, with some striking exceptions, as in hyenas and phalaropes. We may also notice that as the beauty associated with sexual display became less genetic and more behavioral, it has become more female."

Hutchinson felt strongly about the potentially destructive uses of technology. He wrote that it was useless to complain that destructive tendencies were "human nature." Science needed to strive to produce "a technology of love," applicable socially and based on vastly increased knowledge. Although written during the now-ended cold war, this is a widely quoted essay, worth reading in a still violent world of "useless wars" more than twenty years later.

Hutchinson included a few words in this essay aimed at ecologists who were at that time criticizing some of his and Robert MacArthur's ecological work, particularly over the role of competition. Simberloff in particular had used, or misused, in Hutchinson's view, the ideas of Karl Popper and the null hypothesis. (This controversy is discussed below; see the section "Santa Rosalia Was a Goat" in chapter 18.)

It is clear that both Hutchinson's interests and his expertise were wide and deep. His "Marginalia" encompass not only actual marginalia, the depiction of animals on the margins of old manuscripts, but also a host of other topics, including serious scientific concerns. In his first "Marginalia" column in *American Scientist* in 1943 he had announced, "One of the most important [positive] aspects of the intellectual climate of the present time is the increasing tendency of scientific workers to pass the conventional boundaries of their subjects . . . and to borrow from diverse fields information that can be related to the results obtained in their own special investigations." He would endeavor in his columns, Hutchinson continued, to consider papers and reviews that would be of interest to scientists in more than one field. This endeavor would, however, be limited by the intellectual capacities of the writer: "No man today has the knowledge and experience to penetrate the realms of all the sciences. Moreover the interest evoked by any particular work is bound to depend on certain emotional attitudes in the writer and reader. These notes [columns] will reflect the attitude of the philosophic naturalist, rather than that of the engineer."[22]

After a lapse of some years in the "Marginalia" series, Hutchinson had been welcomed back in 1970. Caryl Haskins was then editor-in-chief of *American Scientist*. Hutchinson was to write a 5,000-word "Marginalia" piece every two or three issues, for which he would be given an honorarium of $200, to be raised to $500 the following year if the journal's financial condition made that possible.

His editors wrote, in introducing Hutchinson's article, that this "beloved contributor to *American Scientist*" had produced since 1943 a "notable series of musings, ostensibly on current scientific research but actually taking in the entire universe past, present and future." These "Marginalia" "ranged from dodos and bowerbirds, the colors of the planets, Linear B, the nature of mathematics, the mating habits of bacteria, and the fleeting geometry of the human face."

Hutchinson was a major scientist — and a true polymath.

CHAPTER SEVENTEEN

The Last Years—
From Yale to England

In 1983 both Hutchinson's wife Margaret and his close friend Rebecca
West died. Although Margaret survived Rebecca by a few months, she
was already seriously impaired by Alzheimer's disease. Evelyn wrote
to Yemaiel just after Rebecca's death. He said that Margaret did realize
what had happened when Rebecca's secretary called to tell them but had
forgotten by later in the day. Evelyn "had to go all through the pain again."

During Margaret's decline with Alzheimer's, his doctor had insisted
that Evelyn get out daily for at least an hour in the library or lab, which
he did. He had, nevertheless, carried out most of Margaret's care himself.
He was also currently "bereft," he wrote, of any real help on the fourth
and final volume of the *Treatise on Limnology* because Anna Aschenbach,
his "general factotum, a *very* liberal Quaker," was off in Belgium and Germany doing antinuclear weapons peace work.[1] Anna, though not a biologist, was Hutchinson's able assistant from 1974 until he left for England
in early 1991.

Hutchinson's many young friends at Yale were his great solace. Several of these were biologists, including Janie Wulff and Saran Twombly,
who helped him with Margaret and with volume 4. Others enriched his
life with their shared interests in other fields, particularly art, or visited
him with their small children.

Without Margaret's companionship or Rebecca's letters and visits,
Evelyn had lost his confidants who had shared his daily life and feelings.

Even before Margaret and Rebecca died, they were both ill, Margaret since the 1970s. Although Evelyn and his younger sister Dorothea, in England, had not been particularly close except in childhood, they had occasionally written to each other. It was Dorothea who now became Hutchinson's major correspondent.

Dorothea Hutchinson was a graduate of Newnham College, Cambridge, and had a long career as a psychiatric social worker. She had seen much of English life from the perspective of virtually all the social classes—in a very class-conscious society—in London and elsewhere. In her old age, when this author knew her, she was an interesting and perceptive woman living in Cambridge. Dorothea Hutchinson gave me all of her letters from Evelyn; they chronicle his view of his life in its last decade, from 1980 on.[2]

In January 1980, Evelyn wrote to Dorothea about a very positive review of his recently published partial autobiography, *The Kindly Fruits of the Earth*, written by John Harper, a professor at the University of Bangor in North Wales. This review gave him great pleasure, Hutchinson wrote. He described John Harper to Dorothea as the best living plant ecologist. Harper was one of those who brought much of Hutchinson's work to the attention of other British ecologists.

But Evelyn wrote to his sister at this time primarily about Margaret and her progressing illness (as yet undiagnosed). Evelyn and Margaret had recently had Sunday tea at the master's house in Saybrook, Evelyn's college at Yale. Margaret was physically much better, as noticed by everyone who met her there; by contrast, she had been quite ill the previous year. At the end of Evelyn's letter, however, he added a postscript:

> The above was read by Margaret, but unhappily I have to add more. The sudden great physical improvement is at a great psychological cost. It all happened over the night of 28 Dec 1979. In the morning she seemed much younger, very cheerful but extremely aphasic. . . . A huge system of delusions has developed, so that her life revolves about a number of non-existent or partly non-existent people doing things in the neighborhood.

One such person was a real neighbor, who Margaret imagined had stolen three dollars and change from her bedside table. There were also other problems, requiring him to be with her constantly: "She wandered out[side] holding a heating pad for one of these phantasms, and got lost, though a neighbor found her before I did. . . . Fortunately she seems quite happy and apparently highly intelligent behind the unreality."

There was a further P.P.S. to this letter to Dorothea: "Don't write any-thing about the fourth page to me [the above P.S. about Margaret's ill-ness], even to the lab. *Don't yet say anything to anyone else* [italics in origi-nal]. I think the doctor is very good but says at the moment mainly, wait and see."[3]

But by April 1980 there were new developments to share with Dorothea. He was deeply grateful to her for her last two letters, he wrote, "but there is no need for you to worry. The situation is a very curious one but very far from desperate." He had driven to a nursery and bought some perennials that Margaret had planted "with complete confidence and skill." Today her accompanist had come, and Margaret had sung several parts of the *Marriage of Figaro* "straight through without any hesitation." Her artistic and literary sensibility also appeared "completely untouched, though the rhythms of daily life can go completely astray." She seemed much happier than a year before "when things were beginning to go wrong."

He wrote that he was having no difficulty about running the house. Extraordinary young friends, like Marjorie Garber, whom Hutchinson re-ferred to as an "intellectual daughter," were helping. Marjorie's assistance the preceding August, Evelyn thought, "was largely responsible for Mar-garet's recovery." Another young friend who came to see Margaret was his former graduate student Donna Haraway, with whom he had kept up a constant correspondence. She was back at Yale for a history of science conference and a special meeting of people interested in the history of ecology. Hutchinson himself was then doing some research on Mary Ball, who had carried out early research in entomology, which she had had to publish under her brother's name. Many other subjects, apparently of interest to Dorothea, appear in this letter including discussion about the iconography of apes.[4]

Margaret's illness was not medically diagnosed until summer 1982. In August Hutchinson sent Dorothea a report on Margaret's condition:

> Technically she is diagnosed as having the Alzheimer Syndrome . . . which involves the slow granulation of nerve cells which become non-functional. There seems to be a sort of balance, the unaffected parts being able to compensate to some extent for the losses. This leads to great temporal variability. She is extremely good at compensating in social situations and although she has often, in fact nearly always, no idea what people are talking about, she manages to give them a feeling that she deeply appre-ciates them. . . . She reads a great deal, can tell me practically nothing, in cognitive detail, about what she is reading but manages to convey critical

judgements in a vague and general way but always giving a feeling of perceptiveness, in tone or gesture rather than semantically.

Phoebe [Ellsworth] who was staying with her parents visited two weeks ago with her 14 month old daughter and this was the purest pleasure Margaret had had for months. . . . She gets to church each Sunday but for the past few months has not otherwise been out. . . . Any comment you may want to make please make to the lab.

Love, Evelyn[5]

Hutchinson told this author, when she visited him in 1982, that he had taken Margaret out one afternoon to hear a concert at a friend's house. He said she really seemed to enjoy it while they were there, but afterward she could remember nothing about it.

A Celebration at Cambridge University

Hutchinson had never turned in his scientific papers to Cambridge University to apply for an earned doctorate of science. In his early years he was afraid he might be turned down, as is known from his correspondence with a Cambridge student friend. By the late 1950s his work was so well known and important that a Cambridge doctorate almost surely would have been awarded him had he applied. He did not. In his later years he was proud of having advanced so far in America without an earned doctorate; by that time he had received many American honorary doctorates.

In June 1981 at the age of seventy-eight Hutchinson was awarded an honorary doctor of science degree from the University of Cambridge. He came to England for the award. "It is very sad that Margaret cannot come," he wrote to Rebecca.[6] Margaret was not well enough to go to England with him, but Dorothea, who lived in Cambridge, took good care of him, as did his old friend Dr. Sydney Smith. He also arranged to visit Yemaiel Aris and for Yemaiel and Rebecca West to meet.

A three-page "programme for Professor G. E. Hutchinson" detailed all the events of June 9. For example: "Dine in the Hall of King's College (*Lounge suit and M.A. gown provided by Dr. J. K. Rose.*) The evening will be quite informal; no speeches will be made." On the tenth, there was a full schedule: "Low Mass at 10 a.m., Coffee at 11, Lunch with the Bishop of Ely, a Reception at the Master's Lodge, St. Catherine's College," followed by dinner and coffee. "The Master will propose the Loyal Toast and make a few remarks about the Guest of Honour (*G.E.H. to reply*) [italics in the original]."[7] There are dress notes for these events also.

Finally came the big day, the eleventh of June, beginning at 10 a.m. and not ending until 10 p.m. for the seventy-eight-year-old Hutchinson. He was gowned and introduced and then presented to the queen's husband, Prince Philip, who was the chancellor of Cambridge University. He had arrived by helicopter the night before. All the "honorary graduands" formed an academic procession. The others so honored included Dr. Max Perutz, Nobel Prize winner and the former head of the Molecular Biology Laboratory at Cambridge; the archbishop of Canterbury, Dr. Robert Runcie; the foreign secretary, Lord Carrington; and Dame Helen Gardner, Emeritus Merton Professor of English Literature at Oxford. Only Hutchinson and Perutz were to receive doctor of science degrees. All were duly photographed. After the procession there was music performed by the choirs of two Cambridge colleges and the presentation of "each graduand to the chancellor by the orator in a Latin speech."

In the afternoon the schedule continued with a formal luncheon and then "Miss D [orothea] M. Hutchinson's tea party," followed by dinner at Hutchinson's college, Emmanuel. Italics in Hutchinson's schedule indicate the dress required for these events and some remarks by the master of the college to which Hutchinson would be asked to reply. Friday the twelfth was a more informal day. It included coffee at the home of Evelyn's old Cambridge friend and well-known tropical ecologist Paul Richards and his wife, Anne. But the highlight of that day was a return to Evelyn's (and Grace Pickford's) favorite haunt, Wicken Fen, including a picnic lunch with Professor Sir James Beament and "a party of colleagues and research students." January is not the best time to see the myriad insect and other life of the fen, but surely Hutchinson enjoyed that outing.

On his return, Evelyn wrote thanking Dorothea. "I had a wonderful time and a great deal of its quality was due to you." He had enjoyed every minute of it, particularly the people that Dorothea had invited to tea.[8]

Hutchinson saw Rebecca West on this trip and wrote to her afterward. He was happy that Rebecca and Yemaiel were able to meet. Evelyn had a good visit with Yemaiel as well: "One of the happiest things about my visit was the realization of how much she is our [his and Margaret's] daughter."[9] In a later letter about his honorary doctorate, he enclosed the citation, noting that it was taken largely from his recent partial autobiography, which, he added ironically, ends when he was a young professor at Yale "before reaching anything that possibly could justify an Hon ScD."[10]

On January 3, 1983, many friends and former students gathered for

37a. Hutchinson, with sister-in-law Hannah Hutchinson, after receiving his honorary doctorate of science from Cambridge University in 1981. Courtesy of Dorothea Hutchinson.

37b. Hutchinson after receiving his Cambridge University degree, with sister Dorothea, sister-in-law Hannah, niece Rosemary, and nephew Philip. Courtesy of Dorothea Hutchinson.

another gala event, Evelyn's eightieth birthday party. Margaret was able to come, if not to remember. The dinner menu was printed partly in Latin, e.g.: a salad of Lactuca sativa (with Italian dressing), Gallus domesticus Paragon, and in French: Gateau Hutchinsoniella. The latter was in the form of a marzipan model of a primitive crustacean named *Hutchinsoniella* by his former graduate student, Howard Sanders, who first discovered it. Michelle Press, editor of the *American Scientist* and an old friend of Evelyn's, wrote to Dorothea, who was unable to come:

> Your brother . . . was even more remarkable than usual on that day, spending great time, energy, and consideration on Margaret's comfort and happiness, greeting what seemed to be hundreds of old friends and well-wishers, responding with wit and eloquence to the speeches and toasts in his honor. . . . It was lovely to watch, and I know that everyone there was very moved with affection for him and felt great pride simply in knowing him.[11]

Evelyn himself also described the birthday party to Dorothea: "Margaret did splendidly and enjoyed it as much as I did." He called it a tremendous success, "doing a lot of good in fostering a truly affectionate spirit in a great number of people." Some of these, like Tommy and Yvette Edmondson from Seattle and former graduate student Karen Glaus Porter, from the University of Georgia, came from a long distance away. In all, 205 guests left their signatures. Evelyn could not identify some of these: "A few may have been hungry postgraduate students lured by free wine and chocolate cake," he wrote. A fund had been set up, with the help of Tom Lovejoy, Ruth Patrick, and Dillon Ripley, he told Dorothea, to assist him in completing the fourth volume of his *Treatise on Limnology*.[12]

One more comment about the birthday party appeared in a later letter: "With regard to women doing remarkable things intellectually, it was curious to find that the eminent people, here before about 1957 who had survived [several of his graduate students died young] and came to the party were all men. After that date, when Alison Jolly was here, the women [graduate students] predominated, with about half a dozen of extraordinary ability."[13] Most of these women, now professors themselves, had returned for his birthday party.

Rebecca West was ill and did not come to Evelyn's January birthday party. Although Margaret did attend, her condition had worsened by then. She was apathetic most of the time and had almost stopped read-

ing. Hutchinson was "nurse, cook and ladies maid" as related earlier. He
did continue to get help from several friends so that he could get to his
office and do some writing. Margaret hung on until August, long enough
for their golden wedding anniversary.

Evelyn continued to chronicle his concerns and honors to Dorothea.
In February 1984 he wrote that he had been awarded the prestigious
Daniel Giraud Eliot Medal of the National Academy of Sciences. He decided to go to the annual academy meeting in Washington, D.C., in April
on his way west to lecture and to see Phoebe Ellsworth and her husband
Sam and daughter Sasha in California. "About five places in the West
want lectures which means I can pay for all the travelling from the [Eliot]
honoraria. All that it needs is to have something to say." There were quiet
weeks ahead and he was hoping to get a lot of writing done.

An art historian friend, Ramona Austin, whom he had previously described to Dorothea as of Bahamian descent and "café au lait" rather
than black, was visiting Yale. They had "a couple of good talks." He sent
Dorothea a Polaroid photo of Ramona but wanted it back for his Yale archives. He related that he had given a small dinner party and "was quite
bowled over when Ramona quoted half a dozen lines of a Saxon devo-
tional poem in the original Old English."[14] These anecdotes are of interest
in terms of Hutchinson's marriage to another light-skinned black woman,
Anne Twitty Goldsby, the following year.

In April Evelyn attended his first seder, to which he was invited by his
former student, well-known population ecologist Larry Slobodkin, professor at SUNY Stonybrook, and his wife Tamara. It took place at the Slo-
bodkins' Long Island home, and it was a "spiritual adventure." Hutchinson described it to Dorothea:

> The Passover celebration was not what I expected but very interesting and
> at times most moving. Larry says the current rite is hellenistic, 1st century
> B.C.–A.D. Alexandria. . . . There is an immense amount of ritual involving
> horseradish (the bitter herb of the Bible). . . . Some of the prayers give a
> completely Christian point of view (or rather represent an identical posi-
> tion) and there is a passage where Moses is quoted as telling the children
> of Israel that they must not rejoice in the drowning of the Egyptians but
> be genuinely sorry for them. The whole ritual is clearly designed to im-
> press the children with the continuity of Jewish tradition. . . . At the end
> they sang, "Who knows twelve? Twelve are the tribes of Israel, Seven are
> the stars in Joseph's dream . . . ," which I think I did know but had for-
> gotten so it was a big familiar surprise.[15]

Evelyn and Dorothea were brought up in the Anglican Church. Evelyn was always interested in religious rituals, whether Anglican ones, Catholic rituals in Italy, or Buddhist in Ladakh. He seems not to have had much previous experience of Jewish holiday ritual. He had also been to the High Episcopal services in New Haven during Easter week. He found religious ritual comforting as well as of anthropological interest.

Hutchinson was busy that Easter season preparing lectures he was to give at the University of Colorado and at Stanford University on his western tour. That journey west was a very satisfying one for him, especially in terms of visiting former graduate students and friends. He saw Saran Twombly, whom he always called his "post-ultimate graduate student," in Colorado. He gave two lectures at the University of Colorado and three at Stanford. There he also saw former students and participated with Phoebe Ellsworth in discussions with the students in her psychology course. On his Seattle stop he visited the Edmondsons and talked to and with University of Washington students; he commented that they all learned a lot.

In June he was elected a corresponding member of the Vienna Academy of Sciences. When writing to Dorothea about this honor, he mentioned his new friend, Princess Mary de Rachewiltz. "She is Ezra Pound's daughter, of whom Phoebe Ellsworth once said, 'If you were looking for the ideal human being you would probably not think of starting with the illegitimate daughter of a fascist poet, but you would be quite wrong.'" Mary de Rachewiltz was married to a Polish prince but was spending part of the year at Yale going through her father's papers. She and Hutchinson had become good friends during her stay.[16]

Evelyn described his social life, which included hosting several visiting women scholars. He had dinner with Lynn Margulis, well known for work on symbiotic events in evolution: "Perhaps the most imaginative biologist. . . . We went to a restaurant with a nice outside terrace and had an exciting talk into the summer night." She had just written a book for the "scientifically minded general public," which Hutchinson had sent to the next generation of English relatives, including Yemaiel's daughter who was just entering Cambridge University. He had himself recently made dinner at home for visiting Marjorie Garber, "who helped so much in the early part of Margaret's illness."[17]

Evelyn was able to attend plays and concerts in New Haven for the first time since Margaret's serious illness had begun. He described several such events, including a wonderful production of *The Tempest*. He had been to a concert of medieval and Renaissance dance music on a fine June evening. Evelyn was also able to do much more traveling, both in the United States and to England. He wrote to Dorothea just after the first anniversary of Margaret's death that he would be coming to England on October 4, 1984, and had several nonfamily events planned. He would stay until November 5, and Dorothea could make "any reasonable family suggestions."[18] On these trips he saw Yemaiel and became, for the first time, a good friend to his four nephews, Francis, Andrew, Paul, and Philip, and to his niece, Rosemary, and to their families. These friendships continued and were especially important at the end of his life when Yemaiel and her husband, Ben Aris, and Francis Hutchinson and his wife, Joyce, took him into their homes.

One event of this year, 1984, was a commission by his former student and longtime correspondent Myrdene Anderson to have her artist friend, Christina Olson Spiesel, paint Hutchinson's portrait for Saybrook College, Hutchinson's college at Yale. Christina came to know Evelyn very well in the course of many sittings. The painting also contained the Ladakh masks, from his 1932 Yale North India Expedition, which were always there with him in the lab. She wanted to portray him in the act of working in his lab, not as an effigy, she told me.

By the new year of 1985 Hutchinson was leading a very full life, seeing many old friends and going to parties. He wrote about his friend Dorcas MacClintock, who had published on African animals with wonderful cutout shadow illustrations by Ugo Mochi. Dorcas was herself an animal lover. "Her house was a refuge for stray raccoons; Margaret and I always called it the 'peaceable kingdom' with every animal lying down with every other."

Former Yale postgraduate student Pamela Parker visited him. She was an expert on Australian animals and spent half her time working for the Australian government and the other half at the Brookfield Zoo in Chicago. She told him about some fascinating research with dolphins, which were getting "psychological training," learning to recognize symbols or ideograms for particular types of rewards. They were taught the concept of a reward for any unusual type of behavior; they then invented new behaviors. Hutchinson wrote to Dorothea that he hoped to go to the Brookfield Zoo after Easter to see for himself. "I suspect that chimpanzees could behave in an equivalent way. I was naturally particularly interested as it

38. G. Evelyn Hutchinson with Christina Spiesel (left) and
Myrdene Anderson. Spiesel's portrait of Hutchinson with
masks from Ladakh is on the wall behind them. Photograph
by Charles Erickson. Courtesy of Yale University Office of
Public Information. Manuscripts and Archives,
Yale University Library.

all raises the question as to whether a high intellect cannot develop un-
aided in a liquid medium," he wrote. Hutchinson had an intense interest
in animal behavior. Many of the students on whose graduate committees
he sat had worked with mammals, from lemurs to jackals and to the Lapland reindeer that Myrdene Anderson studied.

There was also a special exhibition at the Yale University Art Gallery of folding screens by artists of different eras, to which he had taken a number of good friends, including Pamela Parker, Phoebe Ellsworth, and Rudi Strickler, who was a "remarkable investigator of planktonic crustaceans." But he wrote that in spite of all these distractions he was getting back to work on his *Treatise*. Volume 4 was yet to be completed. His life sounded very busy and not especially lonely.[19]

But life-changing events were soon to occur in Evelyn's personal life (see chapter 11). On March 25, 1985, Evelyn married Anne Twitty Goldsby. She is pictured with many other friends at his birthday celebration in January. Dorothea received a photo of her new sister-in-law and an account of how Easter week was divided between his High Episcopal church and Anne's black church, where a service was followed by lunch with some of her friends. Evelyn was clearly interested in the black community in a rather anthropological way. The entrée to that community that Anne provided was one of her attractions.

Their New Haven friends seemed to approve of their marriage. Life was hectic; Evelyn wrote, "Everyone wants to see us."[20] They had made reservations to come to England at the beginning of June. Evelyn was admitted to the Zoological Society on June 11, with his wife and his "daughter" (Yemaiel) in attendance.

Evelyn and Anne visited most of his family in England in June and July of 1985, starting with Yemaiel, her husband Ben, and Dorothea, and stopping to see Evelyn's niece, Rosemary, and her husband, Peter Dodd, in Newcastle toward the end of the visit. Rosemary told me (the author) that when they extended the invitation to Evelyn and Anne to visit them, Dorothea was quite concerned about how things would go. It was indeed a strange visit. First of all there was a mix-up with their suitcases; Evelyn had picked up someone else's suitcase at the railroad station. The missing one was full of clothes that Anne had just bought on their visit to the Isle of Wight, where Evelyn's sister-in-law, Hannah, lived, as did her son Francis and his wife, Joyce. Joyce had a boutique, where Anne had bought the clothes.

"Anne retired to bed straight away," Rosemary related, and did not reappear until the following evening's tea. Meanwhile Evelyn was running up and down stairs trying to see if Anne was all right. The missing suitcase was by then found. However, "vodka bottles were coming down the stairs, so I knew what was going on," Rosemary said. As the Dodds did not drink vodka or keep it in the house, Evelyn and Anne must have brought the

39. Evelyn Hutchinson celebrating his birthday in his office with friends in January 1985. From left to right: (seated) Stephanie Singer, Evelyn Hutchinson, Anne Twitty Goldsby (later Hutchinson), Anna Aschenbach; (standing) Dorothea Rudnick, David Furth, Stan Rachootin, Christina Spiesel, Charles Remington, Janie Wulff, Willard Hartman. Courtesy of Yale University Office of Public Information. Manuscripts and Archives, Yale University Library.

bottles with them. It seems clear that Evelyn knew that Anne was drinking. Peter Dodd said kindly that he put it down to Anne's nervousness at meeting the family. Vodka looks clear, like water, and Dorothea, whom they visited earlier, insisted that water was what Anne had been drinking. Later, when Dorothea visited New Haven at Christmas and Anne was out of the room, Dorothea put her finger in Anne's glass. She discovered, and finally admitted to Rosemary, what she had denied previously. It actually was vodka, not water.

Nevertheless, on this summer visit to Newcastle Anne finally came down to tea at Rosemary's and Peter's house and was "friendly and chatty" and not drunk. Peter said she looked presentable, and Rosemary added that she thought Anne a quite attractive woman.[21] Anne's serious problems with alcohol did not surface until several years later. Evelyn, on some level, certainly knew; on another level he denied the problems. Anne's children were also in denial and never used the word "alcoholic" about their mother.

After their English visit they returned to New Haven, and all was going well. They had settled down, according to Evelyn, to a "normal home life which is very delightful. I am getting going on the book [volume 4 of the *Treatise on Limnology*] which gives me much pleasure."[22]

Evelyn wrote that Anne had started writing a book about high school science education, an endeavor that interested Evelyn and with which he was helping her. In August they went to Detroit for Anne's nephew's wedding. Hutchinson's anthropological view reappeared: "Most interesting and enjoyable is my meeting so many kinds of people whom I should never otherwise know. His bride is a veterinarian, as were two of her very beautiful bridesmaids." He was amused to find that for the first time in his life the fact that he came from Cambridge, England, took on new meaning. It meant to the bride, who worked with horses, the big veterinary center outside Cambridge, which was considered the world center of equine medicine. "All the race course officials in Detroit were at the wedding," Evelyn wrote. Hutchinson also found himself, at the wedding, "in the company of several quite fundamentalist black gentlemen discussing the theology of Lot's wife." They very politely asked his opinion of this matter. His reply is unknown.[23]

The *Washington Post* Interview

Toward the end of the summer of 1985, Evelyn's former student Tom Lovejoy brought a reporter, Henry Mitchell, to dinner. A photographer turned up too. The results of this interview appeared in the *Washington Post*. "Occasionally he didn't 'write it right' but it is a most warm piece which gives all our friends pleasure," Evelyn wrote to Dorothea. Other newspapers reprinted it; Hutchinson had to deny the newspapers' report that he was the "father of Ecology." As noted earlier, he always insisted that honor belonged to Darwin.[24]

Henry Mitchell wrote in that article about Hutchinson's science; for example, how lakes change over eons of time. But he wrote about his other interests as well. "I knew that Hutchinson talked about everything under the sun. . . . He spoke that night about dogs, and Neapolitan folklore, the cathedral of Ely and the costume of a Spanish Madonna, the mosaics of Monreale [in Sicily] (which he rather dislikes), the old days in Cambridge, and about homosexuality." On this subject Anne joined the conversation: "There is so much prejudice against homosexuality that nobody should add to it. . . . You know humans are so complex, you cannot take them apart as you can analyze things in a laboratory." She then picked up a poem of Auden's and read it to them. "You can study all the influences on a person but still the essence eludes you. As the poem says, science cannot quite dissect these human things, any more than you can get the square root of a sonnet."[25]

Both Evelyn and Anne were working hard on their books that fall. He reported that his volume 4 of the *Treatise* was "three quarters complete and much of the remainder is laid out in my head so I really have a good chance of completing the volume, which is most encouraging."[26] All of this progress occurred in spite of a hurricane, in which an oak tree was blown onto their car and electricity was out for longer than in any previous New Haven meteorological event.

Toward the end of 1985 Hutchinson was experiencing some new problems. Evelyn wrote Dorothea, explaining his neglected correspondence to her: he was suffering from a condition called "orthostatic hypotension." He described the immediate symptoms: "If you get out of bed in a hurry to answer the telephone you fall onto some sharp object. The first time I experienced this, the object was a chain that broke the base of my nose, the second time a rib that cracked." His book by now had grown to a thousand pages of typescript, but he added, rather sadly, "I am encouraged by being able to do this as in some ways I know my mind is less young than it was."[27]

After the Christmas season he went to a mathematical lecture given by a friend. He was worried beforehand, but afterward he wrote, "My difficulty with elementary arithmetic, which my friend Phoebe Ellsworth says I can get from a desk calculator, does not imply loss of much less elementary thoughts which you can't get in such a way. I am greatly encouraged.[28]

He later wrote that he was feeling all the benefits of his marriage to Anne. "She also gives me the capacity to see the details of her [black] variant of American culture in a way which can be both loving and critically observational, so what some people may find a bit surprising is turning out to be triumphantly successful." In addition, Anne was very good at giving parties, which often included food she had cooked (a change from Margaret) and musical events in their home. Although written to Dorothea, these encouraging words were probably spread around the family, some of whom were certainly among the "surprised."

Anne and Evelyn had given a Christmas party and another for his birthday. "This appears to give an enormous amount of pleasure," he wrote, "partly because Anne was so extraordinarily good at entertaining and because we seem to have an extraordinary number of friends." They both felt it very beneficial to mix up the races, "an area that is still inadequately practiced and one which, I suspect, is increasingly important."[29]

On January 20, 1986 Grace Pickford, Evelyn's first wife, had died in Ohio, where she had moved after her Yale retirement (see chapter 11). Ironically, Evelyn became very involved with working on her scientific obituary for both Hiram College and for Newnham College, Grace's college at Cambridge, stressing the extraordinary range of her zoological research. Although he and Grace were colleagues at Yale, for many, many years after their divorce, his feelings about her remained ambivalent; he never wished to talk about her to interviewers, including myself. Yet he sent the memoir he compiled to those of Grace's friends still alive whom he was able to locate.

Hutchinson's Kyoto Prize

In July 1986 Hutchinson received three telephone calls from Japan to inform him that he was a winner of the very first Kyoto award given by the Inamori Foundation. There is, of course, no Nobel Prize in Ecology, but the Kyoto award is sometimes described as the "Japanese Nobel Prize," and it involves a similar amount of money. These funds proved extremely helpful to the Hutchinsons. As Evelyn wrote to his sister, "The material aspects are certainly very nice, giving us a much more comfortable retirement."[30] Before he won this prize Evelyn was not nearly as wealthy as his family in England had imagined. Afterward he shared much of the prize money with members of his English family, including the youngest generation.

Evelyn was somewhat worried about his forty-five-minute Kyoto lecture. "I am sure I can do [it] in a way that will start up some good, but it will need very careful thought." He wrote, "It was . . . the kind of challenge that . . . I know Anne likes me to have—in moderation." He had to have his lecture written by August, however, and wrote 5,000 words one afternoon in their New Haven garden!

He was not free to discuss the Kyoto Prize as yet with their Yale friends. But when Dorothea confessed that she had not known about the Kyoto

40. Hutchinson with his wife Anne in England celebrating the
450th anniversary of Emmanuel College, Cambridge University.
Courtesy of Dorothea Hutchinson.

Prize, Evelyn related an interesting incident about it: the head of the editorial staff of Yale University Press had met a woman colleague, just back
from Europe, who had bought a copy of the London *Daily Telegraph* at
the Zurich railroad station. The paper announced that the Japanese award
had been given to a man whose name sounded familiar to her. The editor
immediately identified his close friend, Evelyn Hutchinson. Word was
spreading around the world.

Evelyn was looking forward to the trip. He wrote, "Kyoto is the most
beautiful city in Japan. The inhabitants all get drunk when the cherry
blossoms open, but we shall not have to cope with that in November." By

October he and Anne were getting ready to leave for Japan. Friends were inundating them with guidebooks. Dorothea had just had an adventurous trip to Spain, a country Evelyn knew not at all. "I am still as personally ignorant as when I wrote the first page of the *Clear Mirror* [fifty years before] except for three hours off a ship in Barcelona."[31] He would not have known Japan either if it had not been for the prize.

The next month Dorothea received a post card with Japanese stamps from Anne:

> Evelyn and I are enjoying an exciting experience with the people of Japan. The Foundation has done everything to ensure our pleasure. Eve did marvelous on his lectures and people have come from all over Japan to hear and see him.
> Love, Anne.[32]

Unfortunately, whatever Evelyn wrote to Dorothea from Kyoto, that most beautiful city, has been lost.

The talks he gave there, both his opening remarks and his more formal prize lecture, are, however, of considerable interest. He said in his opening remarks, more than forty years after the atomic bombing of Japan: "Here I stand as a member of a nation that perpetrated the greatest acts of violence ever let loose on another nation. There is little constructive I can say save to remark on the really extraordinary fact that we are here together as friends. It is a greeting of such friendship that I want to bring you; this I regard perhaps as the most important aspect of my celebration."[33]

He then went on to discuss his upbringing in the British academic tradition at the University of Cambridge. Both his father and uncle were scientists, professors, and masters of Cambridge colleges. From childhood he had met "casually" other scientific investigators, including three of Darwin's sons. He told the story of how he and his younger brother, then aged eleven and nine, had switched off the dining room lights while Sir George Darwin and wife were dining with their parents (see chapter 2). He told about their locking the dining room door, turning out the lights at the main switchboard, and throwing the key in the garden. "The subsequent parts of the story are repressed," he added. Quite a delinquent beginning for the elderly scientist who had just won such a major award for his work in evolutionary ecology! His early career from Naples to South Africa to Yale was covered in about two minutes, including thanks to those whose kindness and insight made it possible, Lancelot Hogben

41. Hutchinson in Japan in 1986 giving his Kyoto Award
speech. Courtesy of Dorothea Hutchinson.

and Ross Harrison. In the final paragraph of his "Introductory Remarks,"
he focused on his further education at Macy Foundation conferences in
New York from 1944 to 1953, citing among his "educators" there Norbert
Wiener, John von Neumann, Margaret Mead, and Gregory Bateson.

His formal Kyoto Prize lecture was titled "Keep Walking." In it he
talked about five matters concerning human evolution "perhaps of considerable importance to our species. There is clearly a long but very interesting walk for anyone who wants to embark on it."

This speech covered a lot of ground, from k- and r-selection (which he
explained) and its relation to human demography and politics, to speculation on human "helpers and homosexuals" in relation to kin selection.

The last section of his speech, entitled "Knowing What One Is Talking About," involved what Hutchinson considered the most important kind of "practical question with which we are faced in dealing with the inter-action of evolutionary and ecological biology with the practical affairs of daily life." He listed what he felt all societies want—freedom from war, hunger, and disease and respect for all the kinds of people living in it, as well as the "full satisfaction of spiritual wants."

For such a society, Hutchinson continued, one primary requirement must be that we know what we are talking about, thus the importance of education, particularly in science. "At the present time [1986] almost every political decision contains some reference to a partially scientific problem." He was concerned that proposed solutions be carried out in such a way that major ecological changes not prevent the acquisition of new knowledge. He pointed to the contributions of students of animal behavior, the ethologists working on many mammal species, including whales: "I feel strongly about the preservation of whales. . . . Their songs appear to be the most complicated animal productions . . . available for study." (Japan was still sanctioning the hunting of whales.)

In his final remarks he addressed a question that nearly twenty years later became a major subject of public controversy, genetically engineered organisms. His prophetic comments included:

> The extraordinary advances made in molecular genetics . . . suggest that a directive kind of change, already active in some plants and protists, will ultimately be available throughout the living world. We shall not only have artificial selection but also artificially directed variation and thus a whole artificial evolutionary process. Whether we shall be wise enough to handle this may be problematical.

That all sounds very familiar today, but his following comment is perhaps more surprising: "Direct interference with evolution, human or otherwise, will be possible in some parts of a culture in which politi-cally powerful people, notably the fundamentalists of North America, can maintain that the evolutionary process does not exist." Moreover, Hutchinson, himself a religious man, felt that the "approach of the fun-damentalists involved a position as antipathetic to religion as it is to sci-ence." No new scientific insights in this speech, perhaps, but it presents a long walk through Hutchinson's personal thoughts on science and society at the age of nearly eighty-four.

Hutchinson felt that his trip to Japan had been "infinitely worthwhile." Tom Lovejoy told him that the interview that the emperor had given him was one of the longest scientific audiences of his long reign. Hutchinson also thought the "whole story of his retention as a constitutional monarch opposed to war was very interesting." Although the emperor's interest in science was well known, he is not generally thought of as an antiwar ruler in the United States. But because the Allies during World War II had commanded that nothing inside the wall of the imperial park should be bombed, "in a most peculiar way the Emperor could be turned into a non-militarist head of state [after the war], backed by the majority of the people."[34]

Domestic and Political Affairs

It is clear that Anne and her "Eve" enjoyed the trip, and also that at his age and somewhat frail, Evelyn could not have traveled to Japan on his own, without Anne. Phoebe Ellsworth, whom they visited on their way home, discussed their visit at a later date. She wrote they were happy but "not a perfect fit . . . like a humorous 'Odd Couple,' both reveling in the hugeness of their differences." Evelyn told Phoebe that their marriage was in part like a course in "Race Relations 101."[35]

Evelyn's 1986 Christmas card was a photograph of a brass rubbing of Sir Nicholas Hauberk (constable of Flint Castle, 1396–1407). The brass was from 1407, but the rubbing was by Townsend Rich, who had given his rubbings to the Yale Center for British Art. Hutchinson sent these cards of medieval knights to Japanese friends since he thought the equivalent feudal culture in Japan not dissimilar.[36]

The year 1987 began with a party at their home for Evelyn's eighty-fourth birthday. They were inviting people not as a social obligation, he wrote, but only "because we are really fond of them." Evelyn was very busy answering letters that had arrived while they were away. Most of these were letters asking for recommendations! They came until the very end of his life. There were also thank you letters from members of the family in England, including Yemaiel, for the checks Evelyn sent from his Kyoto Prize. There was even a check for his sister-in-law Hannah's beautiful garden on the Isle of Wight, a National Trust garden. Another portion of the prize fund was used for repairs and improvements on Evelyn's house, and by Anne herself, mostly for clothes. She had expensive tastes.

The only luxury Hutchinson permitted himself from the Kyoto money was a Cambridge University doctor of science gown. He wrote that he would leave it to Emmanuel College, his Cambridge College, for "the next time they have an impecunious Sc.D." He asked Dorothea to go to a shop on a certain corner in Cambridge to ask about a secondhand gown, "bonnet," and hood.

On the political scene, he thought that United States–Soviet relations were perhaps about to thaw. He was trying to "get some helpful communications with the Russian scientists—the time seems ripe." Between poetic descriptions of a blizzard with the sun shining on snow is a political comment on the Reagan administration: "I don't know what impression of this American government you are getting, but on the spot it appears to be unthinking idiocy."[37]

A few weeks later Dorothea received another critique of the Reagan administration in a letter that noted the "appalling mess our president has made of his office." Hutchinson seems rarely to have discussed American politics with his Yale colleagues, but Dorothea received his deeply held convictions. The National Medal of Science (bestowed by the president of the United States), which he had turned down during the Nixon administration, was awarded to him posthumously when George H. W. Bush was president; his nephew Francis Hutchinson came to Washington and accepted it for him. It is not known whether Hutchinson himself would have accepted it.

Hutchinson's health began to fail in early 1987; he had several blackouts and falls. In a letter he wrote that he was undergoing physical therapy but was not yet fully recovered from the most recent fall and was in pain. He mentioned the mail that kept coming from the English relatives who received Kyoto Prize checks. "We have now received letters from all the great-nephews and nieces which has given us great pleasure, I think the greatest pleasure we have received from the award."[38]

The Yale commencement in June was a great success. Hutchinson's gown had arrived from England and was the only British doctoral gown in the procession that year. "Sadly, my friend Fred Pottle, who was the greatest living authority on Johnson and Boswell and who always came in an Oxford D. Litt. gown and hood, died two weeks before the ceremony."

Evelyn and Anne continued to travel. They were in Washington, D.C., for the annual National Academy of Sciences meeting. Evelyn was still involved in NAS affairs, writing to Dorothea, "We did have a delightful time in Washington and I have been somewhat occupied with matters connected with the reorganization [of NAS]."

In June Hutchinson wrote to his sister that they were about to set off for Georgia to visit Anne's family, her sister in Atlanta and her mother in Westminster. "I shall undoubtedly discover how much I don't know about my adopted country. Anne, being both passionate and extremely fair, will be a perfect guide." This turned out to be a very interesting "anthropological" trip for Evelyn, as well as an ecological one with a visit to colleagues at the University of Georgia. They also planned to go to England again and would "very likely include some France—Chartres is probably the place that I most want to see anywhere, and I know that Anne would like rather more of Paris than she has had" as well as a "taste of Scotland."[39]

At this time Hutchinson was also looking backward to World War II. He was not directly involved in the war since he was overage and did not become an American citizen until well into the war years. But he was puzzled as to why people had seemed to take the attitude during the war that he should continue his teaching and not be disturbed. He later discovered that this attitude had to do with the kind of scientific training he was giving graduate students, a number of whom, as young officers, had worked on important war projects for the U.S. Navy at Woods Hole. These included Tommy Edmondson, Gordon Riley, and Ed Deevey. He wrote, "I am told that in the war when there were almost no male students and the capacity of women had not been discovered, people used to whisper, 'Is there no chance of getting someone else from Hutchinson?'" He was also looking back and feeling "particular pleasure in four of my Ph.D. students being ultimately elected to the National Academy, the equivalent of the Royal Society."[40]

For Dorothea's eightieth birthday that year there was a big party in Cambridge. Anne and Evelyn sent her a necklace they had purchased in Kyoto. A birthday card with Monet's water lilies contained a letter from Evelyn: "These water lilies and the present Anne has chosen bring deep birthday love. I seem to remember years ago playing trains in the garden on a birthday evening. How we have changed and yet how much we are the same people."

Back in New Haven, Anne was shining as hostess, including a dinner for eight for a "very distinguished Japanese scientist." Then there was a party for fourteen, including Karen Glaus Porter, his former student, and her husband, Jim. Phoebe Ellsworth and her husband were there, too. Anne had a good sense of color and design. There were colorful cloth napkins, crystal, and flowers at these parties, as well as excellent food. Evelyn said about Anne, "The net result is that people just love her." Less cheerful news was that he had had a prostate test followed by the wrong antibiotic, which was making him blind. This problem was corrected, but other medical problems continued to plague him.[41]

Mary Catherine Bateson, the daughter of Evelyn's old friends Gregory Bateson and Margaret Mead, visited him. He had known her since her childhood. He wrote to Dorothea about the Batesons, who were "strict atheists, but their father, William Bateson, the post-Mendelian father of modern genetics . . . insisted that they study the Bible so as not to be empty-headed atheists." When Gregory Bateson died in 1980 he left an unfinished book, to be called *Angels Fear to Tread*, out of which Mary Catherine later constructed a joint book of hers and her father's. Hutchinson had just read it and found it "extraordinary."[42]

Evelyn felt he had to accept an offer to review the book and was finding it "fascinating and very difficult." He wanted Dorothea to say a prayer for his completing a different job, the fourth volume of the *Treatise on Limnology*. "Love from us in a fascinating world in which someone said of himself, 'I'm not retired, just tired.'"[43]

Evelyn now had an extended family in New Haven—Anne's grown children, Tracey and Michael Twitty; various other members of Anne's local family; and also Anne's small grandson, Michael's son. Evelyn was delighted with him. This was the first small child to be a frequent guest in his house since Yemaiel's stay during World War II.

On visits to England he had played with his brother Leslie's children on the Isle of Wight. Rosemary also remembered Evelyn taking her small daughter Anna "about four or five, and Frances Mary, her brother Francis's daughter, out with a big butterfly net and catching butterflies and looking at moths up on the Downs on the Isle of Wight one summer."[44] Hutchinson's wife Margaret was alive then, but she had stayed on the mainland while Evelyn came to visit the island. He continued his interest in Rosemary's and his nephews' children, sending them books according to their interests, finding special mathematics books for Frances Mary when she was older, and introducing Philip's young son, Harry, to natural history on a later trip to England. It was perhaps the greatest sadness of Evelyn's life that he had had no children of his own.

Christmas 1987 found Anne doing a "heroic job" of the Christmas presents and decorations while he was in a "rather somnolent condition," though his doctor found him all right for his age, nearly eighty-five. Medical tests had shown him to have a major vitamin B12 deficiency, which led to both extreme tiredness and confusion. He was given injections to correct this. In addition Anne took him, as well as her two children, to the island of Aruba in the Caribbean. Evelyn described the island to Dorothea as "an extraordinary island, still part of Holland with an odd language, part Spanish, part Portuguese, part Dutch, part English, wide beaches of coral sand and very clear water."[45]

The department gave Evelyn a special birthday party for his eighty-fifth, complete with a seminar on the senses of bees and an elegant lunch at the Lawn Club put on by his former student Tom Lovejoy, then working on conservation problems in Brazil for the Smithsonian. The guests included National Academy of Sciences biochemist Maxine Singer, "very much concerned with the whole field from molecules to tropical forests," and former graduate student Patrick Finnerty, "a very brilliant man, mathematics to tropical forests." In addition there was Hutchinson's friend, the celebrated Yale historian Jaroslav Pelikan. Evelyn commented that "it was all very interesting but at times very amusing, too."[46]

More illness followed, but in May 1988 he and Anne went to Washington, D.C., for the National Academy of Sciences annual meeting, where he saw many old friends and former graduate students. The meeting was overshadowed by the AIDS epidemic. A lot of work was being done, he reported, on the epidemiology of AIDS, with mathematical approaches similar to those used in ecology. Robert May, who was just at that time leaving Princeton for Oxford University, gave a brilliant paper. A major mathematical ecologist, May had worked with Hutchinson's student Robert MacArthur at Princeton before MacArthur's early death.

September 1988 found Evelyn and Anne back in New Haven after a return trip to Aruba. Anne liked that "curious volcanic and coral island" in the West Indies, and Evelyn found it a good place to write. They just missed Hurricane Gilbert, which passed over the island. It was preceded, he wrote, "by an enormous swarm of the sexual flying phase of ants, not yet identified."[47] He brought back much material for Yale's Peabody Museum.

Animal behavior remained an interest he shared with Dorothea. For example, he wrote: "I have been much intrigued by the new work on the social life of many seagulls and terns. After mating the females assort themselves into lesbian pairs. The males live a socially inactive life and all the work is done by the females. I cannot think what the advantage of this [is] but I am fascinated by certain human analogs."[48]

Hutchinson continued to share his interest in—or dismay about—American politics with Dorothea: "I have wondered much during the past few weeks what on earth you make of the American way of choosing—or perhaps purchasing—a president." It appeared recently, he wrote, that Reagan believed that trees cause atmospheric pollution and so should be cut down. About the upcoming debate between the two presidential candidates, Hutchinson commented, "It will be curious to see what lines they take, if *any*." He also commented on the ongoing cold war. He had received a letter from a Russian who elucidated a footnote in Hutchinson's *Introduction to Population Ecology*. "It is nice," Hutchinson wrote, "to have such a friendly letter from a place that is now widely regarded here as the middle of the bottom circle of Inferno."

Life seemed to be going reasonably well for Hutchinson. In his letters he commented on Dorothea's work with Hospice and discussed his own lectures and writings, notably his 1986 essay, "What Is Science For?" Anne continued to make a home for him, nurse him when he was ill, and entertain his old friends and the new ones they made together. She was more welcoming to his family than Margaret had been. Yemaiel had recently visited with her husband and son.

Ed Deevey, one of Hutchinson's first graduate students and a very well-known ecologist, died suddenly, and Hutchinson was asked to write a long obituary. "Sad but necessary to do something good. Thank god I can still write," he concluded. The 1988 Christmas season included social events with Anne's people, enabling Evelyn to "see how things are done in a rather different style, but not as different as I or she had thought." He wrote to Dorothea, "I remember the black hymn books that mother had in the top part of the piano, and how the black hero's [sic] of liberation were hero's to everyone with a liberal type of mind."[49]

In early 1989, just after his eighty-sixth birthday, Hutchinson was given a good medical report from his doctor. He was very relieved since he had a large number of things to do, including writing a preface to a book about Wicken Fen, the favorite collecting spot of his Cambridge days. But 1989 marked the end of his letters to Dorothea; his handwriting had become progressively more difficult to read over the past two years. It also probably marked the end of any happiness in his marriage to Anne.

As he grew older, Evelyn was highly involved in remembering his past life. One of his late letters to Dorothea read:

> I have had at various times very strong feeling about what a splendid family we all were. This has always been coming and going in my mind but became very intense when I saw the enlargement of that photograph of Father hanging outside the reception hall in Kyoto. . . . Have you been able to get a written transcript of your tape about Mother? This is the source material of history and I do not see that the good part should not be preserved as carefully as the bad.[50]

Dorothea, Hutchinson's sister, correspondent, and source for much information about him for this author, lived until 2001. At the age of ninety-four she was still going to Save the Children Fund (SCF) functions in Cambridge. She had supported that organization, for which her (and Evelyn's) nephew Andrew Hutchinson worked, with energy and dedication all around the world for many years. It was Dorothea who had established the "Friends of SCF" group in Cambridge. She had trained as a psychiatric social worker at the London School of Economics (LSE) just before World War II and had worked in London during the wartime bombing. Later she was instrumental in running the Cambridge Child Guidance Clinic and at one time teaching the Mental Health Course at the LSE. She was known as undauntable, and when she needed your help, she was "a lady one did not say no to."[51]

When asked how she would like to spend her ninetieth birthday, Dorothea recalled her expertise at punting on the Cam while at Newnham College, Cambridge. A punting trip for Dorothea on the Cam it was—on her birthday—from Cambridge to Grantchester.

Dorothea was a wonderful hostess to Evelyn in Cambridge after Margaret's death and later to both Evelyn and his third wife Anne. Evelyn's letters to Dorothea are a valuable resource. It is extremely unlikely that her letters to Evelyn in their last years were destroyed; Evelyn kept everything. But thus far they have not been found.

"An Interracial Marriage Is Hard Enough without Alcohol"

What happened in the very last year of Hutchinson's life is something of a Rashomon story. There are several different versions. I (the author) arrived on the scene in September 1990 for a sabbatical year at Yale, working on the history of ecology under the auspices of the Section of the History of Medicine and Life Sciences at the Yale School of Medicine. I had planned to interview Hutchinson, to work in his archives, and to learn about his life from his former students, his colleagues, friends, and family. I interviewed a great many of these people, including Hutchinson himself, but even among those who were actually on the scene, the accounts vary.

It is clear that both Anne and Evelyn had been declining physically. In summer 1989 Evelyn had a series of what were probably small strokes (TIAs) and was losing his short-term memory. Anne was probably suffering from liver damage due to her heavy alcohol consumption. A close friend of Evelyn's, psychiatrist Stanley Leavy, reported having this conversation with Anne in late summer 1990. Anne was just getting into her car to drive. "I [Leavy] said, 'Have you been very ill?' She was just wasted—terribly. She said, 'Oh no, no, I'm fine.' That's what she always said, even when she was manifestly bedridden. She had failed for a long time." Leavy's wife, Margaret, added "You came home that day and said, 'That woman is going to die.'"[52]

Six months later Anne was hospitalized. Her daughter Tracey refused to believe that her mother was truly ill. Stan Leavy said that her family members had chosen to go along with Anne and had closed their eyes to what was happening. But Evelyn's friends were worried about what was happening to him. No one was taking care of him, which he clearly needed.

The Leavys were sure that Anne was seriously ill. They had known her a long time, before she had any real connection with Evelyn. They described a delightful New Year's Eve party at the home of Arthur Galston. Anne was then married to biologist Richard Goldsby, the master of a Yale college. Stan Leavy described Anne as a "vivacious, sparkling person who was fun to meet. I remember dancing. We danced that evening. I never saw her again until she showed up as Evelyn's fiancée. I was coming home . . . and I looked down Whitney Avenue near Canner Street and saw Evelyn walking, almost skipping hand in hand with this woman and I couldn't believe my eyes. . . . I thought I had hallucinations."

When I asked Leavy whether he thought she was good for Evelyn, he answered, "No question about that. His enthusiasm for her could not have been feigned. She was a sprightly interesting person. She took great interest in me because of my Jewish origin, which she rightly thought was kind of an affinity for oppressed races, which she made quite plain that she felt herself to be. She never showed anything but friendliness toward us."

Margaret Leavy agreed, "We enjoyed the parties at their house, which were fascinating." The mixed-race parties were "fantastically interesting," she said. Notable people like Willie Ruff, the jazz musician, appeared. When Evelyn first introduced Anne at Saybrook College, she was dressed gorgeously, the Leavys said. Stan added: "He presented her dressed in all her finery. I think at that time he was not covered with gravy stains as later on, and he said, 'I want you all to know Anne who has consented to look after me the rest of her life.' *Her* life! As if it were prophetic, or as if he were going to be immortal. We were all too polite to comment on it, until afterwards, but it was a startling thing."

This author met Anne only once. I was invited to the Hutchinsons' house in the spring of 1990 to talk to Evelyn about plans for my sabbatical at Yale. I rang the doorbell, and two black women appeared. One was Anne and one was, I thought, the housekeeper. I wasn't sure which was which. I had heard that Anne was very well dressed, but she was in pink tights and bedroom slippers. Anne helped me over this embarrassment, and was very nice to me, very welcoming. I had no idea she was ill.

The Leavys gradually became aware that Anne was an alcoholic. There had been rumors, and Stan said he could smell the alcohol. And by then he could see the neglected state Evelyn was in. "A woman more than thirty years younger should have some concerns about a man's appearance," Leavy said.

At that time Stan and Evelyn often walked home together after the Sunday service at the Episcopal church, which Leavy also attended. This took a long time as Hutchinson was by then very frail. Stan said that even late in Anne's life, when she was no longer giving Evelyn any care, he would express kindly thoughts about her. Evelyn told Stan about Anne's connection with very distinguished people in the black middle class and added that he could give a very good course in anthropology on the black and white family. Stan said, "His mind was always receptive to anything new going on, and to come firsthand into the black world was to him a thrilling experience."

It is important to keep in mind these positive aspects of their marriage because its last days were so sad. Hutchinson himself had earlier spoken of the "bad days" with Anne to his friends, but he declined to describe them. Members of his English family, Paul Hutchinson and his wife Marie, visited and stayed in the house. They later described to me how bad conditions had become.[53]

By September 1990 a group of Evelyn's friends and neighbors were meeting to talk about their concerns. They included Anna Aschenbach, Evelyn's longtime assistant; Christina Spiesel, who had painted his portrait and had become a close friend; his neighbor Margaret Colliton; and Stan and Margaret Leavy. Margaret Leavy said, "We just met together out of desperation because Evelyn was not getting fed, and nobody was allowed in the house by Anne."

The group met as a committee and considered what to do. They agreed to write a letter to a close friend of Anne's and then to the family in England, addressed to Dorothea. Finally they decided that some member of Evelyn's family must come. Evelyn's oldest nephew, Francis, did come over from England, but by that time Anne was in the hospital.

Anna Aschenbach sent a long letter to all the Hutchinsons and gave a copy to me. She wrote about the whole period in which she was his assistant, starting in 1976, and about Evelyn's marriage to Anne. She described the current situation clearly:

> An interracial intergenerational marriage is hard enough without alcohol. . . . It took courage for Evelyn, who never complained, to tell me and others how bad things were for him at home. This alerted me to Anne's illness. A group of us tried to work things out with Anne's children . . . saying there were two ill people there. . . . We persevered in our efforts for his welfare, my part being to answer his request to find some intellectual companionship for him every day since Anne rarely came out of her room.[54]

Anna succeeded extremely well in regard to Evelyn's request. Many people came to talk to him, particularly about science, including Yale professors Charles Remington and Willard Hartman, and also Stan Leavy. I (the author) had to make special appointments to talk to him. Evelyn still came into the office, though I occasionally went to see him at his house. He now had an electric teapot for the office, replacing his long-used Bunsen burner; he still had his favorite Bordeaux cookies for the daily teas. Evelyn had his good and his bad days intellectually during the fall of 1990. Sometimes he was completely lucid for a two-hour talk, sometimes not.

Anne finally went to the hospital of her own accord but with the psychological help of her close friend Jean Bowen on October 19, 1990.

Anna Aschenbach had contacted Jean in September to ask for her help. Jean had been Anne's friend since childhood, through college and the civil rights movement, and had a position in the New Haven city administration that related to day care centers. Anne's grown children from her first marriage, Tracey and Michael Twitty, moved into the Hutchinson house and took responsibility for Evelyn's care. This was not easy; his needs were getting more acute.

The Leavys, who visited Anne in the hospital soon after she got there, said she seemed to look better. She left the hospital for the nursing home to recuperate but died in December. Her death was truly unexpected; everyone was in shock. Her children talked about her condition not as alcoholism but as liver failure, which probably was the ultimate cause of her death. Anne's funeral, or memorial service, was in her own downtown New Haven church. Many of her family and members of the congregation came, as well as Evelyn, his nephew Francis with his wife Joyce, and some of the Yale community. I was there as well, and it seemed to me, during the funeral, that Evelyn was in a complete state of shock. He had to be helped out of the room. Evelyn later told family members that he did not think Anne had had a proper funeral. There was music appropriate to the Christmas season, but surprisingly little was said about Anne as a person or about her work, although the minister had known her for a long time.

The committee of friends that were devoted to helping Evelyn was relieved when members of the Hutchinson family came from England to assess the situation and decide what to do. The house was in a terrible state of neglect. Members of the Twitty family were still there when Rosemary Dodd and her husband Peter came to stay in the house. They did not deny the help that the Twittys had been to Evelyn but felt that it was difficult for him to live in the house. Rosemary recalled one day during their stay: "There were times when it was very difficult to cope with . . . and one day it was just like Piccadilly Circus." They had taken Evelyn to church, returning to find the house full of Twittys and relations and children and Anne's little barking dogs and the noise level horrendous. "There was a radio going in one room, a tele going in another. There was a video recorder going in yet another. We had to bring this old crumpled gentleman into this Piccadilly Circus."[55]

Rosemary and Peter had originally thought it might be better for Evelyn to stay in New Haven rather than to bring him back to England as Francis had decided, but they changed their minds when they saw the situation. In spite of Evelyn's many friends at Yale and in the community, there was really no one who could care for him full time. As a result, Evelyn did go back to England in January and remained there. His sister Dorothea, Francis and his wife Joyce, and Yemaiel and husband Ben were to care for him there.

The Hutchinson house had to be cleaned out. Andrew Hutchinson, another nephew, arrived from England. He, with the help of this author, cleaned out the books and papers, some of which went to England, some to Yale. The remaining items, though not the house itself, went to the Twitty family. The house was eventually sold.

For several years while still in New Haven Hutchinson had been having transitory ischemic attacks (TIAs), often resulting in falls and sometimes in broken bones and bruises. He was aware of these intellectual lapses. When I (the author) was interviewing Hutchinson in fall 1990, he would start to tell me a story about a recent event and would get lost in the middle, much to his dismay. But some days, even at that late date, he would talk very lucidly about some aspect of science that interested him. I would bring letters from his extensive archives at the Yale University Library and ask him about the people and events mentioned in them. He could recall a collecting trip to the Hebrides in the 1920s with perfect clarity. One day we had a fascinating long conversation at his house, but on other days as Anna reported, he retreated into silence or fell asleep in the office. He did manage to finish almost all the writing on volume 4 of the *Treatise on Limnology,* but it took his editor, Yvette Edmondson, Anna Aschenbach, and a whole team of devoted biologists and other friends to finish the drawings and references and many other aspects of that book after he went back to England.

Just before Hutchinson left for England a stream of friends and colleagues, including several beautiful women, came to the house to say good-bye. I was there, too. Sadly, most of us never saw him again.

After Evelyn returned to England, his nephew Francis Hutchinson and Joyce sent an update to the family and others: "Evelyn arrived in London on January 7th with Peter and Rosemary Dodd, very tired after a flight that was delayed for five hours. Yemaiel and her husband, Ben Aris, met them and took Evelyn back to their London home. He was well taken care of by Yemaiel and Ben, an actor by profession, and by a live-in nurse named Sonia, who was kind, sensitive, intelligent and very importantly— attractive."

On the weekend Dorothea came to London from Cambridge and all of them had dinner at Yemaiel's house. "Evelyn was obviously thrilled to be sitting next to Do [his sister Dorothea] and chatted to her throughout the meal." He apparently was in good physical and mental shape. But a week later Evelyn was not well and was taken to the hospital for tests. They thought he might have had a minor heart attack. Although he was still in the hospital, Joyce wrote, "I can assure all his friends that Evelyn has been surrounded by love and constant attention since his arrival, and I think a feeling of coming home has been quite a strong emotion."[56]

Yemaiel said that when he first arrived at their home he was very tired but "mentally he was fine; we would have long conversations." On January 14 they took him to the British Museum of Natural History and met with a young insect curator, who tried, without success, to find a particular water bug, a corixid, that Evelyn wanted to see. Evelyn, in a wheelchair, was taken to see the new ecology exhibit, intended for popular consumption, but was not favorably impressed by recent changes in the museum in that direction, particularly at the expense of curators for the collections.

Yemaiel and Ben thought he would never be strong enough to really get back on his own, but they wanted him to make his own decisions. But then he became ill. He appeared blue, and they thought he had a heart problem. After the hospital tests they took him to an expensive private clinic in Wimbledon. He rallied but "he was painfully aware that he wasn't thinking straight." Both Yemaiel's and Ben's mothers had suffered from senile dementia. "Sometimes they think wonderfully and everything is lucid, and sometimes not enough oxygen is getting to the brain."[57]

After his return from the Wimbledon clinic, Evelyn went to live with Francis and Joyce in the upstairs room in their London house on Battersea High Street. He and Joyce became good friends. Evelyn became very much interested in Joyce's work with professional young women. "Evelyn said the most important thing in life is tenderness. . . . I would ask questions about his feelings: 'Can you remember how you felt when that was happening?' Because in his book, *The Kindly Fruits,* there's nothing about how he felt about things." They had to have a nurse to stay with him, and every new nurse read him his autobiography. "He enjoyed it; he loved it. . . . It meant that the nurse knew aspects of his life."[58] Francis wrote that Evelyn's condition was very changeable during his stay. Some of the time he was "very lucid and very much his own amusing self," but at other times confusion reigned and memory went. "Fortunately Evelyn's humor and politeness came through," which made taking care of him much easier and less stressful.

Early in March they took him to visit his sister Dorothea in Cambridge. He thoroughly enjoyed this visit, especially being wheeled around the streets and "popping into colleges to revive memories."[59] He also lunched at High Table at his own old college, Emmanuel. The family decided to try a month's stay in Cambridge at a nearby residential home so that he could see more of Dorothea and his Cambridge friends. During this time David Furth, a young entomologist friend who had known Hutchinson at Yale and was currently working in London, came to Cambridge. He took Evelyn to his beloved Wicken Fen in a wheelchair and photographed him there. See page 157.

The stay in Cambridge was cut short by a bad chest infection, which took him to Addenbrooke's Hospital there. The infection was cured but his confusion was not. It was thought better that he should be back in the family. Francis and Joyce picked him up in late April and took him back to London with them. At that point it was decided that he would stay in England permanently with the family to care for him, and Evelyn, who had originally wanted to return to New Haven, was happy with that decision. But it was not to be for long.

Old Friends

Hutchinson did have visitors during his last stay in London. Alison Jolly, a former Yale graduate student and an expert on the lemurs of Madagascar, came to see him. They were good friends. She has written several books about lemurs, including her early *Lemur Behavior: A Madagascar Field Study*.[60] Hutchinson kept up with former Yale graduate students, even with those for whom he was not the major professor. He often wrote to them about their work. Alison was one of these. She, like Hutchinson, is concerned about preserving biodiversity, especially in Madagascar. While teaching at Sussex University in England, she recently wrote about Hutchinson in *Nature*:

> Hutchinson's ecological theories were widely acclaimed. His clutch of honorary doctorates culminated in the scarlet robes and Latin oration of the University of Cambridge. . . . But Hutchinson's students were chronically bemused. . . . He set us impossible questions: Imagine an utterly

isolated island in a constant environment. How big would it have to be to permit indefinitely ongoing evolution?[61]

Ruth Patrick, an old and dear friend and scientific colleague, came to see him at the hospital two days before he died. Patrick, an important researcher of the ecology of diatoms and of lakes and rivers even in her nineties, had first met Hutchinson in her twenties. She had written to him at Yale in 1938 from the Philadelphia Academy of Natural Sciences, where she was an associate curator: "Dear Dr. Hutchinson, I have read with much interest your paper in the October issue of the *Transactions of the Connecticut Academy of Arts and Sciences*. As I have done quite a bit of work on the diatoms of lake beds, . . . I would appreciate a reprint of your paper."[62]

Hutchinson sent the paper and became interested in her innovative diatom research. They later worked together on lake paleoecology and remained close friends and scientific colleagues for more than sixty years. Ruth Patrick was, like Hutchinson and several of his students, a member of the National Academy of Sciences. I (this author) met her there and later interviewed her in her office at the Philadelphia Academy of Sciences. She had just returned in her high boots from a river research trip. She was well into her eighties. Ruth Patrick was also instrumental in raising funds for the completion of Hutchinson's fourth volume of the *Treatise on Limnology*.

Hutchinson received many letters after his return to England. Many friends, including former students and other scientists, wished him well. There is a letter from Professor Shoju Horie from Japan that Evelyn answered from Francis's and Joyce's house on Battersea High Street: "Very many thanks for your letter of 30 December 1990. I am at the moment attempting to put the last finishing touches to my book in England. . . . My final volume will be the end of my career, and yours will be the beginning of a new approach. I am delighted to know that you are doing this."[63]

Another of Hutchinson's friends, S. Dillon Ripley, wrote to him at the end of March 1991, when Hutchinson was staying near Dorothea in Dee House in Cambridge. Hutchinson and Ripley were old and good friends; they met during Ripley's undergraduate days and were later faculty colleagues at Yale before Ripley moved to head the Smithsonian Institution. It is an interesting letter about Ripley's cruise with his wife, Mary, in the Indian Ocean, and their good luck on being able to land on the island of Aldabra.

Ripley's letter to Evelyn, detailing Ripley's part in the "saving" of Aldabra, is quite extraordinary, as also noted by Joyce Hutchinson. Although
it goes on for several more pages, particularly about bird preservation in
the Seychelles, it seems to have been written for posterity. At the very end
of the letter Ripley expressed his hopes that Hutchinson's lodgings would
be "happy and successful" and sent affection "from us all." Dillon Ripley
was apparently entirely unaware that his old friend had only a short time
to live.[64]

It was to be a short time indeed. Joyce and Francis truly cared about
him, but they were saddened by his state of mind and body. Joyce said,
"We loved him dearly and we didn't want him to die, but every time he
had a chest infection, less oxygen went to the brain and each time the
brain was a little more damaged. The only time you could hold his attention was when you were talking to him, involving him in conversation
directly." "Could you do that right to the end?" I asked. "Absolutely."[65]
Hutchinson's long and full life ended on May 17, 1991, at age eighty-eight.

There were obituaries in the London papers and a funeral to which the
family came. Tom Lovejoy, one of Evelyn's last graduate students and a
close friend since his student days, came for the funeral from the United
States on very short notice. He stayed with Francis and Joyce. A week
afterward, Tom wrote them from Washington: "You took a stranger into
your home, made him part of your family . . . and made a moment I knew
had to come but hoped would never, easier and more beautiful than I
could have believed." He added that by the end of his stay, which was only
thirty-one hours in all, he "knew why you in particular had looked after
Evelyn, and indeed how suitable and wonderful that must have been for
him."[66]

It was indeed a tribute to Francis and Joyce Hutchinson, as well as to
Yemaiel and Ben, that Evelyn was taken into their homes and cared for
so well. Evelyn's niece, Rosemary, and his nephews, including Francis,
Andrew, and Paul, all helped. They had not known him well before Margaret died in 1983. In that sense, all, apart from Yemaiel and Dorothea,
were new friends. But new friends with a real sense of family; they cherished Evelyn. Many others of his friends in America and elsewhere would
have wanted to be there for Hutchinson's funeral but had to wait for the
memorial services in New Haven the following fall. A great many of his
former students, colleagues, and friends returned at that time to Yale,
where kind words, remembrances, and Hutchinson exhibits abounded.

Many more letters arrived in London, some to Evelyn that arrived too late, but mainly condolence letters to Francis and family. I quote only one, from a very well-known ecologist, John Cairns: "Immediately upon her return from England, Ruth Patrick passed on the shocking news that G. Evelyn Hutchinson had died. Evelyn was one of the people I admired most in the profession. . . . Although this is a time for expressing sympathy to you and to reflect on the loss of one of the great scholars of all times, it is also an occasion to celebrate a rich and meaningful life."[67]

CHAPTER EIGHTEEN

Concluding Remarks:
Hutchinson's Legacy in Ecology

Evelyn Hutchinson, one of the twentieth century's most notable scientific polymaths, was a polymath within his chosen field of ecology as well. His work appears in nearly every chapter of Robert McIntosh's book *The Background of Ecology: Concept and Theory*.[1] Hutchinson, as noted earlier, when called the "Father of Ecology," always insisted that that title belonged to Charles Darwin. Darwin does indeed appear many times in McIntosh's book, but without Hutchinson's ecological breadth. Hutchinson appears in the chapters entitled "Limnology," "Animal Community Dynamics," "Quantitative Community Ecology," "Population Ecology," "Theoretical Ecology," "Competition and Equilibrium," "Ecosystem Ecology," "Ecologists as Philosophers," "Ecological Theory and Evolution," and "Theoretical Mathematical Ecology." Hutchinson is not present in the chapter about terrestrial plant ecology, but he did some research even in this area, on club mosses.[2] If we were to include his students' research, we could add additional ecological fields, for example, marine ecology and behavioral ecology.

As far back as 1940 Hutchinson was the leader in changing current concepts of ecology. Frederic Clements and Victor Shelford, leading lights of the 1930s in plant and animal ecology, respectively, together produced a text, *Bio-ecology*, with which they had hoped to integrate these two then largely separate fields.[3] To quote McIntosh, "This effort had little success, for reasons detailed by Hutchinson."[4] In reviewing for the journal *Ecology*, Hutchinson wrote:

> The general principles that are supposed to emerge are, however, mainly classificatory . . . [and] appear as a set of rules for the construction of a language. This may be inevitable in the present state of the science; it is, however, uncertain that [this] language of ecology will ever become a universal ecological tongue. . . . In dealing with competition and biological populations, the largely alternative language of mathematics is neglected.[5]

Hutchinson pointed out the "total neglect of certain very important approaches to the subject" by these two ecologists, including the metabolism of communities; photosynthesis is hardly mentioned in the book. "This neglect of the biogeochemical approach is due in part to the authors' insistence that the community and the environment must be separated and should not be considered as forming part of the same ecological unit." He continued, "The same fundamental attitude leads to a dissociation of the pelagic and benthonic communities in the sea as separate biomes."

All of this sounds obvious now in any discussion of ecosystems, but it was not so apparent in 1940. It was, in large part, Hutchinson and his students who took up this agenda in the 1940s and 1950s and turned ecology from a largely descriptive into a theoretical and experimental field of science, including the use of the language of mathematics. Of course, Hutchinson had his forebears. As noted earlier, he credited Charles Elton's book *Animal Ecology* and its discussion of trophic levels as a major influence early in his career.[6] Clements had carried out experiments in plant ecology. But it was Hutchinson who brought the mathematical ideas of Volterra, Lotka, and Gause to American ecologists.[7] It was Hutchinson and some of his early graduate students who incorporated biogeochemistry, including the cycling of carbon dioxide and other nutrients in the ecosystem, into ecology.

Hutchinson and his postdoctoral student Raymond Lindeman revolutionized and quantified ideas about transfer of energy in food webs, initiating ecosystem ecology. Lindeman's work, though now thought of as being in the mainstream of ecology, was originally conceived as belonging to the then quite separate field of limnology since his doctoral research was carried out on a bog lake in Minnesota.

Limnology was the field in which Hutchinson was originally best known, particularly in the early decades of his research. After arriving at Yale in 1928, he published on South African pans, Linsley Pond and other Connecticut lakes, high-altitude lakes of Ladakh (North India), and the arid lakes of Nevada. These studies were interwoven with his and his graduate students' biogeochemical studies and with ecological theory. After the publication of the first volume of his *Treatise on Limnology,* he was probably the best-known American limnologist; three other volumes, the last one published posthumously, followed.[8] Limnology by then had become a highly quantitative science closely allied to and even part of the broader field of ecology.

Mathematical and statistical methods, which Clements and Shelford had had "grave doubts" about applying to the complex causes and effects found in ecology, were developed by Hutchinson, and by Gordon Riley and others of his students, especially Robert MacArthur. In the 1950s and 1960s they were applied to population ecology by many ecologists. A new ecology, more theoretical, more quantitative, more experimental, had been born, all trends that continue today, in spite of, or perhaps because of, considerable controversy among ecologists.

Two fields that Hutchinson was instrumental in starting, systems ecology and radiation ecology (see chapter 9), became "big science," especially when a great deal of federal AEC funding became available. One student of Hutchinson's, H. T. Odum, did his Ph.D. research in biogeochemistry but later became a major researcher on a variety of big ecosystems projects. The International Biological Program (IBP) in the 1960s and early 1970s literally went out on a limb, enclosing whole trees, sometimes even whole forests, whose metabolism was measured in the IBP productivity studies.

Hutchinson was wary of becoming involved in big science involving many investigators and vast sums of money. He preferred smaller projects in which he and his students could be in control. In any case he always managed, from his earliest days as a teacher of graduate students at Yale, to find funding both for himself and for his students.

Hutchinson's multidimensional niche concept was introduced in his famous 1957 paper "Concluding Remarks," to which the title of this chapter refers (see also chapter 14). It is probably his most important paper because of its influence on several generations of ecologists and their research. Although it was Rachel Carson who, more than anyone else, brought the term "ecology" to the attention of the more general American public, it was Hutchinson's view that ecological niches can be quantified. His concepts of niche diversification in a great variety of communities and ecosystems moved ecology rapidly ahead. It also spawned a great many Ph.D. theses and ecological papers. These included such mathematical methods as competition equations and equations for determining niche breadth and overlap, statistical methods like ordination, and quantitative analysis of experimental manipulation.

Hutchinson's 1959 paper "Homage to Santa Rosalia or Why Are There So Many Kinds of Animals?" (see chapter 14) spurred much interest in understanding biodiversity.[9] It is not possible to watch a superbly photographed episode of the television program *Nature*, particularly one of the tropical rain forest, without being bombarded with ideas of niche diversification and of the dependence of the great variety of animals, including insects, on the great variety of plants, ideas put forth in this paper.

Both the understanding and the preservation of the earth's biodiversity have been longtime themes of Hutchinson and also, notably, of E. O. Wilson. Both were ardent promoters of the preservation of biodiversity; Wilson continues to carry on this work.[10]

The word "niche," whose rhyme (in English) with "itch" we once had to impress upon our ecology students, is now blithely used in all kinds of popular presentations. Niche concepts existed in ecological literature long before Hutchinson's 1957 paper. Hutchinson himself, in a characteristically erudite page-long footnote in his 1978 *Introduction to Population Ecology*, tracked the earliest ecological use of the word "niche" back to Roswell Johnson.[11] But it was Hutchinson who launched the quantitative understanding of the concept and its subsequent productive career. Indeed, ecologist Simon Levin, in his 1999 book *Fragile Dominion*, called Hutchinson "the most important figure in American ecology." He also lauded Hutchinson as teacher of the following generation of ecologists.

Hutchinson was devastated by the death of his brilliant student Robert MacArthur at only forty-two. As we have seen, MacArthur wrote insightfully about Hutchinson's influence as his teacher. Hutchinson went from MacArthur's mentor to his codeveloper of ecological theory; they wrote joint papers. But MacArthur went beyond Hutchinson in the development of mathematical ecology, especially population ecology. E. O. Wilson and Hutchinson in their *Biographical Memoir* about MacArthur for the National Academy of Sciences wrote that MacArthur, who quite early in his career had an almost unique status as a "mathematician naturalist," eventually reformulated many of the "parameters of ecology, biogeography and genetics into a common framework of fundamental theory," all leading to the unification of population biology.[12]

Martin Cody and Jared Diamond in their 1975 MacArthur memorial volume wrote that the remarkable era of ecology began in the 1950s and changed ecology into a "structured predictive science that combined powerful quantitative theories with the recognition of widespread patterns in nature."[13]

Hutchinson wrote his own paper in this commemorative volume, entitled "Variations on a Theme by MacArthur." Here he quoted one of MacArthur's last papers: "Scientists are perennially aware that it is best not to trust theory until it is confirmed by evidence. It is equally true, as Eddington [and Hutchinson also] pointed out, that it is best not to put much stock in facts until they have been confirmed by theory."[14]

MacArthur had written that ecology was now in a positive position, but that both the facts and the theories had serious inadequacies, which were stumbling blocks to progress. Hutchinson, in his paper, chose to illustrate these problems in several areas of ecological theory, including that of competitive exclusion. Here, he thought, facts and theory had worked out well, but in the area of, for example, population cycles, that had not yet happened. It was the task of the theoretical ecologist, Hutchinson wrote, to investigate "all possible models," whereas the field or experimental ecologist's task was to find out which of these models or possibilities are realized in nature. Hutchinson, however, warned that although many ecology students of the 1970s were able to handle the mathematical aspects of the subject, ecological education must also provide a "wide and quite deep understanding of organisms." It was in this connection that Hutchinson made his classic remark, "Robert MacArthur really knew his warblers."

MacArthur's research had its roots in the natural history of birds. From this, MacArthur was able to "crystallize observations into neat though simple mathematical packages" and, in turn, to inspire a new generation of young ecologists.[15]

"Santa Rosalia Was a Goat"

Controversies over mathematical ecology became vitriolic in the late 1970s. Roger Lewin summarized these controversies in a 1983 *Science* article entitled, "Santa Rosalia Was a Goat." This title of course referred to

Hutchinson's classic paper "Homage to Santa Rosalia or Why Are There So Many Kinds of Animals?" Roger Lewin wrote:

> Evelyn Hutchinson set the fuse to the explosion of competition theory in 1959 when he published this paper. The shrine of Santa Rosalia is next to the Salerno (Sicily) pond in which Hutchinson observed the two similar but separate species of corixid water bug which were a starting point for his paper. Santa Rosalia's bones, reputed to have great curative powers, later turned out to be those of a goat![16]

Both MacArthur and Hutchinson viewed competition, as well as mechanisms for the avoidance of competition, whether in warblers or in water bugs, as essential processes in the development of community structure. Lewin described the debate challenging the importance of competition in the organization of natural communities as a debate "as acerbic and acrimonious as any that has stirred the combative instincts of academia." Daniel Simberloff of Florida State University, a major player in this debate, had declared that much of ecological theory (as espoused by Hutchinson and MacArthur) was a great waste of time for a whole generation of ecologists. Simberloff, ironically, had had MacArthur on his doctoral committee.

Jonathan Roughgarden of Stanford University, on the other hand, worried that this extreme antagonism in the rhetoric about theory had caused a lot of bitterness and would perhaps discourage students from learning the mathematics that is necessary in the study of ecology.

The debate was not entirely about whether competition is an important process in community development; that debate is an ongoing one. The truly acrimonious debate was more about how to do science, a subject Hutchinson discussed in "What Is Science For?" (see chapter 16).

Simberloff, and some though not all of his colleagues at Florida State, espoused the Karl Popper approach to scientific hypotheses. Briefly, ecologists should not test hypotheses by looking for data that is consistent with the theories but by hypothesis falsification. This led these Popperians to much use of the null hypothesis, which has led to "a great blossoming" of null models in the ecological literature. If the ecological patterns observed, they insisted, are not different from community patterns that can be explained by random associations, then the observed patterns do not need biological explanations. According to Lewin, Simberloff and colleagues relentlessly proselytized the Popperian view of science.

The comments of Michael Gilpin and of Jared Diamond on Simberloff critiques and null models were not so oblique. In fact Lewin called the Diamond-Simberloff arguments (in print!) "more than a little intemperate by the standards of most scientific literature." He reported that these two ecologists no longer communicated directly.[17]

Some critiques were well taken. MacArthur disavowed his own broken stick model for species distributions, but other ecologists continued to use it. Simberloff was also probably correct that Hutchinson's 1.3 rule (referring to relative sizes, for example, of the feeding apparatus of coexisting corixids or the beaks of Galapagos finches) for avoiding competition had outlived its usefulness. Simberloff and William Boeklen in a 1981 paper, "Santa Rosalia Reconsidered," talked about the "tenacity of the tyranny" of such ideas, presumably referring to the idea that the saint's bones even after being known to be those of a goat were still thought to have the same curative powers. Mixing animal metaphors, Simberloff and Boeklen called Hutchinson's 1:3 rule "always a red herring."[18] In any case the "broken stick" and the "red herring" engendered a great deal of ecological research, much of it valuable.

Most of the debate participants were able to present their views in an *American Naturalist* symposium. Arguments and accusations continued to surface. Roughgarden and others accused Simberloff, Strong, and Edward Connor in their null models of viewing the world as devoid of competition and devoid of structure. To this Simberloff replied, in part, that he never said that competition wasn't important in generating patterns in nature.[19]

To some, such as leading mathematical ecologist Robert May, the problem seemed to emphasize the need not for null models but for alternative explanations for the patterns observed in community structure. May, Colwell, and other well-known ecologists did point out positive effects of the debate in terms of ecological methods. As Thomas Schoener commented, it had caused ecologists to question their assumptions and examine their data-handling procedures. But John Terborgh, another Princeton ecologist, warned that we should apply null models with circumspection. Otherwise we may submerge important biological patterns in a "statistical mirage."

Hutchinson said much the same in different words. Although he didn't mention the Florida State ecologists by name, he referred to "a misunderstanding and perhaps a too great enthusiasm for the ideas of Sir

Karl Popper" which leads to the impression that the "purpose of scientific research is to falsify the other fellow's hypothesis." This would establish so many negations that it may lead to "a universal null hypothesis that everything is due to chance."[20]

The Current Debate

Hutchinson continued to publish after 1983, his eighty-first year, and finished the writing of the fourth volume of his *Treatise on Limnology* shortly before his death in 1991. But what of his legacy today, at the beginning of the twenty-first century?

One of the current debates in ecological theory is a somewhat different one, but it also involves niche theory and is clearly relevant to Hutchinson's legacy. It concerns the ways in which ecological communities develop and maintain themselves. A 1999 book by Stephen Hubbell, *The Unified Neutral Theory of Biodiversity and Biogeography*, has been important in this debate.[21] An ecological symposium at the 2005 International Botanical Congress in Vienna addressed the question: how we can best account for biodiversity and species composition of ecological communities?[22] Biodiversity is defined by Hubbell (and others) as species richness as well as the relative abundance of the species involved, their commonness or rarity. The alternative paradigms in this particular symposium addressed these processes in terrestrial plant communities. Hutchinson's ecological theories, particularly in relation to the niche, are certainly germane.

The two paradigms, which the symposium sought to integrate, are the niche-assembly theory and the dispersal-assembly theory. The first paradigm views communities as groups of interacting species whose presence and relative abundance depend on "assembly rules" for the development of these natural communities. These are largely based on the ecological niches of the species involved. Papers from the 1970s by MacArthur, Levin, and Diamond, and a 1999 paper by Weier and Keddy are cited as supporting this view.[23] This view was largely developed from Hutchinson's niche model.

The alternative "dispersal-assembly perspective" was developed in large part by Hubbell and asserts, according to him, "that communities are open, nonequilibrium assemblages of species largely thrown together by chance, history and random dispersal." Stochastic local extinction or "ecological drift" is also important. It is a neutral theory not in the sense of "nothing going on" but because it assumes the ecological equivalence of all individuals of all species in a trophically defined community; for example, trees in a forest or zooplankton grazing in a lake. Thus not only are species equivalent, as in some previous island biogeography theory, but, in the forest example, individual trees are also equivalent to each other. "Neutrality" applies only to one trophic level at a time. It might be noted here that these ideas had long been around in ecology in a less extreme form in the concept of guilds.[24]

In the symposium, research from diverse geographical areas— Norway, Estonia, and Panama, as well as the United States and Canada— addressed different aspects of these two theories. Research titles included "Competition, Colonization Trade-Offs and Seed Limitation in Tropical Forests" and "Competition, Colonization, or Herbivory: Which Explains Patterns of Dominance in Herbaceous Communities?" and "Roles of Chance and Adaptation in the Assembly of Fern Communities."[25] A study of Panamanian forests, for example, put the question of maintenance of Panamanian forest diversity in terms of competition-colonization trade-offs and how they affect the species ability to arrive at sites and to win sites once they arrive.[26]

One of Hutchinson's important legacies (and Hubbell's as well) is all the research that has been carried out in response to their theories. The Panamanian study is an amazing example of the effect of theory in generating interesting and extensive ecological research. The authors used eighteen years of seed rain data to quantify seed production, dispersal distance, and seed limitation for eighty different tree species. They also had extensive data on seedling dynamics for evaluating species competition. They found large differences between tree species in their seed production as well as differences in dispersal distances. Seed size was found to mediate a trade-off between seed production and competitive ability among the tree species. It therefore appears that in this tropical forest not all species (or individuals) are equivalent, contradicting one of Hubbell's hypotheses.

In a study of American communities in which goldenrod (*Solidago*) species are dominant but coexist with numerous long-lived herbaceous species, W. F. Carson and A. Baumert sought a "unified and mechanistic explanation" of their dominance. Was it competition, colonization, or resistance to herbivory?[27] The results thus far of a large-scale field test indicate that goldenrods are the best resource competitors as well as the best colonizers, in addition to being unpalatable to herbivores.

Hubbell himself does not deny the importance of niche factors altogether: "Although there is undeniable evidence for niche differentiation among real species in many trophic guilds, this differentiation is not at all essential for coexistence on time scales usually discussed by ecologists."[28]

Hubbell is especially interested in long time scales in which extinction is the rule, not the exception. "Why the presumption of the indefinite coexistence of species? Much evidence supports the opposite conclusion: all species are transient and ultimately go extinct." Hutchinson would not have disagreed altogether. He worried, in fact, about the fate of the human species.

"The Paradox of the Plankton" and "Copepodology for the Ornithologist"

Hutchinson would have enjoyed the symposium; he was always in favor of new testable theories. Niche factors and competition, including Hutchinson's fundamental versus realized niche concept, are still proving important in understanding community structure and species richness, or alpha diversity, of local communities. Hutchinson himself, however, agreed with some of Hubbell's views. He cited factors other than competition in several of his papers, including "The Paradox of the Plankton" (1961), in which essentially equivalent species in an aquatic environment are separated only temporally, as well as in his earlier paper, "Copepodology for the Ornithologist" (1951). During the ten years between these two papers, Hutchinson published three books and forty-six papers; it was a very fertile period.

In "Paradox of the Plankton" Hutchinson showed that competitive exclusion could be avoided in nature, and how even for relatively short ecological time periods in certain ecosystems, nonequilibrium rather than equilibrium conditions could be the rule. Factors that were discussed for the plankton of freshwater communities included mutualism, commensalism, and rapid environmental changes even over a single year cycle.[29]

Ecologists Robert H. Whittaker and Simon Levin in a summary of one section of their edited book, *Niche Theory and Applications*, wrote about this paper:

> In place of our own summation for this section we borrow from a master of exposition, G. E. Hutchinson. . . . As developed by Hutchinson the competitive exclusion principle becomes the concept about which further investigations are organized. Competitive exclusion is a fundamental force, as is gravity. That airplanes can fly does not negate the law of gravity. That species can coexist does not negate the principle of competitive exclusion. In an elegant article, "The Paradox of the Plankton," Hutchinson points out that in many natural situations, competitive exclusion may never have a chance to run its course because of the rapid rate at which the environment changes.[30]

In "Copepodology for the Ornithologist" Hutchinson discussed another method by which organisms escape competitive exclusion. Copepods are tiny crustaceans, not likely to be known to birdwatchers. Hutchinson had discovered in his review of copepod research for the first volume of his *Treatise on Limnology* striking similarities between phenomena of evolutionary significance in copepods and those found in vertebrates, such as birds. He did not think his *Treatise on Limnology* would be read by ornithologists but prepared this paper in the journal *Ecology* to stimulate research by them and other terrestrial ecologists.

In any ecological succession, Hutchinson wrote, any member of the series may be destined to become locally extinct. "Fugitive species" may escape competition with other species capable of entering a particular niche by being "forever on the move," dispersing and reestablishing themselves in another locality as a new, often temporary, niche opens. The original fugitive might become locally extinct, but its descendents could survive in a temporarily available niche in a different locality. Small fluctuations in the environment provide a refuge where they could escape competition. Hutchinson concluded that the common phenomenon of coexistence of many rare species in freshwater ecosystems may be explained by the strategies of such fugitive species. He provided some probable examples from several parts of the world.[31]

Terrestrial ecological research was indeed stimulated. Fugitive species, also called "opportunistic" in terms of birds or "shuttle" species in plants, avoid competitive exclusion. They may become locally extinct but move on to newly opened habitats. This phenomenon has been well explored for many kinds of organisms, including terrestrial plants, since

Hutchinson's paper.[32] Hutchinson postulated good powers of dispersal for such species; dispersal ability has become increasingly important in explaining community establishment and organization since his time.

In conclusion, paradigms do change over time in ecology, but Hutchinson's contributions to ecological theory continue to be important in contemporary ecological research.

Hutchinson's Legacy as Teacher

Stephen Jay Gould called Evelyn Hutchinson, during Hutchinson's later years, the "world's greatest ecologist."[33] When Simon Levin characterized Hutchinson, soon after his death, as the most important figure in American ecology, he wrote, referring in part to the work of Hutchinson's graduate students:

> The gap between population biology and ecosystem science remained too wide for too long. Indeed it was not theoretical work but stunning empirical successes that began to derive a merger of the best of both disciplines. Evelyn Hutchinson saw no sharp boundaries in his work on lakes and the influence of his approach could be seen clearly in the best work of population biologists and ecosystem scientist alike. Lake Washington became a model for how to restore a polluted ecosystem.[34]

The "stunning empirical success" in reversing the pollution of Lake Washington in Seattle was due to the work of Hutchinson student W. T. (Tommy) Edmondson and Edmondson's graduate students. That branch of Hutchinson's intellectual tree extends to four generations, as does the Hutchinson-to-MacArthur branch of the tree.[35] Simon Levin wrote further:

> MacArthur was a student of G. Evelyn Hutchinson, whose role as a mentor of brilliant young ecologists remains unsurpassed. . . . His style of training graduate students . . . was to encourage, reinforce and get out of the way, as Slobodkin [another noted ecologist and Hutchinson student] said. He also chose his students well, and his incredible legacy is testimony to the success of his philosophy of mentoring.[36]

Hutchinson's graduate students themselves commented much earlier on his teaching. MacArthur described Hutchinson's teaching methods early in his own professional career (see chapter 14). In addition to recalling the sign on Hutchinson's office door that read, "Never discourage a

student for you are sure to succeed," it is worth repeating MacArthur's remark: "I believe he has influenced me (and perhaps some of his other students) the most by his genuine enthusiasm about the small, novel ideas
we may have."[37] Others of his graduate students whom I interviewed said
the same.

Howard T. Odum, the graduate student who became a well-known
systems ecologist, wrote, quite independently, about Hutchinson:

> It is difficult for one person to see the whole of a many dimensional phe-
> nomenon. . . . An overriding characteristic of Hutchinson is his imagina-
> tive ability to see important implications in small detailed matters and
> thus become genuinely excited. . . . This absolute love of knowledge . . . is
> ideal as an environment for the student who is getting his first results or
> for an American-type student raised in a materialistic society where such
> worship of the intellectual is rare. . . . Hutchinson teaches with superbly
> clear lectures leveled at the top of the class but with numerous anecdotes
> which keep the rest of the class at least along in spirit. . . . Hutchinson
> balances a complete personal humility with a scorching criticism of sci-
> entific work in the literature, which serves to teach the student by ex-
> ample the difference between good and bad work. . . . He is immensely
> kind but also immensely immersed. Science and ideas are his interest,
> not scientific politics. . . . He is as impartial to personalities and creeds of
> his students as is conceivably possible.[38]

Odum added that at the end of his doctoral students' education,
Hutchinson sent forth ecologists who "strive to match him at least in their
special lines, with an enormous esprit de corps. Yet each time one thinks
one has him appraised, a new contribution more brilliant than before appears, leaving one aghast where one did not even know he was working."

Larry Slobodkin, the innovative population ecologist and another
Hutchinson graduate student, received his degree the same year as Odum,
1951. He had a rather different view of Hutchinson's personality: "Evelyn
is extremely shy and tends to avoid informal contact with strangers. This,
combined with his rather strong accent and erudition, make him seem
aloof at first contact. He will, however, provide ideas and information
with a remarkable free hand to students and colleagues."[39] This assessment is of interest as it perhaps indicates an evolution in Hutchinson's
personality from the late 1940s to the 1960s, as revealed in Denny Cooke's
experience related in chapter 1 and my own in the early 1970s—we were
both "strangers."

In explaining the success of the Hutchinson "school" of ecology, Slobodkin noted its "freshness of approach. Also, the fact of diversity of interests among the students forces each one to maintain a breadth of development that is not as common as it should be in ecology." He commented on the pervasiveness of Hutchinson's influence, demonstrated by "an examination of any recent volume of any ecological journal." He pointed out that (at the time he wrote this) most of Hutchinson's students were not yet forty and that Hutchinson himself was only in his midfifties. All of the above assessments except Levin's were written in 1958, in Hutchinson's midcareer, before the publication of "Homage to Santa Rosalia."

In 1999 Larry Slobodkin and this author published an article about Hutchinson, including his role as teacher:

> His general policy was to carefully avoid excessively close supervision of his students or a regimented curriculum; he felt these to be major weaknesses of American higher education. . . . An obvious feature is how different the students are one from the other and yet how uniform they are in their conviction that Evelyn helped each of them. Anthropologists, entomologists, limnologists, geologists, primatologists all felt that he impelled them along their path. How he did that was not at first glance obvious as even his amazing erudition was not universal. . . . In its most positive sense the feeling of being part of an intellectual elite energized the activities of his students and later of their students through the generations. For many former graduate students, Hutchinson was a constant ghost. Throughout the decades, and there were at least five decades of Hutchinson graduate students, wherever they found, saw, or read something they thought Evelyn might enjoy they wrote or phoned New Haven.[40]

Some of the graduate students of the later decades did not feel they had received equal time and attention from Hutchinson. Whereas earlier students could walk into his office, for later students, especially while his research associate Ursula Cowgill was working with him and "barring the door," appointments were needed. Cowgill in turn felt that he needed protection from constant interruption, especially while he was writing. Some of the male students felt that the women graduate students were getting more attention than they were. Perhaps some of them were. A great many students of both sexes and all eras, however, kept in touch with Hutchinson after leaving Yale.

Women Graduate Students

Undergraduates at Yale were almost entirely white and male throughout most of Hutchinson's career. There were no women among his graduate students until the late 1960s and 1970s, although earlier he was on the committees of and "strongly influenced" several women, including Jane van Zandt Brower, who worked on butterflies, and Alison Jolly, who, as noted, worked on lemurs. Donna Haraway, Maxine Watson, Karen Glaus Porter, and Linda Carlson were among later women doctoral students. Saran Twombly was a strongly influenced doctoral student after Hutchinson's official retirement. Many of these women students have had outstanding professional careers. Hutchinson wrote in 1986: "Not far from my retirement . . . quite remarkable women began to turn up. I would mention most particularly Karen Glaus Porter who is probably one of the two or three best limnologists now working and Maxine Watson who has done beautiful work in fusing the approaches of animal and plant ecology."[41] They and their research were included in an article by a former president of the Ecological Society of America on the "first modern wave" of women ecologists.[42] In the above-cited letter Hutchinson also mentioned "the appearance of a very brilliant woman late in my didactic career, in the person of primate ethologist Alison Jolly."

Hutchinson's positive views of women in science were probably formed as an undergraduate at Cambridge University. Many of Hutchinson's brightest student peers there were women, several of whom went on to distinguished careers in zoology. Hutchinson argued against restrictions on women graduate students at Yale, including women as laboratory instructors. Women in other fields at Yale became Hutchinson's close friends, usually through his Yale college or the Elizabethan Club, of which he was a longtime member. One set of his many published papers went to Yale; the other two went to Margery Garber, Shakespearean scholar at Harvard, and to Phoebe Ellsworth, professor of psychology at the University of Michigan, both former Yale colleagues. Many other women scholars who were not at Yale, from Margaret Mead and Ruth Patrick early on to Lynn Margulis, Janet Browne, and Sharon Kingsland, got help, encouragement, and advice on their books from Hutchinson over many decades. The Hutchinson archives at Yale contain copies of many of his thoughtful letters about their work.

Enthusiastic Recommendations

Hutchinson wrote hundreds of recommendations for his graduate and postdoctoral students. He was often asked for recommendations for university positions, Fulbrights, National Academy of Sciences candidates, and other awards. He kept copies of these letters. They are almost uniformly enthusiastic. Faculty members at two universities reported to this author that in Hutchinson's latter days, his glowing letters for whichever graduate student had applied for a position were no longer taken at face value. This was probably wise. But one earlier graduate student failed his Yale exams. Nevertheless, Hutchinson recommended him to a famous colleague in England to complete his Ph.D.; he indeed did so.

Hutchinson also wrote recommendations for outstanding ecologists who had not been his students. In 1981 Hutchinson was asked to evaluate Robert May's work for a MacArthur Foundation fellows program. He recommended May highly, especially for his "outstanding originality and imagination" as well as for the importance of his research. Hutchinson said he found it difficult to answer questions involving comparisons, explaining: "The kind of people you are interested in have to be unique. Bob May is in this class. In the most scientific kind of ecology, I can think of three much younger people (under 35) whom I instinctively look up to; they are less mathematical than Bob May but I think they all show the same very rare qualities. . . . Of these four, incidentally, three are women."[43] He did not name them but did say that these four had all been professionally close to him. At least three of his women graduate students have indeed had outstanding research careers.

Not all of Hutchinson's students or associates were as brilliant as Robert MacArthur, Robert May, or the women referred to above. Probably not all the young ecologists he recommended were worthy of the top positions he recommended them for. Regardless, he had an extraordinary way of encouraging his own students and those in related fields who asked for his guidance, a knack for helping them to explore their own new ideas and to try them out. He sometimes had, as several related, more faith in their abilities than did the students themselves.

And what about Evelyn himself? Did he ever have doubts about his own abilities? One anonymous reviewer wanted this author to "convey the ways in which Hutchinson's extraordinary, complex, driven, perhaps tortured . . . character affected what he and others did." I have to admit that after interviewing more than fifty people—Hutchinson himself, Hutchinson's students, colleagues, friends and family members, I have found Evelyn Hutchinson extraordinary and complex, but I have not found a "driven, perhaps tortured" man. He did indeed have a few academic enemies at Yale and possibly elsewhere, but it would be hard to find a twentieth-century scientist who had so many close friends; he had a major talent for friendship as well as for ecology.

As to the possibly "driven" man, I saw no evidence that he had any such effect on others, although I have closely observed, though not personally experienced, that effect of major professors on their graduate students. There is some evidence from his letters to his sister Dorothea that he was always trying to live up to what he felt his father expected of him as a scientist, and that even in his old age he had some doubts as to whether he had succeeded.

Hutchinson wrote (in 1986) about his views on the Ph.D. degree at the time of his Cambridge graduation in 1925: "The Ph.D. was not firmly established as a British institution. It was possible to take one but it was solely a trap to catch Americans before they spent their money in Germany."[44] Later, but fairly early in his Yale career, he had self-doubts about sending his published works to Cambridge University for a doctorate degree. One of his Cambridge University classmates had indeed attempted that route and had been turned down; Hutchinson never sent his papers.

In addition to his concern about his father's approbation, a case could be made for the rather negative effects on his psyche of his overpowering mother. He wrote something of this himself to both Dorothea and in his personal letters to women friends, but neither he nor any of these correspondents would have considered his relationship with his mother as "tortured." The two words most often used about Evelyn Hutchinson are "creative" and "kind."

Hutchinson was not uncritical of either scientific papers, to which many of his students attest, or particular scientists. Negative comments about people do appear in his letters to Rebecca West, to his wife Margaret, and to his sister Dorothea, but rarely in print. He certainly had a negative view of longtime Zoology Department chairman John Nicholas, one that he uncharacteristically put in writing in *Kindly Fruits of the Earth*. Nicholas, he wrote, "later in life became too much engrossed with worldly power to continue as an effective scientist." Other faculty members shared these views during Nicholas's later years as department chair.[45]

Two of Evelyn's older Yale colleagues did not wish to be interviewed for this book. One of them, I was told, was unhappy about the power Hutchinson held in his long tenure as director of graduate studies in the Zoology Department. Hutchinson probably did have too high a proportion of Yale Zoology Department graduate students by the 1950s. But the ecology graduate students surely chose to work with Hutchinson because he was the most famous Yale faculty member in this field, not because of his power as director of graduate studies. Clem Markert, chair of the combined zoology and botany departments, removed Hutchinson from his long tenure in that position.

The Molecular Wars at Yale

Clement Markert came to Yale in 1965 to chair the new combined Biology Department. He described himself as a developmental geneticist interested in the role of genes in development. His interests were not in organismal biology or ecology, but in molecular biology. He brought in many new people during his tenure as chairman. Markert thought there were too many of Hutchinson's former graduate students on the faculty, a view shared by some of the other biology faculty members who were not themselves ecologists or evolutionary biologists. In any case, Markert "encouraged" or caused, in one way or another, almost all of Hutchinson's former graduate students who became faculty members to leave Yale. Hutchinson himself had negative views of Markert and the decisions he made according to several biology and forestry faculty members I interviewed; some of their views of Markert's actions as chair were quite vitriolic.

In addition to the Hutchinson group of faculty members who left for other institutions, several women, who had done excellent research at Yale on their own research grants but did not have faculty positions, also left. Markert had made both laboratory space and the ability to apply for further research grants difficult. He did, however, make Grace Pickford Yale's first female full professor shortly before her retirement.

Markert claimed to have no negative feelings about Hutchinson himself, though he admitted to hastening the departure of most of his ecology group. Only Willard Hartman, who was primarily at the Peabody Museum, and Hutchinson himself remained at Yale. Two of the ecologists who left Yale during this period developed their own groups of productive graduate students at other universities.

I interviewed Clem Markert at great length for this book.[46] Markert was doing important innovative research when he was hired as department chair. He told me that since he was hired from outside Yale he did not owe anything to anyone currently in the department. He did not deny the changes he made in the department; he felt they were needed. He certainly favored molecular biology. He declared to me that he could not understand why the questions asked by ecologists were important. He also said, however, that he had a clear directive from Kingman Brewster, then provost but soon to be Yale's president, to modernize biology at Yale, particularly in terms of the increasing importance of molecular biology.

I did not find any documents to substantiate the Kingman Brewster directive, but there is corroborative evidence in a recent book by Geoffrey Kabaservice, a Yale Ph.D. who lectured in the History Department.[47] In his book *The Guardians*, about Kingman Brewster and his circle at Yale, Kabaservice reported that Brewster "diverted resources toward the sciences on a scale not seen during the Griswold era." Griswold was the previous president of Yale, but Brewster, as provost, was already setting policies during Griswold's final illness. One of Kingman Brewster's noted accomplishments during his own presidency was the creation of a new Department of Molecular Biophysics and Biochemistry "over considerable opposition." Neither Markert nor Hutchinson is mentioned in Kabaservice's book, however. Kingman Brewster was among Hutchinson's many, many correspondents, but no correspondence on this subject has been found.

Hutchinson student and Yale faculty member John Brooks left Yale when Markert warned him that he would not be promoted to full professor. He spent the rest of his career at the National Science Foundation. Johnny Brooks was not only a former student and fellow faculty member, but also a friend. He felt bitter that Evelyn, who had supported him earlier, had not supported him in his difficulties with Markert. But Evelyn had indeed written an excellent letter in favor of Brooks's promotion. (As noted above, Hutchinson kept copies of almost all the letters of recommendation he wrote; this one, too, is available in his Yale archives.)

In addition to Hutchinson's graduate students who became faculty members, there were many former Yale students in fields other than biology among both the Yale University faculty and the administration, undoubtedly too many. This practice had a long history at Yale. Kingman Brewster both noted and changed this system.

The 1960s were certainly the new era of molecular biology at Yale, Harvard, and elsewhere. Ecology at Yale hit a low point during and after Markert's tenure as chair and thereafter. It did not recover in the Biology Department until quite recent years, although there were well-known ecologists in the Yale School of Forestry and Environmental Science. Some senior ecologists who were offered positions in the Biology Department turned them down. One of these was Hubbell, whose important and controversial work is described above. Other young ecologists came to Yale for a short time and then moved on to Cornell and elsewhere.

After his retirement in 1971 at the age of sixty-eight Hutchinson still came to department meetings but had little power over decisions that were made. Margaret Davis, a well-known researcher of climate and vegetation change using pollen analysis, was hired in 1973 as Hutchinson's successor, or so it was assumed. Some of the remaining ecology graduate students felt that someone closer to Hutchinson's areas of ecology should have been hired. Davis told me in an interview that she did not view herself as Hutchinson's successor: "No one could succeed Hutchinson. I thought I was Deevey's successor. I thought I could keep up with Deevey."[48] They worked in similar fields; Deevey was one of those who had left Yale.

"Markert and I had a big fight in a faculty meeting. The fight was about whether the department should hire another ecologist." Davis felt that she was the only faculty member who could be described as an ecologist, though there were evolutionary biologists. She said that her area, vegetation change and paleontology, was inadequate to represent the whole field: "They needed a population ecologist, which was certainly the 'with it' field. They definitely needed more strength in this area to train students properly. They debated the question, [some arguing that] maybe we shouldn't have a graduate program in ecology." Davis said she would not stay at Yale if there were no graduate program in ecology. Markert was no longer chair when Davis was hired; Tim Goldsmith was the new chairman. But Markert, in Davis's eyes, still had "extraordinary power." She continued:

> Markert and I had this debate. It was as if we were on a battlefield. Here we were standing opposite each other and he had all his assistant professors and associate professors standing behind his coattails and I looked

around and the ecologists and evolutionists had disappeared from the scene and were standing behind bushes waiting to see who was going to win. . . . He had a very forceful personality and I think many people were afraid of him. He was still a very influential person in the department.

Davis won this debate, at least in part. It was agreed that two assistant professors, though no senior person, would be hired in ecology. She had her own group of graduate students who interacted with her and each other but not with the rest of the department. Davis was not happy in the Yale Biology Department in that era. She never did connect with Hutchinson nor with his remaining graduate students. She felt that her own work and ecology in general were being marginalized in the department. After a relatively short tenure she departed to become a department head at the University of Minnesota, where she has had a productive and happy career. Newly hired ecologists, mostly young, came and went, but senior ecologists at Yale prevailed only in the School of Forestry.

Ecology eventually recovered at Yale and has once again, with new hires and new graduate students, become a major program, the Department of Ecology and Evolutionary Biology. In 2003, the department celebrated Evelyn Hutchinson's one hundredth birthday, in his absence of course, with a big party and a symposium including many of today's leading ecologists.

Hutchinson's Legacy as a Writer

Ecology as a scientific discipline had certainly exploded in many directions during Hutchinson's sixty-year career at Yale and continues to do so in the twenty-first century. In the latter half of the twentieth century Evelyn Hutchinson was the most important figure in the invention of modern ecology. His theories are still being tested and sometimes overturned; that is what happens in any good science, and it is what he would have wanted. What he wrote on ecology, limnology, and a whole spectrum of related subjects is some of the best scientific writing produced in the twentieth century; much will endure and inspire future students. Hutchinson's intellectual family tree has been compiled only up to 1971. By now it contains at least two more generations of Hutchinson's students' students. His intellectual descendents include many of today's leading ecologists. His role as a teacher, in person and through his writings, is a highly important and lasting one, and one he himself highly valued. Many who were not his own graduate students, including this author, were influenced by his writings.

Hutchinson's first book, *The Clear Mirror,* is a work both literary and scientific. He wrote it in 1936 about his research trip to Ladakh when he was not yet thirty. Subsequently he became greatly admired as a writer, particularly as an essayist, on a great array of subjects. One well-known ecologist who was not his student wrote about his writing: "An incisive theoretician as well as a stunning empiricist, Hutchinson brought these talents to bear on his writing, whose substance and imagery captured the reader's attention and impelled him or her to want to study the subject further."[49]

I leave to Hutchinson the concluding remarks on his own writing. At age eighty-five, in a letter dated only "Autumn 1988" he described the "autumnal colouring, brilliant reds and an unusually good bright orange." He then noted that he had recently reread several of the smaller books by his uncle Arthur Shipley with delight, and recalled how much they had influenced him when they first came out. He concluded: "I certainly got to know from him a belief that if one wrote well enough one could write anything that interested one."[50] Almost everything interested Evelyn Hutchinson.

Finally, the question has been asked: if Evelyn Hutchinson was so brilliant and invented so much of modern ecology, why did he not win a Nobel Prize? The easy answer to that question is that there is no Nobel Prize in ecology—nor in limnology. Hutchinson did win the Kyoto Prize for his work in ecology, in his eighties. This prize, often known as the "Japanese Nobel Prize," provided him and his family with a small fortune.

There are other more philosophical answers to this question. Dan Livingstone wrote: "Hutchinson had a first order knowledge of everything that seemed important in the world. That was very wonderful, in an age when most people thought that sort of mastery had gone out with Leonardo and the Renaissance. . . . He recognized the importance of the first *Pneumococcus* transformation experiments before most geneticists did."

Those were the experiments that showed that DNA, not protein, was the genetic material. Hutchinson followed X-ray crystallography closely and, Livingstone said, "might have taken off from Avery et al. and got to the double helix before Watson and Crick." This is fantasy, but Livingstone's point was that Hutchinson did not wish to focus his whole attention on a "few small but vital problems where he could make a direct and personal advance." Livingstone concluded:

> It is vital that a few people in each generation should hold up the waving banner of ideas, and that the ability to do so is even scarcer than the ability that will, in a few fields, lead to the sort of recognition that a Nobel Prize confers.[51]

Hutchinson, the idea man, the inventor of new hypotheses and new fields of ecology, certainly waved that banner for future generations of ecologists.

Notes

Chapter 1. A Man So Various

1. Thomas H. Huxley to Sir John Simon in Bibby (1960), *T. H. Huxley*, 1.

2. A potto is a small African primate or in Hutchinson's words "the attractive prosimian *Perodictus potto.*" Hutchinson (1979), *Kindly Fruits*, 3.

3. Hagen (1992), *Entangled Bank*; Golley (1993), *History of the Ecosystem Concept in Ecology*; Kingsland (1985), *Modeling Nature*; Mills (1989), *Biological Oceanography*.

4. McIntosh (1985), *Background of Ecology*.

5. Hutchinson and Bowen (1947), "Direct Demonstration of the Phosphorus Cycle in a Small Lake."

6. Clements and Shelford (1939), *Bio-Ecology*; Hutchinson (1940), "Review of *Bio-Ecology.*"

7. Slobodkin and Slack (1999), "George Evelyn Hutchinson."

8. In Y. Edmondson (1971), *Limnology and Oceanography*, fig. 7.

9. Mills (1989), *Biological Oceanography*, 259–260.

10. Hutchinson (1957), *Treatise on Limnology*, vol. 1.

11. Hutchinson (1936), *Clear Mirror*.

12. Hutchinson (1965), *Ecological Theater and the Evolutionary Play*.

13. Riley, introduction in Y. Edmondson (1971), *Limnology and Oceanography*.

14. Slack (2003), "Are Research Schools Necessary?"

15. G. Evelyn Hutchinson to George J. Caranosos, M.D., University of Florida, January 1985. Hutchinson Papers, Manuscripts and Archives, Yale University Library (hereafter Hutchinson Papers, MSSA, Yale University Library).

16. See the bibliography for a list of many of Hutchinson's published books and papers.

17. Thomas Lovejoy, Evelyn Hutchinson memoir (2000). In author's possession. Quotes in following paragraphs are from this memoir.

18. Saran Twombly from Pellanza, Italy, to G. Evelyn Hutchinson, September 3, 1983. Hutchinson Papers, MSSA, Yale University Library.

19. G. Dennis Cooke e-mail correspondence with Nancy G. Slack, June 3 and June 13, 2009. Cooke, a limnologist, is the lead author of *Restoration and Management of Lakes and Reservoirs* (2005).

20. G. Evelyn Hutchinson from Gloucester, Eng., to Karen Glaus Porter, July 1969. In author's possession.

21. Letters between W. T. Edmondson, University of Washington, and H. Guyford Stever, Carnegie Commission on Science, Technology, and Government, 1991, and from G. Evelyn Hutchinson to H. Guyford Stever, July 27, 1973. In author's possession.

Chapter 2. "The Circumstances of My Upbringing"

1. Mills (1989), *Biological Oceanography.*

2. Hutchinson (1979), *Kindly Fruits.*

3. *London Times,* obituary, December 1937. "Arthur Hutchinson, O.B.E., F.R.S., 71 (1866–1937)"; Hutchinson (1979), *Kindly Fruits;* Dorothea Hutchinson interview with Nancy G. Slack, July 25, 1991, Cambridge, Eng. G. Evelyn Hutchinson Biographical Interviews RU9621, Oral History Collection, Smithsonian Institution Archives (hereafter Smithsonian Institution).

4. *London Times,* obituary, December 1937.

5. Dorothea Hutchinson interview with Nancy G. Slack, July 25, 1991, Cambridge, Eng. She explained: "When we were children we were quite hard up because in those days professors had no retiring age. The professor was 84, a bachelor living in Trinity, and did absolutely nothing in the department. Father ran the department, did all the lecturing. . . . He wasn't even a lecturer, only a demonstrator."

6. G. Evelyn Hutchinson to Dorothea Hutchinson, November 12, 1984, December 24, 1987, and February 1988. Hutchinson Papers, MSSA, Yale University Library.

In the 1987 letter to Dorothea, Evelyn wrote, "Do you have any recollections of anything Father said about my scientific work. Did it puzzle him?" He also wanted know whether his father had introduced him to Walter Campbell Smith. He had recently written to Smith thanking him for the very great part he played in introducing Hutchinson to many of Smith's colleagues at the Natural History Museum (South Kensington), who helped him in his profession: "I can only hope that I have been as successful with the extraordinary group of young ecological people whom I have tried to help get started. Of course I particularly treasure your remark that father was the best teacher of mineralogy living when you were a student. I have consciously tried to model my teaching career on what you said about him." G. Evelyn Hutchinson to Walter Campbell Smith, December 29, 1987. Hutchinson Papers, MSSA, Yale University Library.

7. G. Evelyn Hutchinson, dictated to hospital nurse Megan, in London, April 1991. Notes in author's possession.

8. Dorothea Hutchinson interview, July 25, 1991. Dorothea also had fond memories of Cumberland walks with her father. "I love maps," she said. "My father introduced me to geologic maps. Quite often he told me about . . . geological formations on country walks when we went to Cumberland."

9. Hutchinson letter to editor of *Nature*, December 14, 1937. Hutchinson Papers, MSSA, Yale University Library.

10. Willard Hartman interview with Nancy G. Slack, May 1991, New Haven.

11. Hutchinson (1979), *Kindly Fruits*, 11–15; Dorothea Hutchinson interview with Nancy G. Slack, Cambridge, Eng., 1993. It is definitely known that Shipley subsequently became a hotel keeper in Hartley-Wintney in Hampshire. E. William Shipley of St. Martin's in the Fields married Catherine Demezy of Hartley-Wintney in 1782. These were Evelyn's great-great-grandparents. Sir Arthur Shipley, like Evelyn's father, became master of a Cambridge College, Christ's College. When the master's lodge at Christ's was being restored, mummies of four black rats were found. His uncle introduced him to this rat, to its role in bubonic plague, and to its replacement by the brown rat. Evelyn considered this discussion his first introduction to interspecific competition. Uncle Arthur was an important influence on the "embryo" zoologist.

12. Dorothea Hutchinson to Nancy G. Slack, January 29, 1992. In author's possession.

13. Rowbotham (1987), *Friends of Alice Wheeldon*.

14. Pankhurst (1932), *Home Front*.

15. G. Evelyn Hutchinson to Mary Catherine Bateson, daughter of Margaret Mead and Gregory Bateson, n.d. (1980s). Hutchinson Papers, MSSA, Yale University Library.

16. E. D. Hutchinson (1936), *Creative Sex*.

17. Van de Velde (1928), *Ideal Marriage*, 232.

18. Stopes (1918), *Married Love*.

19. For an in-depth discussion of Stopes's books, see Jackson (1994), *Real Facts of Life*, 129–158.

20. Jackson (1994), *Real Facts of Life*, 162–163.

21. E. D. Hutchinson (1936), *Creative Sex*.

22. Hutchinson (1962), "Cambridge Remembered."

23. Hutchinson (1979), *Kindly Fruits*, 1; G. Evelyn Hutchinson to Dorothea Hutchinson, May 9, 1980. Hutchinson Papers, MSSA, Yale University Library.

24. Hutchinson (1962), "Cambridge Remembered." G. Evelyn Hutchinson, Introductory Remarks for Kyoto Prize, Inamori Foundation, November 1986. In possession of author.

25. Hutchinson (1979), *Kindly Fruits*, 19.

26. Christina Innes Morley to Nancy G. Slack, May 17, 1995. In author's possession.

27. Dorothea Hutchinson interview, July 25, 1991. Smithsonian Institution.

28. Hutchinson (1962), "Cambridge Remembered"; Hutchinson (1979), *Kindly Fruits*, 30–31, 39.

29. Dorothea Hutchinson interview, July 25, 1991. Smithsonian Institution.

30. Hutchinson (1979), *Kindly Fruits*, 52.

31. Hickson (1995), *Poisoned Bowl*; Farrar (1962), *St. Winfred or the World of School*; Dorothea Hutchinson interview, July 25, 1991. Smithsonian Institution.

32. Hutchinson (1979), *Kindly Fruits*, 55–57. Hammick could "so excite a group of schoolboys about this subject in a single lecture that the feeling has lasted sixty years in one of them."

33. Hutchinson saved even his very early correspondence, now in his papers at Manuscripts and Archives, Yale University Library. There are letters from Butler beginning in 1920.

34. Hutchinson (1918), "Swimming Grasshopper."

35. Hutchinson (1979), *Kindly Fruits*, 134.

36. Butler (1923), *Biology of the British Hemiptera-Heteroptera*. E. A. Butler to G. Evelyn Hutchinson, August 28, 1920; September 17, 1920; May 2, 1921; August 2, 1921. Hutchinson Papers, MSSA, Yale University Library.

37. F. H. Day to G. Evelyn Hutchinson, August 15, 1920. Hutchinson Papers, MSSA, Yale University Library.

38. Winifred E. Brenchley to G. Evelyn Hutchinson, January 28, 1922. Hutchinson Papers, MSSA, Yale University Library.

39. Gresham's School, Natural History Society annual reports for 1919, 1920, and 1921.

40. Hutchinson (1979), *Kindly Fruits*, 72. The book's index refers to this paragraph under "chemical ecology."

41. Hutchinson (1979), *Kindly Fruits*, 73; Macan (1968), "Tarns on Dufton Fell." Hutchinson found a related species, new to science, in a high-altitude lake in Ladakh, India, in 1932.

42. Hutchinson (1977), "Influence of the New World on Natural History."

43. Hutchinson to George J. Caranosos, January 1985. Hutchinson Papers, MSSA, Yale University Library.

Chapter 3. Becoming a Zoologist

1. Howarth (1978), *Cambridge between Two Wars*.

2. Raine (1975), *Land Unknown*, 17.

3. G. Evelyn Hutchinson to Jean H. Langenheim, April 1, 1986. Hutchinson Papers, MSSA, Yale University Library.

4. Howarth (1978), *Cambridge between Two Wars*, 36–41, 46. *Cambridge Review* quoted in Howarth, 36–37.

5. Raine (1975), *Land Unknown*, 20–25. As a secondary school graduate, she did not feel at home with the public school women at Girton.

6. Ibid.

7. Hutchinson (1979), *Kindly Fruits*, 76. Throughout many of his Yale years, Hutchinson boiled eggs at home for visiting scientists and made tea on a Bunsen burner in his lab. The latter was still the case in 1990 when the author was a visiting scholar at Yale and contributed an electric teakettle to Hutchinson's office.

8. Howarth (1978), *Cambridge between Two Wars,* 26.

9. Hutchinson (1979), *Kindly Fruits,* 76–77, 80–81.

10. Anna Bidder interviews with Nancy G. Slack, July 25, 1991, and August 2, 1993, Cambridge, Eng. Smithsonian Institution.

11. G. Evelyn Hutchinson interviews with Nancy G. Slack, November 1990, New Haven. Smithsonian Institution.

12. E. A. Butler to G. Evelyn Hutchinson, June 9, 1922, and September 4, 1922; August 28, August 29, and October 6, 1923; and October 20, 1924. Hutchinson Papers, MSSA, Yale University Library. Hutchinson was already aware of the importance of archives, however. At the end of the Butler file of letters is a note reading, "In case of mishap to myself all Butler correspondence should go with the collections in the insect room, Cambridge Museum of Zoology." The letters went with him to Yale instead.

13. Anna Bidder interview with Nancy G. Slack, August 2, 1993. Smithsonian Institution. Bidder related that since she had been born in Cambridge, gone to school in Cambridge, and lived in what might be called a Cambridge University house, her parents thought she should leave there and spend one year at University College, London. At this excellent science institution in 1921, women students could receive degrees, unlike at Cambridge. Bidder had an exciting year there, but then she returned to Cambridge to do her zoology degree. "I don't think it occurred to either of my parents that they could possibly consider anything but Cambridge." After finishing her degree, Anna studied in Basle, Switzerland, and also at the marine stations in Plymouth and in Naples, and then returned to Cambridge to do research. When her mother died, Anna had to take over care of the household, as Hutchinson's mother had had to do. She, like Evaline Shipley (Hutchinson), had an older sister who was married with children. Anna managed to continue her research: "When I took over keeping house after she died I thought I would rather put it [research] in second place than not have it." Anna Bidder later did research on swimming and buoyancy control in the pearly nautilus.

14. Hutchinson (1979), *Kindly Fruits,* 89, 144.

15. Hutchinson (1962), "Cambridge Remembered."

16. For the establishment of biochemistry at Cambridge University, see Kohler (1978), "Walter Fletcher, F. G. Hopkins and the Dunn Institute of Biochemistry."

17. Joseph Needham interview with Nancy G. Slack, July 25, 1991. Smithsonian Institution. When the author interviewed Needham in 1991, he had his own institute in Cambridge, the East Asian Library and Research Center.

For Needham's account of his own career, see Henry Holorenshaw (pseudonym for Needham), "The Making of an Honorary Taoist," 1–20, in *Changing Perspectives in the History of Science,* ed. Robert Young and Mikulás Teich (London: Heinemann, 1973). For another account of Needham's life and work, especially in China, see Simon Winchester, *The Man Who Loved China* (New York: Harper Perennial, 2007).

18. Joseph Needham interview with Nancy G. Slack, July 25, 1991. Harrison was Yale's famous embryologist who had invented tissue culture and much of modern embryology. He was later a great positive influence on Hutchinson.

19. Howarth (1978), *Cambridge between Two Wars,* 50–57, 98–101, 134–135. For

an analysis of the founding and function of the Balfour Biological Laboratory for Women, see Richmond (1997), "Lab of One's Own."

20. G. Evelyn Hutchinson to Jean L. H. Langenheim, April 1, 1986. Hutchinson Papers, MSSA, Yale University Library.

21. G. Evelyn Hutchinson to John A. Wilkinson, the Secretary, Yale University, May 13, 1983. Hutchinson Papers, MSSA, Yale University Library.

22. Howarth (1978), *Cambridge between Two Wars*, 53–57.

23. Koestler (1971), *Case of the Midwife Toad*.

24. Hutchinson (1962), "Enchanted Voyage." These interests later evolved into a lecture that included cowrie shells, Tyrian purple from sea snails, and other "magical" sea organisms, which Hutchinson gave several decades later at Woods Hole for the dedication of the Laboratory of Oceanography. It began with Shakespeare's *The Tempest*, included T. S. Eliot's *The Waste Land*, and in general contained too much sex, magic, and religion for some of the oceanographers present.

25. Penelope Jenkin interview with Nancy G. Slack, August 4, 1993, in North Wales. Smithsonian Institution.

26. Hutchinson (1979), *Kindly Fruits*, 116–123, 134–135, 146–147.

27. Penelope Jenkin interview with Nancy G. Slack, August 4, 1993. Quotes in this section about the expedition are from this interview.

28. E. A. Butler to G. Evelyn Hutchinson, September 5, 1925. Hutchinson Papers, MSSA, Yale University Library.

Chapter 4. From Naples to South Africa

1. E. A. Butler to G. Evelyn Hutchinson, September 5, 1925. Hutchinson Papers, MSSA, Yale University Library.

2. Kofoid (1910), *Biological Stations of Europe*, 7–32. Kofoid, of the University of California, Berkeley, wrote extensively about the Naples Zoological Station, which he visited in 1909.

3. Ibid., 11.

4. Anton Dohrn to Doctor Ida Hyde, April 20, 1897. In Kofoid, *Biological Stations of Europe*, 15.

5. G. Evelyn Hutchinson to "Dear Sir" at Naples Station, April 25, 1925, and letter to "Miss G. Evelyn Hutchinson" from Reinhard Dohrn, April 30, 1998. Stazione Zoologica archives.

Christiane Groeben, archivist of the Stazione Zoologica, kindly sent the author the list of all the scientists working at the station in 1926. Hutchinson is listed there as "G. Evelyn Hutchinson." Christiane Groeben to Nancy G. Slack, November 17, 1998. Letter and list in author's possession.

6. Charles Resser to G. Evelyn Hutchinson in relation to a Smithsonian paper of Hutchinson's about the Burgess Shale, 1930. "According to unanimous usage in America your name as you have written it would be mistaken by all as that of a woman." Hutchinson Papers, MSSA, Yale University Library.

7. G. Evelyn Hutchinson to "Dr. Dohrn" (Reinhard Dohrn) August 2 and September 25, 1925. Stazione Zoologica archives.

8. Hogben (1924), *Pigmentary Effector System*. Hogben (1927), *Comparative Physiology of Internal Secretion*.

9. Winifred Allen to G. Evelyn Hutchinson, "early" 1927, quoted in Hutchinson (1979), *Kindly Fruits*, 149–150.

10. Winifred Allen to G. Evelyn Hutchinson, September 26, 1926. Hutchinson Papers, MSSA, Yale University Library.

11. Hutchinson (1979), *Kindly Fruits*, 177–180.

12. G. Evelyn Hutchinson to Jean H. Langenheim, April 1, 1986. Hutchinson Papers, MSSA, Yale University Library.

13. No such correspondence from Pickford to Hutchinson has yet been found. Pickford destroyed many of her personal papers, specifically including her letters home from South Africa. Winifred Allen wrote a letter, and this one is extant, to Hutchinson in South Africa. Allen had seen his mother and reported that she had heard from Evelyn only once since he and Grace had married. Winifred Allen to G. Evelyn Hutchinson, September 26, 1926. Hutchinson Papers, MSSA, Yale University Library.

14. G. Evelyn Hutchinson to Jean H. Langenheim, April 1, 1986. Hutchinson Papers, MSSA, Yale University Library.

15. Hutchinson (1932), "On Corixidae from Uganda."

16. Hutchinson (1929), "Revision of the Notonectidae and Corixidae of South Africa."

17. Hutchinson (1979), *Kindly Fruits*, 202–206.

18. Ibid., 202–206.

19. Hutchinson, Pickford, and Schuurman (1929), "Inland Waters of South Africa." Hutchinson, Pickford, and Schuurman (1932), "Contribution to the Hydrobiology of Pans and Other Inland Waters of South Africa."

20. Thienemann (1925), *Die Binnengewasser Mitteleuropas*.

21. Harvey (1928), *Biological Chemistry and Physics of Sea Water*.

22. The actual term "limnology" had appeared in 1892. Francois Alphonse Forel used this term. It was later expanded to include other bodies of fresh water, such as rivers, and hydrography as well as biology. Hydrography is the part of limnology concerned with the physics and chemistry of water, the shapes of lake basins, temperature stratification, etc. The term "limnology" (*Limnologie* in German) was used by Einar Naumann and August Thienemann in 1922 to mean the science of fresh water as a whole. They were proposing an international association of limnologists. It is unlikely that Hutchinson had seen that paper since he was introduced to Thienemann's work only after leaving South Africa. For further information, see Golley (1993), *History of the Ecosystem Concept in Ecology;* Naumann and Thienemann (1922), "Vorschlag zur Grundung einer internationalen Vereinigung für theoretische und angewandte Limnologie."

23. Hutchinson (1979), *Kindly Fruits*, 203.

24. Penelope Jenkin interview with Nancy G. Slack, August 4, 1993. Later comments from Jenkin in this section are from this interview.

25. G. Evelyn Hutchinson to Jean H. Langenheim, April 1, 1986. Hutchinson Papers, MSSA, Yale University Library.

26. Hutchinson (1979), *Kindly Fruits*, 208.

27. Hutchinson, quote from Fantham's obituary: "McGill students mourned the loss of a friend." *Kindly Fruits* (1979), 212. Max Dunbar interview with Nancy Slack, June 20, 1993: "He [Fantham] had a very bad temper and an unpleasant way of treating his subordinates in the department. . . . They were very glad to get rid of him."

28. Lancelot Hogben to Ross G. Harrison, February 2, 1928. Ross G. Harrison Papers, Manuscripts and Archives, Yale University Library (henceforth Harrison Papers, MSSA, Yale University Library).

29. L. L. Woodruff to Lancelot Hogben, February 29, 1928. Harrison Papers, MSSA, Yale University Library.

30. G. Evelyn Hutchinson cable to L. L. Woodruff, March 31, 1928. G. Evelyn Hutchinson letter and photo to L. L. Woodruff, April 4, 1928. Harrison Papers, MSSA, Yale University Library.

31. L. A. Borradaile to Yale University, the Graduate School, confidential statement concerning G. E. Hutchinson, April 23, 1928, stamped May 8, 1928; G. Stanley Gardiner to L. L. Woodruff, April 24, 1928, and G. Stanley Gardiner to Yale University, the Graduate School, confidential statement concerning G. Evelyn Hutchinson, April 23, 1928; all in Harrison Papers, MSSA, Yale University Library.

32. Lancelot Hogben to Yale University Osborn Zoological Laboratory, confidential report on Mr. E. Hutchinson, M.A., March 26, 1928. Harrison Papers, MSSA, Yale University Library.

33. G. Evelyn Hutchinson application for Sterling Fellowships for Research, Graduate School, Yale University, stamped May 8, 1928, probably sent April 4, 1928. Harrison Papers, MSSA, Yale University Library.

34. G. Evelyn Hutchinson cable to L. L. Woodruff, May 11, 1928, and letter, May 13, 1928. Harrison Papers, MSSA, Yale University Library.

35. L. L. Woodruff to G. E. Hutchinson, May 17 and June 18, 1928. Harrison Papers, MSSA, Yale University Library.

36. G. Evelyn Hutchinson to L. L. Woodruff, June 20, 1928, Harrison Papers, MSSA, Yale University Library.

37. John Nicholas to R. G. Harrison, July 16, 1928, Harrison Papers, MSSA, Yale University Library.

Chapter 5. First Years at Yale

1. Citation from Hutchinson's honorary degree at Princeton University, 1961, quoted by Margaret Hutchinson in letter to Dorothea Hutchinson, June 15, 1961.

2. Hutchinson (1979), *Kindly Fruits*, 224, 234.

3. Ibid., 229-230.

4. Hutchinson (1980), *Alexander Petrunkevitch*.

5. Hutchinson (1987), "Keep Walking."

6. J. S. Nicholas to Ross G. Harrison, July 16, 1928. Harrison Papers, MSSA, Yale University Library.

7. Hutchinson (1979), *Kindly Fruits*, 214.

8. For further information and insights on Ross Harrison's life and work, see Abercrombie (1961), "Ross Granville Harrison." Maienschein (1983), "Experimental Biology in Transition."

9. Twitty (1966), *Of Scientists and Salamanders.*

10. Ibid., 9–10, 17, 19–21.

11. Ibid., 39.

12. Ibid., 51–53. There were many Americans at the institute in Berlin at that time. Twitty escorted one, Professor Mary Stuart MacDougall, on a tour of Berlin's night spots, probably not exactly the form of culture Harrison had envisioned. Twitty left Germany for Stanford in 1932, before Hitler's rise but not before other American students had witnessed the political upheaval in the making.

13. "Professor Ross Granville Harrison," short biography written by a student in an undergraduate embryology course Harrison taught (no name or date; probably written in the 1930s). Harrison Papers, MSSA, Yale University Library.

14. Hutchinson, "Introductory Remarks," Kyoto Prize (Inamori Foundation), November 14, 1986.

15. Ross G. Harrison to "To whom it may concern," July 22, 1930, letter of recommendation for G. Evelyn Hutchinson. Harrison Papers, MSSA, Yale University Library.

16. Ross G. Harrison to L. L. Woodruff, January 15, 1944. Harrison Papers, MSSA, Yale University Library.

17. John S. Nicholas to Charles Seymour, November 22, 1945. J. S. Nicholas Papers, Manuscripts and Archives, Yale University Library.

18. Hutchinson (1979), *Kindly Fruits*, 234.

19. Ibid., 240–242.

Chapter 6. Chief Biologist

1. De Terra (1932), Memorandum for an Expedition to the Himalaya and Karakorum. Harrison Papers, MSSA, Yale University Library.

2. Ibid.

3. Ibid.

4. George Parmly Day, Treasurer, Yale University, to Professor Ross G. Harrison, April 14, 1931. Harrison Papers, MSSA, Yale University Library.

5. Ibid.

6. Ross G. Harrison to George P. Day, April 17, 1931. Harrison Papers, MSSA, Yale University Library.

7. George Parmly Day, Treasurer, Yale University, to Professor Ross G. Harrison, April 18, 1931. Harrison Papers, MSSA, Yale University Library.

8. Hutchinson (1931), Program of Research That Should Be Undertaken by a

Freshwater Biologist Accompanying Dr. de Terra's Expedition to Kashmir. Hutchinson Papers, MSSA, Yale University Library. Information in the following paragraphs is drawn from this proposal.

9. Authorized account by the Yale North India Expedition. Geological and biological research in Kashmir, May 1932. Harrison Papers, MSSA, Yale University Library.

10. Hellmut de Terra to President J. R. Angell, Yale University, May 11, 1932. Harrison Papers, MSSA, Yale University Library.

11. Hutchinson to Grace E. Pickford, August 25, 1932. Hutchinson Papers, MSSA, Yale University Library.

12. G. Evelyn Hutchinson to Ross G. Harrison, August 10, 1932. Harrison Papers, MSSA, Yale University Library.

13. Ross G. Harrison to President J. R. Angell, Yale University, October 1, 1932. Harrison Papers, MSSA, Yale University Library.

14. Hellmut de Terra to President J. R. Angell, Yale University, October 7, 1932. Harrison Papers, MSSA, Yale University Library.

15. Hellmut de Terra, authorized report from the leader of the Yale North India Expedition from Srinagar, Kashmir, October 6, 1932. Harrison Papers, MSSA, Yale University Library.

16. Ibid.

17. G. Evelyn Hutchinson to Ross G. Harrison, October 8, 1932. Harrison Papers, MSSA, Yale University Library.

18. Ross G. Harrison to G. Evelyn Hutchinson, October 8, 1932. Harrison Papers, MSSA, Yale University Library.

19. President J. R. Angell, Yale University, to Ross G. Harrison, October 6, 1932. Harrison Papers, MSSA, Yale University Library. Harrison wrote to President Angell as soon as he received Hutchinson's letter, written on August 10 (see note 12, above) but not received at Yale until the end of September. He quoted most of Hutchinson's letter as to the distribution of the collections. Harrison had already consulted with Professors Ball and R. S. Lull (mammalian paleontologist and museum director) and with Petrunkevitch, all of whom agreed that "Hutchinson's proposal is wise and would recommend that he be permitted to make such disposition of the material as would lead to its most satisfactory working up." A representative collection was to be reserved for the Peabody Museum. Harrison, however, did not feel that he had the final say in this matter and asked for Angell's permission to do this.

20. G. Evelyn Hutchinson to Grace Pickford, November 13, 1932. Hutchinson Papers, MSSA, Yale University Library.

21. Hutchinson (1933), "Limnological Studies at High Altitudes in Ladakh." Hutchinson (1937), "Limnological Studies in Indian Tibet." Hutchinson (1939), "Ecological Observations on the Fishes of Kashmir and Indian Tibet."

22. John Nicholas to Charles Seymour, November 22, 1945. J. S. Nicholas Papers, MSSA, Yale University Library.

23. Many letters from the 1930s in the Hutchinson Papers, MSSA, Yale University Library.

24. Catherine Skinner, talk at G. Evelyn Hutchinson memorial service luncheon, October 19, 1991. Smithsonian Institution.

25. G. Evelyn Hutchinson to M. E. Mosely, October 2, 1933. Hutchinson Papers, MSSA, Yale University Library. In an earlier letter from Hutchinson to Mosely (September 25, 1933, Hutchinson Papers), Hutchinson had written about the difficulty of obtaining funds. He wrote that he was very much averse to having the results scattered in many journals but that this might be necessary. He asked Mosely to write his position on this matter. "By showing your letter to the Yale authorities it might be possible to get something done."

26. G. Evelyn Hutchinson, Instructions to Investigators Examining the Biological Material Collected by the Yale North India Expedition. Hutchinson Papers, MSSA, Yale University Library.

27. G. Evelyn Hutchinson to Malcolm Cameron, October 12, 1933. He also wrote to Cameron (April 25, 1934, Hutchinson Papers) objecting to the species name kasmirensis for a beetle discovered on the expedition. "It seems better to characterize all the trans-Himalayan part . . . as Indian Tibet as Francke, the great archaeologist, has done, thus bringing out the very great differences in landscape and vegetation, once one is over the pass."

28. G. Evelyn Hutchinson to August Thienemann, April 1, 1936. Hutchinson Papers, MSSA, Yale University Library.

29. Hutchinson (1937), "Limnological Studies in Indian Tibet."

30. Hutchinson (1978; repr. of 1936 ed.), *Clear Mirror,* 43.

31. Hutchinson (1978), *Clear Mirror,* back cover.

32. Ibid., 99.

33. Ibid., 119.

34. Ibid., 42.

Chapter 7. The First Crop

1. W. T. Edmondson interview with Nancy G. Slack, April 30, 1991. National Academy of Sciences, Washington, D.C., Smithsonian Institution.

2. The first rotifer was *Scaridium longicaudum.* W. T. Edmondson (1989), "Rotifer Study as a Way of Life."

3. Edmondson interview, April 30, 1991.

4. Edmondson (1989), "Rotifer Study as a Way of Life."

5. Edmondson interview, April 30, 1991.

6. Forbes (1887), "Lake as a Microcosm."

7. Hutchinson (1957), *Treatise on Limnology,* vol. 1.

8. Hutchinson (1979), *Kindly Fruits,* 17–18, note 12.

9. Edmondson (1989), "Rotifer Study as a Way of Life."

10. Edmondson interview, April 30, 1991.

11. Edmondson and Hutchinson (1934), "Yale North India Expedition: Report on Rotatoria."

12. Edmondson interview, April 30, 1991. Hutchinson's replacement at Yale while

he was on the North India Expedition was Richard Bond. He had some collections from Hispaniola. Edmondson did the rotifers and wrote up that paper, which, like Hutchinson's first paper, was published when he was fifteen, while he was still in high school.

13. G. E. Hutchinson interview with Nancy G. Slack, November 1990, at Yale University. Smithsonian Institution.

14. Raymond Rappaport interview with Nancy G. Slack, June 15, 1999, Bar Harbor, Maine. In author's possession.

15. Edmondson interview, April 30, 1991. He stayed in Hutchinson's lab throughout his undergraduate years at Yale. Miss Ross later took him to the American Museum of Natural History to meet F. J. Myers, with whom he had been corresponding at Hutchinson's suggestion. Myers was an independently wealthy honorary research associate, a rotifer expert, and an artist.

16. G. E. Hutchinson to Chairman of the Zoology Department, University of Wisconsin, December 11, 1937. Hutchinson Papers, MSSA, Yale University Library.

17. Edmondson interview, April 30, 1991. Smithsonian Institution.

18. See Hudson and Gosse (1886), *Rotifera or Wheel Amimalcules.*

19. Gordon A. Riley (1984), "Reminiscences of an Oceanographer," unpub. memoir, 166 pages. In author's possession. The quotes from Riley that follow are taken from his memoir.

20. Gordon A. Riley interview with Eric Mills, Dalhousie University, Halifax, Nova Scotia, October 27, 1980. In author's possession.

21. Hutchinson (1982), preface in Wroblewski, ed., *Selected Works of Gordon A. Riley.*

22. Gordon Riley interview with Eric Mills.

23. Riley (1984), "Reminiscences of an Oceanographer."

24. Hutchinson (1982), preface.

25. Gordon Riley interview with Eric Mills. Gordon Riley (1939), "Limnological Studies in Connecticut," *Ecological Monographs* 9:53–94.

26. Edward S. Deevey was a faculty member at Yale, at Dalhousie University, and at the University of Florida; he was the eminent ecologist of the Ecological Society of America in 1982, a member of the National Academy of Sciences, and the author of well over one hundred papers over a fifty year period.

27. Dian Deevey interview with Nancy G. Slack, October 5, 1996, Quechee Lake, Vt. Smithsonian Institution. Dian Deevey was Ed Deevey's second wife. She recounted a conversation she had had with Hutchinson in 1990.

28. Fossil pollen had been studied in Europe for many years, going back to a 1896 paper by Carl A. Weber, a German who produced the first pollen grain percentages from a peat deposit.

29. Edmondson (1997), National Academy of Sciences memoir of Edward S. Deevey, Jr.

30. Hutchinson (1979), *Kindly Fruits,* 240–241.

31. Eleven of Nichols's papers are included in Deevey's dissertation reference list, although none of Nichols's research was in paleoecology. The papers of geologist R. F.

Flint, later an associate of Deevey's at Yale, were also helpful to Deevey. Flint had done research on many aspects of the Pleistocene glacial geology of Connecticut. This work and Flint's interpretations were directly applicable to Deevey's research.

32. Hutchinson, December 7, 1984, comments from a symposium, "Topics in Historical Ecology," quoted by W. T. Edmondson, in NAS mcmoir of Edward S. Deevey, Jr.

33. Deevey (1937), "Pollen from Interglacial Beds in the Pang-gong Valley."

34. Hutchinson response to Deevey proposal. G. Evelyn Hutchinson to Dr. E. S. Deevey, Oceanographic Institution, Woods Hole, Mass., September 26, 1944. Hutchinson Papers, MSSA, Yale University Library.

35. G. Evelyn Hutchinson, Confidential Statement concerning E. S. Deevey. November 10, 1937. Hutchinson Papers, MSSA, Yale University Library.

36. Max Dunbar interview with Nancy G. Slack, June 20, 1993, in Montreal, Canada. Smithsonian Institution. Quotes from Dunbar in this section are from this interview.

Chapter 8. The Old and the New Limnology

1. Hutchinson (1979), *Kindly Fruits*, 70-71.
2. Hutchinson (1919), "*Gerris lateralis asper* Fab."
3. Hutchinson (1979), *Kindly Fruits*, 72-73.
4. Hutchinson (1940), "Revision of the Corixidae of India and Adjacent Regions."
5. Hutchinson (1979), *Kindly Fruits*, 72.
6. Hutchinson (1923, 1924), "Contributions toward a List of the Insect Fauna of the South Ebudes. IV. The Hemiptera (Rynchota)."
7. Hutchinson (1979), *Kindly Fruits*, 89.
8. Butler (1923), *Biology of the British Hemiptera-Heteroptera*.
9. E. A. Butler, Clapham Commons, W. W. 11, to G. Evelyn Hutchinson, June 9, 1922. Hutchinson Papers, MSSA, Yale University Library.
10. Hutchinson, Pickford, and Schuurman (1932), "Contribution to the Hydrobiology of Pans and Other Inland Waters of South Africa." See fig. 1, p. 19, and fig. 2, p. 24.
11. Hutchinson, Pickford, and Schuurman (1929), "Inland Waters of South Africa."
12. Elster (1972), *History of Limnology*.
13. Ibid.
14. Ibid.
15. Mills (1989), *Biological Oceanography*.
16. Welch (1935), *Limnology*.
17. Worthington (1983), *Ecological Century*, 51-52.
18. Ibid., 54-57.
19. Ibid., 67-68.
20. Agassiz (1850), *Lake Superior*.
21. Frey, ed. (1963), *Limnology in North America*.
22. For evaluations of the Birge-Juday era at the University of Wisconsin, see Frey

(1963), "The Birge-Juday Era." In Frey, ed., *Limnology in North America*, 3–54; and Beckel (1987), *Breaking New Waters*.

23. G. Evelyn Hutchinson to E. B. Wilson, January 14, 1938, and E. B. Wilson to G. Evelyn Hutchinson, January 15, 1938. Both in the Hutchinson Papers, MSSA, Yale University Library.

24. Hutchinson, Deevey, and Wollack (1939), "Oxidation-Reduction Potential of Lake Waters."

25. Gould (1989), *Wonderful Life*, 77, and Hutchinson (1930), "Restudy of Some Burgess Shale Fossils."

26. Hutchinson (1932), "Experimental Studies in Ecology. I. Magnesium Tolerance of Daphniidae and Its Ecological Significance."

27. Hutchinson and Pickford (1932), "Limnological Observations on Mountain Lake, Virginia."

28. Hutchinson (1933), "Limnological Studies at High Altitude in Ladakh."

29. Hutchinson (1937), "Limnological Studies in Indian Tibet."

30. Hutchinson (1939), "Ecological Observations on the Fishes of Kashmir and Indian Tibet."

31. Hutchinson (1937), "Contribution to the Limnology of Arid Regions."

32. Hutchinson and Wollack (1940), "Studies on Connecticut Lake Sediments. II. Chemical Analyses of a Core from Linsley Pond, North Branford."

33. Hutchinson (1941), "Limnological Studies in Connecticut. IV. Mechanism of Intermediary Metabolism in Stratified Lakes."

34. Beckel (1987), *Breaking New Waters*. See also Hutchinson, *Kindly Fruits*, 246–248 and note 21, and Mills (1989), *Biological Oceanography*, 265–268, for a discussion of Hutchinson's and Riley's views of hypothesis testing and the difficulties of publishing both Riley's and Lindeman's papers.

35. Lindeman (1942), "The Trophic-Dynamic Aspect of Ecology."

36. Cook (1977), "Raymond Lindeman and the Trophic-Dynamic Concept in Ecology."

37. Beckel (1987), *Breaking New Waters*.

38. G. Evelyn Hutchinson to L. G. M. Baas Becking, Sydney, Australia, January 17, 1951. Hutchinson Papers, MSSA, Yale University Library.

39. Hutchinson (1957, 1967, 1975, 1993), *Treatise on Limnology*, vols. 1–4.

40. Brooks (1946), "Cyclomorphosis in Daphnia. I. An Analysis of *D. retrocurva* and *D. galeata*."

41. Hutchinson (1970), "Chemical Ecology of Three Species of *Myriophyllum*."

42. Cowgill (1974), "Hydrogeochemistry of Linsley Pond. II. The Chemical Composition of the Aquatic Macrophytes."

43. Vitt and Slack (1975), "Analysis of the Vegetation of *Sphagnum*-Dominated Kettle-Hole Bogs in Relation to Environmental Gradients."

44. Patrick (1967), "Invasion Rate, Species Pool and Size of Area on the Structure of the Diatom Community."

45. Hutchinson (1975), Bibliography, *Treatise on Limnology*, vol. 3, 573.

46. Y. Edmondson (1993), foreword to Hutchinson, *Treatise on Limnology*, vol. 4, ix–xi.

47. Hutchinson (1993), preface, *Treatise on Limnology*, vol. 4, xiv.

48. Ibid., xiv–xviii.

49. Hutchinson (1993), *Treatise on Limnology*, vol. 4, 9.

50. Hutchinson (1918), "Swimming Grasshopper."

51. Hutchinson (1993), preface, *Treatise on Limnology*, vol. 4, xvii.

52. Wetzel (1975), *Limnology*, viii.

53. Cowgill and Hutchinson (1964), "Cultural Eutrophication in Lago di Monterosi during Roman Antiquity."

54. Hutchinson (1968), "Vittorio Tonolli, 1913–1967." This article is the source for the next two paragraphs.

55. Hutchinson (1959), "Il Concetto Moderno di Nicchia Ecologica."

56. See bibliography of this book for these papers.

Chapter 9. Radioisotopes

1. Hutchinson and Bowen (1947), "Direct Demonstration of the Phosphorus Cycle in a Small Lake."

2. Hutchinson and Bowen (1950), "Limnological Studies in Connecticut. IX. A Quantitative Radiochemical Study of the Phosphorus Cycle in Linsley Pond."

3. Hutchinson (1932), "Experimental Studies in Ecology. I. The Magnesium Tolerance of Daphniidae and Its Ecological Significance."

4. F. Ronald Hayes, Dalhousie University, to G. Evelyn Hutchinson, January 30, 1950. Daniel Livingstone interview with Nancy G. Slack, February 21, 1999, New York State University at Stonybrook. Smithsonian Institution.

5. Edward S. Deevey to Arthur Hasler, University of Wisconsin, September 25, 1944. Hutchinson Papers, MSSA, Yale University Library.

6. Cook (1977), "Raymond Lindeman and the Trophic-Dynamic Concept in Ecology."

7. Slack and Hagen (1992), "Raymond Lindeman and G. Evelyn Hutchinson." Poster Session, Ecological Society of America annual meeting.

8. G. Evelyn Hutchinson to Edward Deevey, September 26, 1944. Hutchinson Papers, MSSA, Yale University Library.

9. Warren Weaver, Rockefeller Foundation, to Dean William C. Devane, Yale University, June 22, 1950. Dean William Devane to Provost E. S. Furniss, Yale University, December 18, 1950. Hutchinson Papers, MSSA, Yale University Library.

10. W. T. Edmondson interview with Nancy G. Slack, April 30, 1991. Smithsonian Institution.

11. McIntosh (1985), *Background of Ecology*.

12. G. Evelyn Hutchinson to H. H. Goldsmith, Atomic Energy Commission Group, United Nations, October 5, 1948. Hutchinson Papers, MSSA, Yale University Library.

13. L. R. Swart, Chief of Personnel, Brookhaven National Laboratory, to G. Evelyn Hutchinson, August 28, 1947. Hutchinson Papers, MSSA, Yale University Library.

14. Bocking (1997), *Ecologists and Environmental Politics.* Auerbach (1965), "Radionuclide Cycling."

15. Taylor (1988), "Technological Optimism."

16. Wolfe (1967), "Radioecology."

17. Golley (1993), *History of the Ecosystem Concept in Ecology,* 96.

18. Olson (1965), "Equations for Cesium Transfer in a Liriodendron Forest."

19. Olson (1964), "Advances in Radiation Ecology." Olson (1966), "Progress in Radiation Ecology."

20. Witcamp (1966), "Biological Concentrations of Cesium 137 and Strontium 90 in Arctic Food Chains."

21. Golley (1993), *History of the Ecosystem Concept in Ecology,* 73.

22. Slack (2003), "Are Research Schools Necessary?"

Chapter 10. Biogeochemistry

1. Gorham (1991), "Biogeochemistry."

2. G. Evelyn Hutchinson, February 20, 1989, unpub. letter for Foreign Member R.S. [of the Royal Society], London, Mitarbeiter Akademie Wissenschaften Wien. In author's possession.

3. G. Evelyn Hutchinson to Dr. Cecilia H. Paine, Harvard Observatory, January 17, 1933. Hutchinson Papers, MSSA, Yale University Library.

4. G. Evelyn Hutchinson to Dr. Hugh Ramage, Carrow Hill, Norwich, Eng., January 28, 1933. Hutchinson Papers, MSSA, Yale University Library.

5. G. Evelyn Hutchinson (1930), "On the Chemical Ecology of Lake Tanganyika." Hutchinson (1932), "Experimental Studies in Ecology. I. The Magnesium Tolerance of Daphniidae and Its Ecological Significance."

6. Hutchinson and Wollack (1940), "Studies on Connecticut Lake Sediments. II. Chemical Analyses of a Core from Linsley Pond, North Branford." Gorham (1991), "Biogeochemistry."

7. McIntosh (1985), *Background of Ecology.* Hutchinson (1940), Review of *Bio-ecology* (F. E. Clements and V. E. Shelford), *Ecology* 21:267–268.

8. Ibid. Odum (1953), *Fundamentals of Ecology.*

9. Gorham (1991), "Biogeochemistry."

10. Hutchinson (1943), "Thiamin in Lake Waters and Aquatic Organisms." Hutchinson and Setlow (1946), "Limnological Studies in Connecticut. VIII. The Niacin Cycle in a Small Inland Lake."

11. G. Evelyn Hutchinson to Edward Conway, 1945. Hutchinson Papers, MSSA, Yale University Library.

12. In 1949–50 the staff of the American Museum of Natural History is listed as Albert Parr, director; Margaret Mead under anthropology; Ernst Mayr under birds; and G. Evelyn Hutchinson, M.A., as consultant. Memo in Hutchinson Papers, MSSA, Yale University Library.

13. G. Evelyn Hutchinson to Dr. W. A. Albrecht, December 28, 1944. Hutchinson Papers, MSSA, Yale University Library.

14. Hutchinson (1947), "Problems of Oceanic Geochemistry." Hutchinson (1945), "Aluminum in Soils, Plants and Animals."

15. Hutchinson (1950), "Survey of Contemporary Knowledge of Biogeochemistry. III. *Biogeochemistry of Vertebrate Excretion*," preface, xi.

16. Ibid. All references mentioned or quoted can be found in the bibliography of *Biogeochemistry of Vertebrate Excretion*.

17. Ibid., introduction, 1.

18. Ibid., 45. Written by Cieza de León, trans. C. R. Markham.

19. Ibid.

20. Ibid., 49 ff.

21. Ibid., 50, fig. 17, and 55, fig. 18.

22. G. Evelyn Hutchinson to the company, Guanos y Fertilizantes de Mexico, March 29, 1985. Reply from the company to Hutchinson, April 9, 1945. Hutchinson Papers, MSSA, Yale University Library.

23. Gale Blosser to G. Evelyn Hutchinson, February 19, 1950. Hutchinson Papers, MSSA, Yale University Library.

24. G. Evelyn Hutchinson to James F. Chapin, May 9, 1949. Hutchinson Papers, MSSA, Yale University Library.

25. Douglas C. Carroll, Sr., to G. Evelyn Hutchinson, March 5, 1948, and G. Evelyn Hutchinson to Douglas C. Carroll, Sr., March 9, 1948. Hutchinson Papers, MSSA, Yale University Library.

26. Quoted by Hutchinson in *Biogeochemistry of Vertebrate Excretion*, 140, from an anonymous writer of about 1845.

27. Hutchinson (1950), *Biogeochemistry of Vertebrate Excretion*, 162, fig. 46.

28. Ibid., 201.

29. Ibid., 310. The 1154 *Geographie d'Edrisi* was translated by Jaubert and published in Paris in 1836.

30. Ibid., 346. Hutchinson quoted Darwin's 1844 *Geological Observations of the Volcanic Islands Visited during the Voyage of HMS "Beagle."*

31. Ibid., 364–365.

32. Lynge (1934), "General Results of Recent Norwegian Research on Arctic Lichens."

33. Hutchinson (1950), *Biogeochemistry of Vertebrate Excretion*, 371–372.

34. Hutchinson (1947), "Note on the Theory of Competition between Two Social Species."

35. Hutchinson (1950), *Biogeochemistry of Vertebrate Excretion*, 481–482.

36. Murphy (1950), "Review of G. Evelyn Hutchinson, *Biogeochemistry of Vertebrate Excretion* (1950)."

37. N. H. Fisher to Albert Parr, Director, American Museum of Natural History, December 13, 1950. Hutchinson Papers, MSSA, Yale University Library. Luis Gamarra Dulanto to G. Evelyn Hutchinson, May 5, 1951. Hutchinson Papers, MSSA, Yale University Library.

38. Catherine Skinner interview with Nancy G. Slack, April 9, 1992, Yale University. Smithsonian Institution.

39. Cowgill, Hutchinson, and Skinner (1968), "Elementary Composition of *Latimeria chalumnae* Smith." Thomson (1991), *Living Fossil.*

40. Hutchinson, Setlow, and Brooks (1946), "Biochemical Observations on *Asterias forbesi."*

41. E. H. Bradley to G. Evelyn Hutchinson, September 17, 1948. Hutchinson Papers, MSSA, Yale University Library.

42. G. Evelyn Hutchinson to E. H. Bradley, September 24, 1948. Hutchinson Papers, MSSA, Yale University Library.

Chapter 11. The Three Wives and Yemaiel

1. Daniel A. Livingstone, from Duala, Cameroon, to G. Evelyn Hutchinson, May 29, 1985. Hutchinson Papers, MSSA, Yale University Library.

2. Slack (1995), "Continuum of Couples in Ecology and Botany."

3. Hutchinson (1979), *Kindly Fruits.*

4. Biographical information about Grace Pickford comes, in part, from a Newnham College (Cambridge) obituary written by Penelope Jenkin and Anna Bidder for the January 1987 Newnham College newsletter. Penelope Jenkin also talked about Pickford with me: Nancy G. Slack interview with Penelope Jenkin, August 4, 1993, in North Wales. Smithsonian Institution.

5. Richmond (1997), "Lab of One's Own."

6. E. A. Butler to G. Evelyn Hutchinson, September 5, 1925. Hutchinson Papers, MSSA, Yale University Library.

Dorothea Hutchinson and her family received a cable from Evelyn from Cape Town, South Africa, announcing their wedding there. Dorothea also wrote to me (the author): "I'm certain Grace went out to South Africa first. I remember E.[velyn] getting amusing letter from her describing the voyage." In author's possession.

7. Further biographical information about Pickford, especially her scientific work, can be found in Atz (1970), "Grace Pickford Retires"; Ball (1987), "In Memoriam"; Patricia Stocking Brown (1994), "Early Women Ichthyologists."

8. G. Evelyn Hutchinson to Grace E. Pickford, late March 1932, from Kashmir, India. Hutchinson Papers, MSSA, Yale University Library.

9. G. Evelyn Hutchinson to Grace E. Pickford from the Yale North India Expedition, October 13, 1931, and late March 1932. Hutchinson Papers, MSSA, Yale University Library.

10. Joseph Omer-Cooper to G. Evelyn Hutchinson, n.d., 1934. Hutchinson Papers, MSSA, Yale University Library.

11. John Nicholas to Charles Seymour, November 22, 1945. Hutchinson Papers, MSSA, Yale University Library.

12. Margaret Wright interview with Nancy G. Slack, November 29, 1996, Vassar College, Poughkeepsie, N.Y. Smithsonian Institution.

13. Patricia Stocking Brown (1994), "Early Women Ichthyologists."

14. See Pickford's account of her *Fundulus* work in Pickford (1973), "Introductory Remarks." Also Atz (1986), "*Fundulus heteroclitus* in the Laboratory." Atz dedicated this paper to Pickford: "archetypal experimental zoologist, strict and inspiring teacher, mentor and friend. In her hands *Fundulus heteroclitus* (the mummingchog) has illuminated comparative endocrinology."

15. James Barrow interview with Nancy G. Slack, September 21, 1998. In author's possession.

16. Pickford (1973), "Introductory Remarks."

17. Thomson (1991), *Living Fossil.*

18. Pickford and Atz (1957), *Physiology of the Pituitary Gland of Fishes.*

19. Dennis Taylor telephone interview with Nancy G. Slack, October 1, 1998. In author's possession.

20. Pickford (1973), "Introductory Remarks."

21. Ball (1987), "In Memorium."

22. Ibid.

23. Anna Bidder interview with Nancy G. Slack, August 2, 1993, in Cambridge, Eng. Smithsonian Institution.

24. Grace E. Pickford, from Hiram College, Hiram, Ohio, to Hannah Hutchinson, September 29, 1974. In author's possession.

25. Margaret Wright interview with Nancy G. Slack, November 29, 1996. Smithsonian Institution.

26. Anna Bidder interview with Nancy G. Slack, August 2, 1993, in Cambridge, Eng. Smithsonian Institution.

27. G. Evelyn Hutchinson to Dorothea Hutchinson, July 31, 1984. Hutchinson Papers, MSSA, Yale University Library.

28. David Boulton eulogy at Christ Church, New Haven. "In Memoriam Margaret Hutchinson," September 27, 1983. In author's possession.

29. Letters and news clippings about Margaret Hutchinson. In author's possession.

30. Dorothea Hutchinson interview with Nancy G. Slack, July 25, 1991, Cambridge, Eng. Smithsonian Institution.

31. Willard Hartman (and Charles Remington) interview with Nancy G. Slack, May 23, 1991, at Yale University. Smithsonian Institution.

32. Dorothea Hutchinson from Cambridge, Eng., to Nancy G. Slack, May 5, 1995. In author's possession.

33. Dorothea Hutchinson from Cambridge, Eng., to Nancy G. Slack, July 25, 1991. In author's possession.

34. Margaret Seal Hutchinson to Dorothea Hutchinson, June 15, 1961, to Leslie, Hannah, and Dorothea (Hutchinson), June 1, 1961, to Dorothea Hutchinson, August 20, 1977. In author's possession.

35. G. Evelyn Hutchinson to Margaret Hutchinson from Krakow, Poland, August 21 (no year). Hutchinson Papers, MSSA, Yale University Library.

36. Ibid.

37. G. Evelyn Hutchinson to Margaret Hutchinson en route to Italy (n.d.). Hutchinson Papers, MSSA, Yale University Library.

38. G. Evelyn Hutchinson to Margaret Hutchinson from Pisa, Italy, "Saturday" (n.d.). Hutchinson Papers, MSSA, Yale University Library.

39. G. Evelyn Hutchinson to Margaret Hutchinson from Pisa, Italy, "Sunday" (n.d.). Hutchinson Papers, MSSA, Yale University Library.

40. G. Evelyn Hutchinson to Margaret Hutchinson from Florence, Italy, "Tuesday" (n.d.). Hutchinson Papers, MSSA, Yale University Library.

41. G. Evelyn Hutchinson to Margaret Hutchinson from Florence, Italy, "Wednesday" (n.d.). Hutchinson Papers, MSSA, Yale University Library.

42. Ibid.

43. G. Evelyn Hutchinson to Michael Fletscher, Geological Survey, Reston, Va., April 7, 1981. Hutchinson Papers, MSSA, Yale University Library.

44. Letter to G. Evelyn Hutchinson from a friend in Sweden (no name), December 1982. Hutchinson Papers, MSSA, Yale University Library.

45. G. Evelyn Hutchinson to Yemaiel Aris, December 17, 1982. In author's possession.

46. G. Evelyn Hutchinson to Sybil Marcuse, September 12, 1983. In author's possession.

47. Dorothea Hutchinson to Nancy G. Slack, May 5, 1995. In author's possession.

48. Condolence letters from Fred and Peggy Smith and Luigi and Rose Provasoli, 1983. Hutchinson Papers, MSSA, Yale University Library.

49. Edwine Martz to Evelyn Hutchinson, September 1, 1983. Hutchinson Papers, MSSA, Yale University Library.

50. Patrick Finnerty to Evelyn Hutchinson, July 8, 1984. Hutchinson Papers, MSSA, Yale University Library.

51. Letter from "Jerry," of Monks Wood, Broadway, Worcester, Eng., to G. Evelyn Hutchinson, December 1984. Hutchinson Papers, MSSA, Yale University Library.

52. G. Evelyn Hutchinson to Yemaiel Aris, September 6, 1983. In author's possession.

53. Sybil Marcuse interview with Nancy G. Slack, August 17, 1992, San Francisco. Smithsonian Institution.

54. Dorothea Hutchinson interview with Nancy G. Slack, July 25, 1991, and August 1, 1993, Cambridge, Eng. Smithsonian Institution.

55. According to Yemaiel, her parents had never officially married: "They had a long sort of standing relationship but never got married because my father was already married and my mother insisted that she wouldn't break up the family. By the time they actually got together, my mother was in her thirties and my father between fifty and sixty." Yemaiel Aris interview with Nancy G. Slack, July 24, 1981, London. Smithsonian Institution.

56. This and succeeding quotes are from the scrapbook that Gwendle (Sah Oved) kept during her daughter's stay in America. It contains many letters (copied by Sah Oved into the book) about Yemaiel from Evelyn and Margaret Hutchinson, news clip-

pings, copies of Epstein's drawings, and her own thoughts. It is inscribed, "This book is about Yemaiel Oved and is made for her." The scrapbook is in Yemaiel Oved Aris's possession.

57. Yemaiel Aris interview with Nancy G. Slack, July 24, 1991, London. Smithsonian Institution.

58. Letters from Margaret Hutchinson to Yemaiel's mother, Sah Oved, February 2, February 11, and March (n.d.) 1941. In Yemaiel Aris's possession.

59. Further letters from Margaret Hutchinson to Sah Oved in 1943, all collected in the latter's scrapbook. In Yemaiel Aris's possession.

60. Ibid.

61. Letters from G. Evelyn Hutchinson to Sah Oved, 1941 to 1943. Copied into Sah Oved's scrapbook. In Yemaiel Aris's possession.

62. Yemaiel Aris interview with Nancy G. Slack, July 24, 1991. Smithsonian Institution.

63. Newspaper clippings in Sah Oved's scrapbook. In Yemaiel Aris's possession.

64. Yemaiel Aris interview with Nancy G. Slack, July 24, 1991. Smithsonian Institution.

65. Ibid.

66. Yemaiel and Ben Aris interview with Nancy G. Slack, August 7, 1993. Smithsonian Institution.

Ben's family was connected with Hutchinson's: Ben's mother corresponded with the master of Pembroke College, Cambridge, about a portrait she had painted. The master was Arthur Hutchinson, Evelyn's father, who suggested the name of Benjamin or Ben for her son (from the above interview).

67. Ibid.

68. G. Evelyn Hutchinson to Yemaiel Aris, December 17, 1982, and March 28 and September 6, 1983. In author's possession.

69. G. Evelyn Hutchinson to Yemaiel Aris, December 23, 1984, and January 11, 1985. In author's possession.

70. G. Evelyn Hutchinson to Yemaiel Aris, February 8, 1985. In author's possession.

71. G. Evelyn Hutchinson to Dorothea Hutchinson, February 16, 1985. Hutchinson Papers, MSSA, Yale University Library.

72. G. Evelyn Hutchinson to Dorothea Hutchinson, March 25, 1985. Hutchinson Papers, MSSA, Yale University Library.

73. Daniel A. Livingstone, from Duala, Cameroon, to G. Evelyn Hutchinson, May 29, 1985. Hutchinson Papers, MSSA, Yale University Library.

Chapter 12. Good Friends

1. The last months of his life, from January to May 1991, were spent in England. See chapter 17.

2. Hutchinson (1979), *Kindly Fruits.*

3. Ibid., 105–110. Koestler (1971), *Case of the Midwife Toad.*

4. Margaret Mead to Evelyn Hutchinson, April 6, 1948. Hutchinson Papers, MSSA, Yale University Library.

5. G. Evelyn Hutchinson to Sharon Kingsland, October 14, 1980. Hutchinson Papers, MSSA, Yale University Library.

6. Josiah Macy, Jr., Foundation (1950), Instructions for editing manuscripts. Hutchinson Papers, MSSA, Yale University Library.

7. G. Evelyn Hutchinson to W. T. Edmondson, Oceanographic Institution, Woods Hole, Mass., March 18, 1946. Hutchinson Papers, MSSA, Yale University Library.

8. Hutchinson (1948a), "Circular Causal Systems in Ecology." See also Taylor (1988), "Technocratic Optimism."

9. Margaret Mead to Evelyn Hutchinson, February 19, 1948, on Institute for Intercultural Studies letterhead. Hutchinson Papers, MSSA, Yale University Library.

10. Jane Howard (1984), *Margaret Mead*, 156, 272–273 and note.

11. Margaret Mead to Evelyn Hutchinson, January 16, 1948, and Margaret Mead to Evelyn Hutchinson, January 8, 1948. Both in Hutchinson Papers, MSSA, Yale University Library.

12. Mary Catherine Bateson (1984), *With a Daughter's Eye*, 88.

13. Ibid., 94. Hutchinson attended Christ Church, the High Episcopal church in New Haven, Connecticut; his memorial service was held there.

14. Hutchinson (1948b), "On Living in the Biosphere."

15. M. C. Bateson, *With a Daughter's Eye*.

16. Mead (1949), *Male and Female*.

17. G. Evelyn Hutchinson to Margaret Mead, American Museum of Natural History, June 23, 1948. Hutchinson Papers, MSSA, Yale University Library.

18. Margaret Mead to Evelyn Hutchinson, July 16, 1948. Hutchinson Papers, MSSA, Yale University Library.

19. Mead (1949), *Male and Female*, 246.

20. Ibid., 395–396.

21. Margaret Mead to Evelyn Hutchinson, March 19, 1948, and G. Evelyn Hutchinson note to Margaret Mead written on the above letter. Hutchinson Papers, MSSA, Yale University Library.

22. Margaret Mead to Evelyn Hutchinson, February 22, 1951. Hutchinson Papers, MSSA, Yale University Library.

23. G. Evelyn Hutchinson to Margaret Mead, March 3, 1951. Hutchinson Papers, MSSA, Yale University Library.

24. Margaret Mead to Evelyn Hutchinson, January 16, 1948. Hutchinson Papers, MSSA, Yale University Library.

25. G. Evelyn Hutchinson to Margaret Mead, January 23, 1962. Hutchinson Papers, MSSA, Yale University Library.

26. Hutchinson (1959), "Speculative Consideration."

27. Margaret Mead to Evelyn Hutchinson, August 6, 1959. Hutchinson Papers, MSSA, Yale University Library.

28. Freeman (1982), *Making and Unmaking of an Anthropological Myth*.

29. Hutchinson (1950), "Notes on the Functions of a University."

30. M. C. Bateson (1984), *With a Daughter's Eye*, 97.

31. G. Evelyn Hutchinson to Miss Sarah C. Carslake, Director of Admissions, Brearley School, February 24, 1950. Hutchinson Papers, MSSA, Yale University Library.

32. M. C. Bateson (1984), *With a Daughter's Eye*, 69–70.

33. Mark (1984), "Personal Views of an Enigma." Review of Bateson, *With a Daughter's Eye*.

34. Jane Howard (1984), *Margaret Mead*.

Chapter 13. Fond Correspondents

1. G. Evelyn Hutchinson to Yemaiel Aris, March 28, 1983, given to Nancy G. Slack by Yemaiel Aris. In author's possession.

2. Rebecca West to G. Evelyn Hutchinson from Ibstone House, Ibstone, Bucks., Eng., January 17, 1947. Rebecca West Papers, Beinecke Rare Book and Manuscript Library, Yale University (henceforth West Papers, Beinecke Library).

3. Rebecca West telegram to G. Evelyn Hutchinson, February 13, 1947, West Papers, Beinecke Library.

4. Rebecca West to G. Evelyn Hutchinson, April 17, 1947. West Papers, Beinecke Library.

5. Glendinning (1987), *Rebecca West*, 205.

6. Rebecca West to Margaret and Evelyn Hutchinson, July 5, 1947. West Papers, Beinecke Library. Quotes in the following paragraphs are from this letter.

7. Ibid.

8. Rollyson (1996), *Rebecca West*, 28–30.

9. Ibid., 34–35.

10. Rebecca West (1928) in *Time and Tide*, London, quoted in Marcus (1982), *Young Rebecca*, 3–5.

11. Rebecca West (1913), Review of G. B. Shaw's *Androcles and the Lion* in *New Freewoman*, September 15, quoted in Marcus (1982), *Young Rebecca*, 24–25.

12. Rebecca West (1912), in *Freewoman*, July 25, 1912, quoted in Marcus, *Young Rebecca*, 49–52.

13. Rebecca West (1912) in *Freewoman*, September 19, quoted in Marcus (1982), *Young Rebecca*, 64–69.

14. Marcus (1982), *Young Rebecca*, 94–96.

15. Ibid., 351.

16. Quoted by Marcus in *Young Rebecca*, 352.

17. G. Evelyn Hutchinson, private notebook, 1984, with notes about a visit to Durham Cathedral and the Grand Coulee Dam, Beinecke Library, Yale University.

18. Schaffner (1988), "Rebecca West Bio Charts Rare Life," review of Glendinning, *Rebecca West: A Life*.

19. Ibid.

20. West (1941), *Black Lamb and Grey Falcon*.

21. *Time Magazine*, December 8, 1948.

22. Rollyson (1996), *Rebecca West*, 260–261.

23. Rebecca West (1982), *1900*.

24. Rebecca West (1977), *Rebecca West: A Celebration*.

25. Glendinning (1987), *Rebecca West*, 206.

26. G. Evelyn Hutchinson to Rebecca West, January 6, no year [1948?]. G. Evelyn Hutchinson Papers, Special Collections, McFarlin Library, University of Tulsa (henceforth Hutchinson Papers, University of Tulsa).

27. Rebecca West to Margaret and Evelyn Hutchinson from High Wycombe, Bucks., Eng., January 16, 1950. West Papers, Beinecke Library. Thirty-five years of Hutchinson's letters to Rebecca West are not at Yale. After her death in 1983 her heirs sold them to the University of Tulsa (Oklahoma), where this author read a great many of them.

28. G. Evelyn Hutchinson to Rebecca West, April 6, no year [1954]. Hutchinson Papers, University of Tulsa.

29. Rollyson (1996), *Rebecca West*, 290.

30. Hutchinson (1953), "The Dome," in Hutchinson, *The Itinerant Ivory Tower: Scientific and Literary Essays*.

31. Margaret Hutchinson to Rebecca West, December 12, 1957. Hutchinson Papers, University of Tulsa.

32. G. Evelyn Hutchinson to Rebecca West, February 23, 1958. Hutchinson Papers, University of Tulsa.

33. G. Evelyn Hutchinson to Rebecca West, April 15, 1958. Hutchinson Papers, University of Tulsa.

34. G. Evelyn Hutchinson to Rebecca West, November 16, 1958. Hutchinson Papers, University of Tulsa.

35. Margaret Hutchinson to Rebecca West, July 25, 1959. Hutchinson Papers, University of Tulsa.

36. Ibid.

37. G. Evelyn Hutchinson to Rebecca West, September 29, 1959. Hutchinson Papers, University of Tulsa.

38. Rebecca West to Margaret and Evelyn Hutchinson, April 2, 1955. West Papers, Beinecke Library.

39. Rebecca West to Margaret and Evelyn Hutchinson, not dated but "February 1955" notation on letter. West Papers, Beinecke Library.

40. Rebecca West telegram to Hutchinson, October 15, 1955. West Papers, Beinecke Library.

41. Rebecca West to Margaret and Evelyn Hutchinson, January 16, 1956. West Papers, Beinecke Library.

42. Rollyson (1996), *Rebecca West*, 308; West (1957), *Court and the Castle*.

43. Rollyson (1996), *Rebecca West*, 311.

44. Rebecca West to Margaret and Evelyn Hutchinson, February 7, 1959. West Papers, Beinecke Library.

45. Rollyson (1996), *Rebecca West*, 332.

46. Ibid., 337.

47. G. Evelyn Hutchinson to Henry Andrews, March 22, 1961. Hutchinson Papers, University of Tulsa.

48. G. Evelyn Hutchinson to Rebecca West, May 27, 1961. Hutchinson Papers, University of Tulsa.

49. Margaret Hutchinson to Rebecca West, December 30, 1962. Hutchinson Papers, University of Tulsa.

50. Rebecca West to Professor Hutchinson, January 17, 1947. Hutchinson Papers, University of Tulsa.

51. G. Evelyn Hutchinson to Rebecca West, March 5, 1963. Hutchinson Papers, University of Tulsa.

52. G. Evelyn Hutchinson to Henry Andrews, April 8, 1963. Hutchinson Papers, University of Tulsa.

53. S. Dillon Ripley interview with Nancy G. Slack, October 16, 1991, at his home in Litchfield, Conn. Smithsonian Institution.

54. G. Evelyn Hutchinson to Rebecca West, August 11, 1963. Hutchinson Papers, University of Tulsa.

55. Slack (2003), "Are Research Schools Necessary?"; Nancy G. Slack, "The Rise and Fall of Ecology and the Hutchinson Research Group at Yale." In author's possession.

56. Rebecca West to Margaret and Evelyn Hutchinson, May 27, 1970. West Papers, Beinecke Library.

57. G. Evelyn Hutchinson to Rebecca West, October 25, 1971. Hutchinson Papers, University of Tulsa.

58. Rebecca West to Margaret and Evelyn Hutchinson, June 4, 1971. West Papers, Beinecke Library.

59. Edward Deevey to Rebecca West, October 21, 1970. Hutchinson Papers, MSSA, Yale University Library.

60. West (1972), "Evelyn Hutchinson" in Deevey, ed. (1972), *Growth by Intussusception,* 11–16.

61. Leslie Hutchinson to Rebecca West, January 15, 1972. Hutchinson Papers, University of Tulsa.

62. G. Evelyn Hutchinson to Rebecca West, October 13, 1972. Hutchinson Papers, University of Tulsa.

63. G. Evelyn Hutchinson to Rebecca West, November 19, 1972. Hutchinson Papers, University of Tulsa.

64. G. Evelyn Hutchinson to Rebecca West, 1976, undated. Hutchinson Papers, University of Tulsa.

65. Margaret Hutchinson to Rebecca West, October 28, 1978. Hutchinson Papers, University of Tulsa.

66. G. Evelyn Hutchinson to Rebecca West, March 22, 1978. Hutchinson Papers, University of Tulsa.

67. Glendinning (1987), *Rebecca West,* xiii, 205–206.

68. Hutchinson (1979), *Kindly Fruits,* 126. Rebecca West arranged, at Hutchinson's suggestion, to have her letters from H. G. Wells and many other letters and

papers deposited at Yale. After Rebecca's death, Hutchinson contributed his (and his wife Margaret's) letters from Rebecca to the Beinecke Library.

Chapter 14. From the N-dimensional Niche to Santa Rosalia

1. McIntosh (1985), *Background of Ecology*, 277, 280.
2. Platil and Rosenzweig (1979), *Contemporary Quantitative Ecology and Related Econometrics*, preface.
3. For Ph.D. dates and dissertation topics of these and other Hutchinson students to 1971, see Y. Edmondson (1971), "Doctoral Dissertations Completed under the Supervision of G. Evelyn Hutchinson," 169-172.
4. McIntosh (1985), *Background of Ecology*, 190.
5. Hutchinson (1957), "Concluding Remarks."
6. Hutchinson (1944), "Limnological Studies in Connecticut."
7. Hutchinson (1959), "Homage to Santa Rosalia." Quotations in the following paragraphs are from this work.
8. Wilson, ed. (1988), *Biodiversity*.
9. Hutchinson (1957), "Concluding Remarks."
10. Lack (1954), *Natural Regulation of Animal Numbers*; E. P. Odum (1953), *Fundamentals of Ecology*.
11. Hutchinson (1959), "Homage to Santa Rosalia."
12. MacArthur (1955), "Fluctuations of Animal Populations."
13. Brown and Wilson (1956), "Character Displacement."
14. Hutchinson (1959), "Homage to Santa Rosalia."
15. Woodwell and Smith, eds. (1969), *Diversity and Stability in Ecological Systems*, conference at Brookhaven National Laboratory. This was the first major conference attended by the author while a Ph.D. student in ecology. I and also several people interviewed for this book felt that this conference was seminal in our thinking about ecology. The Cold Spring Harbor Symposium more than ten years earlier, where Hutchinson delivered his "Concluding Remarks," may have had a similar effect on an earlier cohort of ecology students.
16. Slobodkin (1961), "Preliminary Ideas for a Predictive Theory of Ecology." Slobodkin (1962), *Growth and Regulation of Animal Populations*.
17. MacArthur (1962), "Growth and Regulation of Animal Populations" (a review).
18. MacArthur (1958), "Population Ecology of Some Warblers of Northeastern Coniferous Forests."
19. Hutchinson (1978), *Introduction to Population Ecology*, 167, fig. 106.
20. Ibid., 168.
21. Watson (1981), "Patterns of Microhabitat Occupation of Six Closely Related Species."
22. Levins (1968), *Evolution in a Changing Environment*.
23. For reviews, see Hutchinson (1980), *Introduction to Population Biology*; also Slack (1990), "Bryophytes and Ecological Niche Theory."

24. May and MacArthur (1972), "Niche Overlap as a Function of Environmental Variability."

25. Ashmole and Ashmole (1967), "Comparative Feeding Ecology of Sea Birds."

26. Diamond (1975), "Assembly of Species Communities."

27. Dayton (1979), "Ecology, a Science and a Religion."

28. Hutchinson (1975), "Variations on a Theme by Robert MacArthur."

29. Lovejoy (1974), "Bird Diversity and Abundance in Amazon Forest Communities."

30. F. M. Scudo and J. R. Ziegler (1978), *The Golden Age of Theoretical Ecology, 1923–1940* (New York: Springer-Verlag).

31. McIntosh (1985), 280. In the late sixties ecologist Michael Rosenzweig, currently at the University of Arizona, taught a group of ecology graduate students, including the author, to read and evaluate these mathematical papers.

32. Krebs (1979), "Small Mammal Ecology."

33. H. T. Odum and E. P. Odum (1955), "Trophic Structure and Productivity of a Windward Coral Reef Community on Eniwetok Atoll." H. T. Odum (1957), "Trophic Structure and Productivity of Silver Springs, Florida."

34. Kingsland (1985), *Modeling Nature.* See, e.g., the following: Schoener (1982), "Controversy over Interspecific Competition"; Simberloff (1982), "Status of Competition Theory in Ecology"; Lewin (1983), "Santa Rosalia Was a Goat."

35. McIntosh (1985), *Background of Ecology,* 281.

36. May (1975), "Patterns of Species Abundance and Diversity."

37. Cody and Diamond, eds. (1975), *Ecology and Evolution of Communities,* vii.

38. Robert H. MacArthur to G. Evelyn Hutchinson, July 9, 1971, from Marlboro, Vt., and 1972 (undated). Both in Hutchinson Papers, MSSA, Yale University Library. MacArthur died November 1, 1972.

39. Elizabeth (Betsy) MacArthur to G. Evelyn Hutchinson, undated, from Marlboro, Vt. Hutchinson Papers, MSSA, Yale University Library.

40. Robert MacArthur, University of Texas, to Bernard Patten at Duke University, February 7, 1958. In author's possession. This letter and those of other Hutchinson students were related to a course taught at Duke by Oosting.

41. Hutchinson and MacArthur (1959a), "Theoretical Ecological Model of Size Distributions among Species of Animals"; Hutchinson and MacArthur (1959b), "On the Theoretical Significance of Aggressive Neglect in Interspecific Competition."

42. G. Evelyn Hutchinson to Gerald Freund, June 4, 1981. Hutchinson Papers, MSSA, Yale University Library.

43. Wilson and Hutchinson (1989), "Robert Helmer MacArthur."

44. Kingsland (1985), *Episodes in the History of Population Biology,* 205.

Chapter 15. Hutchinson the Environmentalist

1. Hutchinson (1943), "Marginalia."

2. Hutchinson (1979), *Kindly Fruits.*

3. Hutchinson (1970), "Biosphere."

4. Hutchinson's Tyler award acceptance speech, 1974. In author's possession.

5. Ibid.

6. G. Evelyn Hutchinson to Raymond Baldwin, Governor of Connecticut, February 19, 1943. Hutchinson Papers, MSSA, Yale University Library.

7. Fairfield Osborn, President, New York Zoological Society, to G. Evelyn Hutchinson, June 19, 1947. Hutchinson Papers, MSSA, Yale University Library. Osborn (1948), *Our Plundered Planet*.

8. H. M. Gray, University of Illinois, Urbana, to Fairfield Osborn, President, New York Zoological Society, May 21, 1948. Hutchinson Papers, MSSA, Yale University Library.

9. Minutes by G. E. Hutchinson of meeting on Yale University Conservation Program, April 25, 1957. Hutchinson Papers, MSSA, Yale University Library.

10. Hutchinson (1949), "Note on Two Aspects of the Geochemistry of Carbon."

11. Hutchinson (1973), "Eutrophication."

12. Clyde Goulden interview with Nancy G. Slack, September 26, 1991, Philadelphia Academy of Sciences. Smithsonian Institution.

13. Philip Crowley, Rice University, Laboratory of Environmental Science and Engineering, Houston, to G. Evelyn Hutchinson, October 8, 1970. Hutchinson Papers, MSSA, Yale University Library.

14. G. Evelyn Hutchinson to Philip Crowley, Rice University, Houston, October 19, 1970. Hutchinson Papers, MSSA, Yale University Library.

15. Donald C. McNaught, State University of New York at Albany, to G. Evelyn Hutchinson, September 21, 1971. Hutchinson Papers, MSSA, Yale University Library.

16. G. Evelyn Hutchinson to Donald C. McNaught, October 5, 1971. Hutchinson Papers, MSSA, Yale University Library.

17. Vitt and Slack (1975), "Analysis of the Vegetation of Sphagnum-Dominated Kettle-Hole Bogs," quoted in Hutchinson (1975), *Treatise on Limnology*, v. 3:58–61.

18. G. Evelyn Hutchinson to Edwin L. Chapin, Jr., August 26, 1969. Hutchinson Papers, MSSA, Yale University Library.

19. U.S. Congress, *Congressional Record*, 95th Cong., 123 (56), March 30, 1977; Hon. Frederick W. Richmond to Dr. E. Hutchinson, May 9, 1977. Both in Hutchinson Papers, MSSA, Yale University Library.

20. Hutchinson (1978), *Introduction to Population Ecology*, 91.

21. Ibid., 95–96, notes 6, 7, 13, 14.

22. Hutchinson (1978), *Introduction to Population Ecology*, 105, note 32.

23. G. Evelyn Hutchinson to Edwin L. Chapin, Jr., August 26, 1969. Hutchinson Papers, MSSA, Yale University Library.

24. Ruth Patrick, Academy of Natural Sciences, Philadelphia, to G. Evelyn Hutchinson, July 29, 1980, with appended biographical sketch by Hutchinson. Hutchinson Papers, MSSA, Yale University Library.

25. Jeremy Holloway, Commonwealth Institute of Entomology, London, to Thomas Lovejoy, World Wildlife Fund, March 10, 1981. Hutchinson Papers, MSSA, Yale University Library.

26. Kenneth R. McKaye, Duke University, to G. Evelyn Hutchinson, March 11, 1982. Hutchinson Papers, MSSA, Yale University Library.

27. G. Evelyn Hutchinson to Thomas E. Lovejoy, February 22, 1980. Hutchinson Papers, MSSA, Yale University Library.

28. Thomas E. Lovejoy, World Wildlife Fund–U.S., to Walter G. Rosen, National Academy of Sciences, July 10, 1982. Hutchinson Papers, MSSA, Yale University Library.

Quammen (1996), *Song of the Dodo*, 445–463. Lovejoy (1979), "Refugia, Refuges and Minimal Critical Size in the Conservation of the Neotropical Herpetofauna." Thomas Lovejoy interview with Nancy G. Slack, April 25, 1994, at the Smithsonian Institution. Smithsonian Institution.

29. Ripley (1975), *Paradox of the Human Condition*, 41.

30. List of charitable contributions by G. Evelyn and Margaret Hutchinson. In author's possession.

31. "The Status of the Ecological Society of America" December 14, 1945, report by the Endowment Committee, Charles C. Adams, Chairman. Hutchinson Papers, MSSA, Yale University Library.

32. Ibid.

33. Minutes of the third annual meeting of the Ecologists Union, December 31, 1947, Ralph W. Dexter, Secretary-Treasurer. Hutchinson Papers, MSSA, Yale University Library.

34. Minutes of the annual meeting of the Ecologists Union, December 28, 1949, "Ecologists Union at the Crossroads," Ralph W. Dexter, Secretary-Treasurer. Hutchinson Papers, MSSA, Yale University Library.

35. "Preservation of 'Living Museums' Sought through Nature Conservancy Bill," memoir of the Ecologists Union "for release at will," undated [1950]. Hutchinson Papers, MSSA, Yale University Library. Congressman Charles E. Bennett of Florida had introduced a bill to establish a Nature Conservancy of the United States. The bill did not provide for any appropriations from the federal government.

36. *AIBS Newsletter* 3 (2), September 1950, published by the American Institute of Biological Sciences, Washington, D.C. Item: "The Ecologists Union Becomes the Nature Conservancy." Hutchinson Papers, MSSA, Yale University Library.

37. G. Evelyn Hutchinson to Raymond Baldwin, Governor of Connecticut, February 19, 1943. Hutchinson Papers, MSSA, Yale University Library.

38. Letter and survey from the Legislative Reference Service of the Library of Congress on the preservation of wilderness areas, sent to the Limnological Society of America, October 22, 1948. Hutchinson Papers, MSSA, Yale University Library.

39. "Living Museums of Primeval America: A Need and an Opportunity." Ecologists Union, May 1950. Hutchinson Papers, MSSA, Yale University Library. E. G. Leigh to Nancy G. Slack, October 19, 2003. In author's possession. Alison Jolly interview with Nancy G. Slack, March 17, 1993. See their publications in the bibliography.

40. Hutchinson (1948), "On Living in the Biosphere." Quotations in the following paragraphs are from this paper.

41. Hutchinson (1949), "Note on Two Aspects of the Geochemistry of Carbon."

42. Sitwell (1925), "Little Ghost Who Died for Love."

43. Morris Miller, Economic Development Institute, Washington, D.C., to G. Evelyn Hutchinson, December 3, 1970. Hutchinson Papers, MSSA, Yale University Library.

44. W. T. Edmondson (1991), "Uses of Ecology."

45. Beamish (1970), *Aldabra Alone*, 48–49. Quoted from the Protest Letter of 1874.

46. Ibid., 31.

47. Ibid.

48. Ibid., 180–181. Quoted from the *New Yorker*.

49. Hutchinson to Rebecca West, March 21, 1967. Hutchinson Papers, University of Tulsa.

50. Huxley (1970), foreword to Beamish, *Aldabra Alone*.

51. G. Evelyn Hutchinson's notes for the Royal Society about Aldabra. In author's possession.

52. Beamish (1970), *Aldabra Alone*, 198–199.

53. G. Evelyn Hutchinson to Rebecca West, November 9, 1959. Hutchinson Papers, University of Tulsa.

54. Arthur Galston interview with Nancy G. Slack, April 25, 1994, at Yale University. Smithsonian Institution. Quotes in the following paragraphs are from this interview.

55. G. Evelyn Hutchinson to Rebecca West, December 16, 1973. Hutchinson Papers, University of Tulsa.

56. G. Evelyn Hutchinson to Neil Chalmers, director, British Museum (Natural History), December 11, 1990, written with the assistance of Nancy G. Slack. Copy in author's possession.

Chapter 16. The Polymath

1. Snow (1958), *The Search*. Snow was a fellow student of Bernal's at Cambridge University. In this novel the character Constantine, about whom these words were actually written, was the well-recognized surrogate for Bernal. See footnote 5, *Search*, for the "sage of science."

2. Friday (1997), *Wicken Fen* (quotation from Hutchinson).

3. Y. Edmondson (1971), *Limnology and Oceanography* issue dedicated to G. Evelyn Hutchinson.

4. Slobodkin and Slack (1999), "George Evelyn Hutchinson."

5. Mattox (2002), *Rosalind Franklin*.

6. Brown (1995), *J. D. Bernal*; Snow (1958), *Search*, 39, 45–46, 62–63.

7. Hutchinson (1953), *Itinerant Ivory Tower*; Hutchinson (1962), *Enchanted Voyage*; Hutchinson (1965), *Ecological Theater and the Evolutionary Play*.

8. Hutchinson (1979), *Kindly Fruits*.

9. Hutchinson (1974), "Attitudes toward Nature in Medieval England."

10. Ibid.

11. The contents of this library, belonging to the earl of Macclesfield, were sold at auction at Sotheby's in 2004 and bought by the J. Paul Getty Museum in Los Angeles. But because the Macclesfield Psalter was considered a national treasure by the British arts minister, it was barred for export and eventually, after a public fundraising campaign, was bought back for the Fitzwilliam.

12. Fitzwilliam Museum (2005), *Macclesfield Psalter.* The text quoted is by Stella Panayotova, Fitzwilliam Museum, Cambridge, Eng.

13. Hutchinson (1978), "Zoological Iconography in the West after A.D. 1200."

14. Ibid., p. 679, fig. 5.

15. Garber (1996), "John and Elizabeth Gould."

16. Hutchinson (1979b), review of Zoltán Kádár (1978), *Survival of Greek Zoological Illuminations in Byzantine Manuscripts.*

17. Kingsland (2010), "The Beauty of the World."

18. Hutchinson (1972), "Long Meg Reconsidered."

19. Hutchinson (1973), "Eutrophication."

20. Clyde Goulden interview with Nancy G. Slack, September 26, 1991, Philadelphia Academy of Sciences. Smithsonian Institution.

21. Hutchinson (1986), "What Is Science For?" Quotations in the following paragraphs come from this article.

22. Hutchinson (1943), "Marginalia."

Chapter 17. The Last Years

1. G. Evelyn Hutchinson to Yemaiel Aris, March 28, 1983. In author's possession.

2. G. Evelyn Hutchinson letters to Dorothea Hutchinson from 1980 to 1989, given to Nancy G. Slack, now in the Hutchinson Papers, MSSA, Yale University Library.

3. G. Evelyn Hutchinson to Dorothea Hutchinson, January 17, 1980. Hutchinson Papers, MSSA, Yale University Library.

4. G. Evelyn Hutchinson to Dorothea Hutchinson, April 22, 1980. Hutchinson Papers, MSSA, Yale University Library.

5. G. Evelyn Hutchinson to Dorothea Hutchinson, August 25, 1982. Hutchinson Papers, MSSA, Yale University Library.

6. G. Evelyn Hutchinson to Rebecca West, May 7, 1981. Hutchinson Papers, University of Tulsa.

7. Programme for Professor G. E. Hutchinson (honorary degree, Cambridge University), Xerox copy provided by Dorothea Hutchinson. Hutchinson Papers, MSSA, Yale University Library.

8. G. Evelyn Hutchinson to Dorothea Hutchinson, June 17, 1981. Hutchinson Papers, MSSA, Yale University Library.

9. G. Evelyn Hutchinson to Rebecca West, July 3, 1981. Hutchinson Papers, University of Tulsa.

10. G. Evelyn Hutchinson to Rebecca West, September 24, 1981. Hutchinson Papers, University of Tulsa.

11. Michelle Press, letter enclosing eightieth birthday invitation to Dorothea Hutchinson, April 5, 1983. Copy in author's possession.

12. G. Evelyn Hutchinson to Dorothea Hutchinson, January 23 and January 29, 1983. Hutchinson Papers, MSSA, Yale University Library.

13. G. Evelyn Hutchinson to Dorothea Hutchinson, February 20, 1983. Hutchinson Papers, MSSA, Yale University Library.

14. G. Evelyn Hutchinson to Dorothea Hutchinson, February 15, 1984. Hutchinson Papers, MSSA, Yale University Library.

15. G. Evelyn Hutchinson to Dorothea Hutchinson, April 22, 1984. Hutchinson Papers, MSSA, Yale University Library.

16. G. Evelyn Hutchinson to Dorothea Hutchinson, June 16, 1984. Hutchinson Papers, MSSA, Yale University Library.

17. G. Evelyn Hutchinson to Dorothea Hutchinson, June 26, 1984. Hutchinson Papers, MSSA, Yale University Library.

18. G. Evelyn Hutchinson to Dorothea Hutchinson, September 1, 1984. Hutchinson Papers, MSSA, Yale University Library.

19. G. Evelyn Hutchinson to Dorothea Hutchinson, January 6, 1985. Hutchinson Papers, MSSA, Yale University Library.

20. G. Evelyn Hutchinson to Dorothea Hutchinson, Easter Eve, 1985. Hutchinson Papers, MSSA, Yale University Library.

21. Rosemary and Peter Dodd interview with Nancy G. Slack, July 29, 1991, Newcastle-on-Tyne. Smithsonian Institution.

22. G. Evelyn Hutchinson to Dorothea Hutchinson, August 1, 1985. Hutchinson Papers, MSSA, Yale University Library.

23. G. Evelyn Hutchinson to Dorothea Hutchinson, September 20, 1985. Hutchinson Papers, MSSA, Yale University Library.

24. G. Evelyn Hutchinson to Dorothea Hutchinson, September 20, 1985. Hutchinson Papers, MSSA, Yale University Library.

25. Henry Mitchell, "Ecologist Evelyn Hutchinson," *Washington Post*, September 8, 1985. Original article in author's possession.

26. G. Evelyn Hutchinson to Dorothea Hutchinson, October 12, 1985. Hutchinson Papers, MSSA, Yale University Library.

27. G. Evelyn Hutchinson to Dorothea Hutchinson, November 11, 1985. Hutchinson Papers, MSSA, Yale University Library.

28. G. Evelyn Hutchinson to Dorothea Hutchinson, January 14, 1986. Hutchinson Papers, MSSA, Yale University Library.

29. G. Evelyn Hutchinson to Dorothea Hutchinson, June 14, 1986. Hutchinson Papers, MSSA, Yale University Library.

30. G. Evelyn Hutchinson to Dorothea Hutchinson, July 10, 1986. Hutchinson Papers, MSSA, Yale University Library. Quotations in following paragraphs are from this letter.

31. G. Evelyn Hutchinson to Dorothea Hutchinson, October 19, 1986. Hutchinson Papers, MSSA, Yale University Library.

32. Anne Hutchinson post card to Dorothea Hutchinson, November 14, 1986. Hutchinson Papers, MSSA, Yale University Library.

33. G. Evelyn Hutchinson, "Introductory Remarks" and "The Lecture of the Kyoto Prize (Basic Science)." In author's possession. The lecture was published as Hutchinson (1987), "Keep Walking."

34. G. Evelyn Hutchinson to Dorothea Hutchinson, December 8, 1986. Hutchinson Papers, MSSA, Yale University Library.

35. Phoebe Ellsworth, University of Michigan, to Nancy Slack, July 24, 1995. In author's possession.

36. G. Evelyn Hutchinson card with rubbing to Dorothea Hutchinson, "late 1986 or early 1987" written on it. Hutchinson Papers, MSSA, Yale University Library.

37. G. Evelyn Hutchinson to Dorothea Hutchinson, February 20, 1987. Hutchinson Papers, MSSA, Yale University Library.

38. G. Evelyn Hutchinson to Dorothea Hutchinson, March 19, 1987. Hutchinson Papers, MSSA, Yale University Library.

39. G. Evelyn Hutchinson to Dorothea Hutchinson, June 1, 1987. Hutchinson Papers, MSSA, Yale University Library.

40. G. Evelyn Hutchinson to Dorothea Hutchinson, June 20, 1987. Hutchinson Papers, MSSA, Yale University Library.

41. G. Evelyn Hutchinson to Dorothea Hutchinson, birthday card, undated. In author's possession. G. Evelyn Hutchinson to Dorothea Hutchinson, June 20, 1987. Hutchinson Papers, MSSA, Yale University Library.

42. Hutchinson (1988), Review of *Angels Fear*.

43. G. Evelyn Hutchinson to Dorothea Hutchinson, November 22, 1987. Hutchinson Papers, MSSA, Yale University Library.

44. Rosemary Dodd interview with Nancy G. Slack, July 29, 1991, Newcastle-on-Tyne. Smithsonian Institution.

45. G. Evelyn Hutchinson to Dorothea Hutchinson, December 24, 1987, and n.d., 1987, from Aruba. Hutchinson Papers, MSSA, Yale University Library.

46. G. Evelyn Hutchinson to Dorothea Hutchinson, February (n.d.) 1988. Hutchinson Papers, MSSA, Yale University Library.

47. G. Evelyn Hutchinson to Dorothea Hutchinson, May 5, 1988. Hutchinson Papers, MSSA, Yale University Library.

48. G. Evelyn Hutchinson to Dorothea Hutchinson, September (n.d.) 1988. Hutchinson Papers, MSSA, Yale University Library.

49. G. Evelyn Hutchinson to Dorothea Hutchinson, December 20, 1988. Hutchinson Papers, MSSA, Yale University Library.

50. G. Evelyn Hutchinson to Dorothea Hutchinson, December 23, 1988. Hutchinson Papers, MSSA, Yale University Library.

51. Frances Evans, "Tributes for 'Daughter of Cambridge,'" *Cambridge Evening News*, October 2001.

52. Stanley and Margaret Leavy interview with Nancy G. Slack, September 15, 1991, New Haven. Smithsonian Institution. Quotations in the following paragraphs are from this interview.

53. Paul Hutchinson and his wife Marie, interview with Nancy G. Slack, July 25, 1991, London. Smithsonian Institution.

54. Anna Aschenbach, New Haven, to the Hutchinson family in England, October 15, 1991. Copy in author's possession.

55. Rosemary and Peter Dodd interview with Nancy G. Slack, July 29 and 30, 1991, Newcastle, Eng. Smithsonian Institution.

56. "Update on Evelyn on Sunday 20th January, 1991," memo from Joyce and Francis Hutchinson, London. In author's possession.

57. Yemaiel Aris interview with Nancy G. Slack, July 24, 1991, London. Smithsonian Institution.

58. Joyce Hutchinson interview with Nancy G. Slack, July 24, 1991, London. Smithsonian Institution.

59. Francis Hutchinson, London, to "Dear Friends," an update on Evelyn Hutchinson [April 12, 1991], sent to the author and to other friends and relatives.

60. Jolly (1966), *Lemur Behavior.*

61. Jolly (2006), "Global Vision."

62. Ruth Patrick to Dr. G. E. Hutchinson, October 7, 1938. Hutchinson Papers, MSSA, Yale University Library.

63. Professor Shoji Horie, Kyoto University, Japan, to G. Evelyn Hutchinson, December 30, 1990, and G. Evelyn Hutchinson to Shoji Horie, from Battersea High Street, London, February 1, 1991. Copy in author's possession.

64. S. Dillon Ripley, Smithsonian Institution, Washington D.C., to G. Evelyn Hutchinson, Cambridge, Eng., March 25, 1991. In author's possession.

65. Joyce Hutchinson interview with Nancy G. Slack, July 24, 1991, London. Smithsonian Institution.

66. Thomas Lovejoy, Smithsonian Institution, Washington D.C., to Francis and Joyce Hutchinson, London, May 29, 1991. Copy in author's possession.

67. John Cairns, Jr., University Distinguished Professor of Environmental Biology, Virginia Polytechnic Institute, to Francis Hutchinson, London, May 23, 1991. Copy in author's possession.

Chapter 18. Concluding Remarks

1. McIntosh (1985), *Background of Ecology.*
2. Hutchinson (1945), "Aluminum in Soils, Plants and Animals."
3. Clements and Shelford (1939), *Bio-Ecology.*
4. McIntosh (1985), *Background of Ecology,* 88.
5. Hutchinson (1940), "Review of *Bio-Ecology* by F. E. Clements and V. E. Shelford."
6. Elton (1927), *Animal Ecology.*
7. Kingsland (1985), *Modeling Nature.*
8. Hutchinson (1957–93), *Treatise on Limnology,* 4 vols.
9. Hutchinson (1959), "Homage to Santa Rosalia."
10. Wilson (1992), *Diversity of Life.*
11. Hutchinson (1978), *Introduction to Population Ecology,* 155.
12. Wilson and Hutchinson (1989), "Robert Helmer MacArthur."

13. Cody and Diamond, eds. (1975), *Ecology and Evolution of Communities*, vii.

14. Hutchinson (1975), "Variations on a Theme by MacArthur." The quote is from MacArthur (1972), "Coexistence of Species."

15. Levin (1999), *Fragile Dominion*, 82.

16. Lewin (1983), "Santa Rosalia Was a Goat." Quotations by Lewin throughout this section are from this article.

17. Ibid.

18. Simberloff and Boeklen (1981), "Santa Rosalia Reconsidered."

19. Roughgarden (1983), "Competition and Theory in Community Ecology." See also Strong, Simberloff, Abele, and Thistle, eds. (1984), *Ecological Issues and the Evidence*, for articles by many of the ecologists cited and the subjects discussed, including null models.

20. Hutchinson (1983), "What Is Science For?"; Egbert G. Leigh to Nancy G. Slack from the Smithsonian Tropical Research Institute (STRI), Panama, October 19, 2003; Leigh (1999), *Tropical Forest Ecology*.

21. Hubbell (2001), *Unified Neutral Theory of Biodiversity and Biogeography*.

22. Deborah Goldberg, Program for the symposium "Integrating the Dispersal-Assembly and Niche-Assembly Paradigms in Plant Community Ecology." XVIII. International Botanical Congress, Vienna, Austria, July 17-23, 2005.

23. Hubbell (2001), *Unified Neutral Theory*, 8.

24. G. Evelyn Hutchinson (1978), *Introduction to Population Ecology*, 166-168, note 35.

25. V. Vandik, University of Bergen, Norway, abstract 3.14.1, "Quantifying the Role of Dispersal, Ecological Drift and Species Sorting in Determining the Diversity of Local Communities." Paper presented at the International Botanical Congress, Vienna, 2005.

26. H. K. Muller-Landau and S. Wright, abstract 3.14.3, "Competition-Colonization Trade-offs and Seed Limitation in Tropical Forests." Paper presented at the International Botanical Congress, Vienna, 2005.

27. W. F. Carson and A. Baumert, abstract 3.14.4. "Competition, Colonization, or Herbivory: Which Explains Patterns of Dominance in Herbaceous Communities?" Paper presented at the International Botanical Congress, Vienna, 2005.

28. Hubbell (2001), *Unified Neutral Theory*, 10; 321-322.

29. Hutchinson (1961), "Paradox of the Plankton."

30. Whittaker and Levin, eds. (1975), *Niche Theory and Application*, 75.

31. Hutchinson (1951), "Copepodology for the Ornithologist," *Ecology* 32:571-577.

32. See, e.g., Slack (1977), *Species Diversity and Community Structure in Bryophytes*.

33. Gould (1989), *Wonderful Life*, 77-78.

34. Levin (1999), *Fragile Dominion*, 82, 109. W. T. Edmondson (1991), *Uses of Ecology*.

35. Y. Edmondson (1971), *Limnology and Oceanography*, 162-163.

36. Levin (1999), *Fragile Dominion*, 82-83.

37. Robert MacArthur, Edward Gray Institute, Oxford University, to Bernard Patten, Duke University, February 7, 1958. In author's possession.

38. Howard T. Odum, University of Texas, Port Arkansas, to Bernard Patten, Duke University, February 7, 1958. In author's possession.

39. Lawrence B. Slobodkin, University of Michigan, to Bernard Patten, Duke University, January 28, 1958. In author's possession.

40. Slobodkin and Slack (1999), "George Evelyn Hutchinson."

41. G. Evelyn Hutchinson to Jean H. Langenheim, April 1, 1986. Hutchinson Papers, MSSA, Yale University Library.

42. Langenheim (1996), "Early History and Progress of Women Ecologists." The author of this book, Nancy G. Slack, also appeared among the later "pioneer" women ecologists in Jean H. Langenheim's paper.

43. G. Evelyn Hutchinson to Gerald Freund, Director, Prize Fellow Program, MacArthur Foundation, June 4, 1981. Hutchinson Papers, MSSA, Yale University Library.

44. G. Evelyn Hutchinson to Jean H. Langenheim, April 1, 1986. Hutchinson Papers, MSSA, Yale University Library.

45. Hutchinson (1979), *Kindly Fruits*, 234.

46. Clem Markert interview with Nancy G. Slack, April 26, 1991, Yale University. Smithsonian Institution.

47. Kabaservice (2004), *Guardians*.

48. Margaret Davis interview with Nancy G. Slack, May 1, 1995, at the University of Minnesota. Smithsonian Institution.

49. Levin (1999), *Fragile Dominion*, 83.

50. G. Evelyn Hutchinson to Dorothea Hutchinson, "Autumn" 1988 and December 19, 1988. Hutchinson Papers, MSSA, Yale University Library.

51. Daniel Livingstone, Duke University, to W. T. Edmondson, University of Washington, December 27, 1991. In author's possession.

Bibliography

Abercrombie, M. 1961. Ross Granville Harrison, 1870–1959, *Biographical Memoirs of the Royal Society*. Vol. 7. London: Royal Society.

Adams, Charles C. 1945. The Status of the Ecological Society of America by the Endowment Committee; Charles C. Adams, Chairman. December 14, 1945.

Agassiz, Louis. 1850. *Lake Superior: Its Physical Character, Vegetation, and Animals* (Narrative by J. E. Cabot). Boston: Gould, Kendall and Lincoln.

American Public Health Association. 1923. *Standard Methods for the Examination of Water and Sewage*, 5th ed. New York.

Ashmole, N. P., and M. J. Ashmole. 1967. Comparative Feeding Ecology of Sea Birds of a Tropical Island. *Bulletin of the Peabody Museum of Natural History* 24:1–131.

Atz, James W. 1970. Grace Pickford Retires. *Discovery* 6:41–43.

———. 1986. *Fundulus heteroclitus* in the Laboratory: A History. *American Zoologist* 26:11–120.

Auerbach, Stanley I. 1965. Radionuclide Cycling: Current Uses and Future Needs. *Health Physics* 11:1355–1361.

Ball, J. N. 1987. In Memoriam, Grace E. Pickford, 1902–1986. *General and Comparative Endocrinology* 65:162–165.

Bateson, Mary Catherine. 1984. *With a Daughter's Eye: A Memoir of Margaret Mead and Gregory Bateson*. New York: William Morrow.

Beamish, Tony. 1970. *Aldabra Alone*. San Francisco: Sierra Club Books.

Beckel, Annamarie L. 1987. Breaking New Waters, a Century of Limnology at the University of Wisconsin. *Transactions of the Wisconsin Academy of Sciences, Arts and Letters*. Special Issue.

Behnke, J. A., ed. 1972. *Challenging Biological Problems*. New York: A.I.B.S., Oxford University Press.

Bibby, Cyril. 1960. *T. H. Huxley: Scientist, Humanist, and Educator*. New York: Horizon Press.

Bocking, Stephen. 1997. *Ecologists and Environmental Politics: A History of Contemporary Ecology*. New Haven: Yale University Press.

Brooks, John L. 1946. Cyclomorphosis in Daphnia. I. An Analysis of D. retrocurva and D. galeata. *Ecological Monograph* 16:409–447.

Brown, Andrew. 1995. *J. D. Bernal: The Sage of Science*. Oxford: Oxford University Press.

Brown, Patricia Stocking. 1994. Early Women Ichthyologists. *Environmental Biology of Fishes* 41:9–30.

Brown, W. L., and E. O. Wilson. 1956. Character Displacement. *Systematic Zoology* 5:49–64.

Butler, E. A. 1923. *A Biology of the British Hemiptera-Heteroptera*. London: Witherby.

Carson, Rachel. 1962. *Silent Spring*. New York: Houghton Mifflin.

Clements, Frederic E., and Victor E. Shelford. 1939. *Bio-Ecology*. New York: John Wiley and Sons.

Cody, Martin L., and Jared M. Diamond, eds. 1975. *Ecology and Evolution of Communities*. Cambridge: Belknap Press.

Cook, Robert E. 1977. Raymond Lindeman and the Trophic Concept in Ecology. *Science* 198:22–26.

Cowgill, Ursula M. 1974. The Hydrogeochemistry of Linsley Pond. II. The Chemical Composition of the Aquatic Macrophytes. *Archiv für Hydrobiologie. Supplementband* 45:1–119.

Cowgill, Ursula M., and G. Evelyn Hutchinson. 1964. Cultural Eutrophication in Lago di Monterosi during Roman Antiquity. *International Association of Theoretical and Applied Limnology Proceedings* 15:644–645.

Cowgill, Ursula M., G. Evelyn Hutchinson, and H. C. W. Skinner. 1968. The Elementary Composition of *Latimeria chalumnae* Smith. *Proceedings of the National Academy of Sciences* 60:456–463.

Dayton, Robert. 1979. Ecology, a Science and a Religion. In *Ecological Processes in Coastal and Marine Ecosystems*, ed. R. J. Livingston. New York: Plenum Press.

Deevey, E. S. 1937. Pollen from Interglacial Beds in the Pang-gong Valley and Its Climatic Interpretation. *American Journal of Sciences* 235:44–56.

———, ed. 1972. Growth by Intussusception: Ecological Essays in Honor of G. Evelyn Hutchinson. *Transactions*, Connecticut Academy of Arts and Sciences 13:16.

Diamond, Jared M. 1975. Assembly of Species Communities. In *Ecology and Evolution of Communities*, ed. M. L. Cody and J. M. Diamond, 342–444. Cambridge: Belknap Press.

Edmondson, W. T. 1989. Rotifer Study as a Way of Life. *Hydrobiologia* 186/187:1–9.

———. 1991. *The Uses of Ecology: Lake Washington and Beyond*. Seattle: University of Washington Press.

Edmondson, W. T., and G. Evelyn Hutchinson. 1934. Yale North India Expedition. Report on Rotatoria. *Memoir, Connecticut Academy of Arts and Sciences* 10:153–186.

Edmondson, Yvette H. 1971. *Limnology and Oceanography* 16, no. 2. Dedicated to G. Evelyn Hutchinson. Reprinted 1991 as vol. 36, no. 3.

———. 1993. Foreword. In G. Evelyn Hutchinson, *A Treatise on Limnology*, vol. 4. New York: John Wiley and Sons.

Elster, Hans Joachim. 1972. History of Limnology. Trans. from German by T. T. Macan. *Mitteilungen: Internationale Vereinigung für theoretische und angewandte Limnologie* 20:7–30.

Elton, Charles S. 1927. *Animal Ecology*. London: Sidgwick and Jackson.

Farrar, W. W. 1962. *St. Winfred or the World of School*. Edinburgh: A. and C. Black.

Fitzwilliam Museum. 2005. *The Macclesfield Psalter*. Cambridge, Eng., Fitzwilliam Museum.

Forbes, S. A. 1887. The Lake as a Microcosm. *Bulletin of the Science Association (Peoria, Illinois)*, 77–87.

Freeman, Derek. 1982. *The Making and Unmaking of an Anthropological Myth*. Cambridge: Harvard University Press.

Frey, David G., ed. 1963. *Limnology in North America*. Madison: University of Wisconsin Press.

Friday, Laurel. 1997. *Wicken Fen: The Making of a Wetland Nature Reserve*. Colchester, Eng.: Harley Books.

Garber, Janet Bell. 1996. John and Elizabeth Gould: Ornithologists and Scientific Illustrators, 1829–1841. In *Creative Couples in the Sciences*, ed. Helena M. Pycior, Nancy G. Slack, and Pnina G. Abir-Am. New Brunswick, N.J.: Rutgers University Press, 87–97.

Glendinning, Victoria. 1987. *Rebecca West: A Life*. New York: Alfred A. Knopf.

Golley, Frank B. 1993. *A History of the Ecosystem Concept in Ecology*. New Haven: Yale University Press.

Gorham, Eville. 1991. Biogeochemistry: Its Origins and Development. *Biogeochemistry* 13:199–230.

Gould, Stephen Jay. 1989. *Wonderful Life: The Burgess Shale and the Nature of History*. New York: W. W. Norton.

Goulden, Clyde E., ed. 1977. *The Changing Scenes in Natural Science*. Philadelphia: Philadelphia Academy of Natural Sciences, Special Publication 12.

Hagen, Joel B. 1992. *The Entangled Bank: On the Origins of Ecosystem Ecology*. New Brunswick, N.J.: Rutgers University Press.

Harvey, H. W. 1928. *The Biological Chemistry and Physics of Sea Water.* Cambridge: Cambridge University Press.

Hickson, Alisdare. 1995. *The Poisoned Bowl: Sex, Repression and the Public School System.* London: Constable.

Hogben, Lancelot. 1924. *The Pigmentary Effector System: A Review of the Physiology of Color Response.* Edinburgh: Oliver Boyd.

———. 1927. *The Comparative Physiology of Internal Secretion.* Cambridge: Cambridge University Press.

Holorenshaw, Henry (pseudonym for Joseph Needham). The Making of an Honorary Taoist. In *Changing Perspectives in the History of Science: Essays in Honor of Joseph Needham,* ed. Mikulás Teich and Robert Young. London: Heinemann, 1–20.

Howard, H. Eliot. 1920. *Territory in Bird Life.* London: J. Murray.

Howard, Jane. 1984. *Margaret Mead: A Life.* New York: Simon and Schuster.

Howarth, T. E. B. 1978. *Cambridge between Two Wars.* London: Collins.

Hubbell, Stephen P. 2001. *The Unified Neutral Theory of Biodiversity and Biogeography.* Princeton: Princeton University Press.

Hudson, C. T., and P. H. Gosse. 1886. *The Rotifera or Wheel Animalcules, Both British and Foreign, I and II.* London: Longmans, Green.

Hutchinson, E. D. 1936. *Creative Sex.* London: George Allen and Unwin.

Hutchinson, G. Evelyn. 1918. A Swimming Grasshopper. *Entomological Record. Journal of Variation* 30:138.

———. 1919. *Gerris lateralis* Fab. *Norfolk Entomologists Monthly Magazine* 55:33.

———. 1923 and 1924. Contributions toward a List of the Insect Fauna of the South Ebudes. *Scottish Naturalist,* 1923, pp. 185–191; 1924, pp. 21–27.

———. 1929. A Revision of the Notonectidae and Corixidae of South Africa. *Annals of the South African Museum* 25:359–474 and plates 27–41.

———. 1930a. On the Chemical Ecology of Lake Tanganyika. *Science* 71:616.

———. 1930b. Restudy of Some Burgess Shale Fossils. *Proceedings of the United States National Museum* 78(2584): Article 11:1–24.

———. 1932a. Experimental Studies in Ecology. I. The Magnesium Tolerance of Daphniidae and Its Ecological Significance. *Internationale Revue der Gesamten Hydrobiologie und Hydrographie* 28:90–108.

———. 1932b. On Corixidae from Uganda. *Stylops* 1:37–40.

———. 1933. Limnological Studies at High Altitudes in Ladakh. *Nature* 132:136–138.

———. 1936. *The Clear Mirror: A Pattern of Life in Goa and in Indian Tibet.* Cambridge: Cambridge University Press. Repr. 1978.

———. 1937a. A Contribution to the Limnology of Arid Regions. *Transactions of the Connecticut Academy of Arts and Sciences* 33:47–132.

———. 1937b. Limnological Studies in Indian Tibet. *Internationale Revue der Gesamten Hydrobiologie und Hydrographie* 35:134–177.

————. 1939. Ecological Observations on the Fishes of Kashmir and Indian Tibet. *Ecological Monographs* 9:142–182.

————. 1940a. Review of *Bio-Ecology* (by F. E. Clements and V. E. Shelford). *Ecology* 21:267–268.

————. 1940b. A Revision of the Corixidae of India and Adjacent Regions. *Transactions of the Connecticut Academy of Arts and Sciences* 33:339–476.

————. 1941. Limnological Studies in Connecticut. IV. Mechanisms of Intermediary Metabolism in Stratified Lakes. *Ecological Monographs* 11:21–60.

————. 1943a. Marginalia. *American Scientist* 31:270.

————. 1943b. Thiamin in Lake Waters and Aquatic Organisms. *Archives of Biochemistry* 2:143–150.

————. 1944. Limnological Studies in Connecticut. VII. A Critical Examination of the Supposed Relationship between Phytoplankton Periodicity and Chemical Changes in Lake Waters. *Ecology* 25:3–26.

————. 1945. Aluminum in Soils, Plants and Animals. *Soil Science* 60:29–40.

————. 1947a. A Note on the Theory of Competition between Two Social Species. *Ecology* 28:319–321.

————. 1947b. The Problems of Oceanic Geochemistry. *Ecological Monographs* 17:299–307.

————. 1948a. Circular Causal Systems in Ecology. *Annals of New York Academy of Sciences* 50:221–246.

————. 1948b. On Living in the Biosphere. *Scientific Monthly* 67:393–398.

————. 1949. A Note on Two Aspects of the Geochemistry of Carbon. *American Journal of Science* 247:27–32.

————. 1950a. Notes on the Function of a University. *American Scientist* 38:127–131.

————. 1950b. Survey of Contemporary Knowledge of Biogeochemistry. III. *The Biogeochemistry of Vertebrate Excretion. Bulletin of the American Museum of Natural History,* no. 96. 554 pp.

————. 1951. Copepodology for the Ornithologist. *Ecology* 32:571–577.

————. 1953. *The Itinerant Ivory Tower: Scientific and Literary Essays.* New Haven: Yale University Press.

————. 1957a. Concluding Remarks. *Cold Spring Harbor Symposium on Quantitative Biology* 22:415–427.

————. 1957b. *A Treatise on Limnology.* vol. 1. *Geography, Physics, and Chemistry.* New York: John Wiley and Sons.

————. 1959a. Il Concetto Moderno di Nicchia Ecologica. *Memorie: Instituto Italiano di Idrobiologia* 11:9–22.

————. 1959b. Homage to Santa Rosalia or Why Are There So Many Kinds of Animals? *American Naturalist* 93:145–159.

————. 1959c. A Speculative Consideration of Certain Possible Forms of Sexual Selection in Man. *American Naturalist* 93:81–91.

————. 1961. The Paradox of the Plankton. *American Naturalist* 95:137–146.

————. 1962a. Cambridge Remembered. In G. E. Hutchinson, *The Enchanted Voyage and Other Studies*, 149–156. New Haven: Yale University Press.

————. 1962b. The Enchanted Voyage. In G. Evelyn Hutchinson, *The Enchanted Voyage and Other Studies*, 1–11. New Haven: Yale University Press.

————. 1964. The Lacustrine Microcosm Reconsidered. *American Scientist* 52:334–341.

————. 1965. *The Ecological Theater and the Evolutionary Play*. New Haven: Yale University Press.

————. 1967. *A Treatise on Limnology*. Vol. 2. *Introduction to Lake Biology and the Limnoplankton*. New York: John Wiley and Sons.

————. 1968: Vittorio Tonolli, 1913–1967. *Archiv für Hydrobiologie* 64:491–495.

————. 1970a. The Biosphere. *Scientific American* 223, no. 3 (September): 45–53.

————. 1970b. The Chemical Ecology of Three Species of *Myriophyllum* (Angiospermae, Haloragacae). *Limnology and Oceanography* 15:1–5.

————. 1972. Long Meg Reconsidered. *American Scientist* 60, pt. 1, pp 24–31; pt. 2, pp. 210–219.

————. 1973. Eutrophication. *American Scientist* 61:269–279.

————. 1974. Attitudes toward Nature in Medieval England: The Alphonso and Bird Psalters. *Isis* 65:5–37.

————. 1975a. *A Treatise on Limnology*. Vol. 3. *Limnological Botany*. New York: John Wiley and Sons.

————. 1975b. Variations on a Theme by Robert MacArthur. In *Ecology and Evolution of Communities*, ed. M. L. Cody and Jared M. Diamond, 492–521. Cambridge: Belknap Press.

————. 1977. The Influence of the New World on Natural History. In *The Changing Scenes in Natural Science*, ed. C. E. Goulden. Philadelphia: Philadelphia Academy of Natural Sciences, Special Publication 12.

————. 1978a. *An Introduction to Population Ecology*. New Haven: Yale University Press.

————. 1978b. Zoological Iconography in the West after A.D. 1200. *American Scientist* 66:675–684.

————. 1979a. *The Kindly Fruits of the Earth: Recollections of an Embryo Ecologist*. New Haven: Yale University Press.

————. 1979b. Review of *Survival of Greek Zoological Illuminations in Byzantine Manuscripts* by Zoltán Kádár, 1978. *Isis* 70:452–453.

————. 1980. *Alexander Petrunkevitch: Biographical Memoir*. Washington, D.C.: National Academy of Sciences.

————. 1982. Preface: Reminiscences and Notes on Some Otherwise Undiscussed Papers. In *Selected Works of Gordon A. Riley*, ed. J. S. Wroblewski. Halifax, Nova Scotia: Dalhousie University Press.

————. 1983. What Is Science For? *American Scientist* 71:639–644.

———. 1987a. The Ecological Niche. *Physiology and Ecology Japan* 24: s03–s07.

———. 1987b. Keep Walking. *Physiological Ecology Japan,* Special Number 24: s81–s87.

———. 1988. Review of Gregory Bateson and M. C. Bateson, *Angels Fear: Toward an Epistemology of the Sacred. American Scientist* 76:285–286.

———. 1993. *A Treatise on Limnology.* Vol. 4. *The Zoobenthos.* New York: John Wiley and Sons.

Hutchinson, G. Evelyn, and Vaughan T. Bowen. 1947. A Direct Demonstration of the Phosphorus Cycle in a Small Lake. *Proceedings of the National Academy of Sciences* 33:148–153.

———. 1950. Limnological Studies in Connecticut. IX. A Quantitative Radiochemical Study of the Phosphorus Cycle in Linsley Pond. *Ecology* 31:194–203.

Hutchinson, G. Evelyn, E. S. Deevey, and A. Wollack. 1939. The Oxidation-Reduction Potential of Lake Waters and Their Ecological Significance. *Proceedings of the National Academy of Sciences* 25:87–90.

Hutchinson, G. Evelyn, and Robert H. MacArthur. 1959a. A Theoretical Ecological Model of Size Distribution among Species of Animals. *American Naturalist* 93:117–125.

———. 1959b. On the Theoretical Significance of Aggressive Neglect in Interspecific Competition. *American Naturalist* 93:133–134.

Hutchinson, G. Evelyn, and G. E. Pickford. 1932. Limnological Observation on Mountain Lake, Virginia. *Internationale Revue der Gesamten Hydrobiologie und Hydrographie* 27:252–264.

Hutchinson, G. E., G. E. Pickford, and J. F. M. Schuurman. 1929. The Inland Waters of South Africa. *Nature* 123:832–834.

———. 1932. A Contribution to the Hydrobiology of Pans and Other Inland Waters of South Africa. *Archiv für Hydrobiologie* 24:1–154.

Hutchinson, G. Evelyn, and Jane K. Setlow. 1946. Limnological Studies in Connecticut. VIII. The Niacin Cycle in a Small Inland Lake. *Ecology* 27:13–22.

Hutchinson, G. Evelyn, Jane K. Setlow, and John L. Brooks. 1946. Biochemical Observations on *Asterias forbesi. Bulletin of the Bingham Oceanographic Collection* 9:44–58.

Hutchinson, G. Evelyn, and Anne Wollack. 1940. Studies on Connecticut Lake Sediments. II. Chemical Analyses of a Core from Linsley Pond, North Branford. *American Journal of Science* 238:493–517.

Huxley, Julian. 1970. Foreword. In *Aldabra Alone* by T. Beamish. San Francisco: Sierra Club Books.

Jackson, Margaret. 1994. *The Real Facts of Life: Feminism and the Politics of Sexuality, 1850–1940.* London: Taylor and Francis.

Jolly, Alison. 1966. *Lemur Behavior: A Madagascar Field Study.* Chicago: University of Chicago Press.

———. 2006. A Global Vision. *Nature* 148.

Jolly, Alison, Philippe Oberle, and Roland Albignac, eds. 1984. *Key Environments: Madagascar.* Oxford: Pergamon Press.

Kabaservice, Geoffrey. 2004. *The Guardians: Kingman Brewster, His Circle, and the Rise of the Liberal Establishment.* New York: Henry Holt.

Kingsland, Sharon E. 1985. *Modeling Nature. Episodes in the History of Population Biology.* Chicago: University of Chicago Press.

———. 2010. The Beauty of the World. In *The Art of Ecology: The Writings of G. Evelyn Hutchinson,* ed. David K. Skelly, David M. Post, and Melinda Smith. New Haven: Yale University Press.

Koestler, Arthur. 1971. *The Case of the Midwife Toad.* New York: Random House.

Kofoid, C. A. 1910. *The Biological Stations of Europe. United States Bureau of Education Bulletin,* no. 4, whole no. 440. Washington, D.C.: Government Printing Office.

Kohler, Robert E. 1978. Walter Fletcher, F. G. Hopkins, and the Dunn Institute of Biochemistry: A Case Study in the Patronage of Science. *Isis* 69:331–355.

Krebs, Charles J. 1979. Small Mammal Ecology. *Science* 203:350–351.

Lack, David. 1954. *The Natural Regulation of Animal Numbers.* Oxford: Clarendon Press.

Langenheim, Jean H. 1996. Early History and Progress of Women Ecologists: Emphasis upon Research Contributions. *Annual Review of Ecology and Systematics* 27:1–53.

Leigh, Egbert G. 1999. *Tropical Forest Ecology: A View from Barro Colorado Island.* New York: Oxford University Press.

Levandowsky, M. 1977. A White Queen Speculation. *Quarterly Review of Biology* 52:383–386.

Levin, Simon A. 1999. *Fragile Dominion: Complexity and the Commons.* Cambridge, Mass.: Perseus Publishing.

Levins, Richard. 1968. *Evolution in a Changing Environment.* Princeton: Princeton University Press.

Lewin, Roger. 1983. Santa Rosalia Was a Goat. *Science* 221:636–639.

Lindeman, Raymond L. 1942. The Trophic-Dynamic Aspect of Ecology. *Ecology* 23:399–418.

Livingston, R. J., ed. 1979. *Ecological Processes in Coastal and Marine Ecosystems.* New York: Plenum Press.

Lovejoy, Thomas E. 1974. Bird Diversity and Abundance in Amazon Forest Communities. *Living Bird* 13:127–191.

———. 1979. Refugia: Refuges and Minimal Critical Size in the Conservation of the Neotropical Herpetofauna. In *The South American Herpetofauna, Evolution and Dispersal,* ed. W. E. Duellman. Museum of Natural History, Monograph 7. Lawrence: University of Kansas Press.

Lynge, B. 1934. General Results of Recent Norwegian Research on Arctic Lichens. *Rhodora* 36:133–171.

Macan, T. T. 1968. The Tarns on Dufton Fell. *Newsletter of the Cumbria Trust for Nature Conservation.* August.

MacArthur, Robert H. 1955. Fluctuations of Animal Populations and a Measure of Community Stability. *Ecology* 35:533–536.

———. 1958. Population Ecology of Some Warblers of Northeastern Coniferous Forests. *Ecology* 39: 599–619.

———. 1962. *The Growth and Regulation of Animal Populations* (a review). *Ecology* 43:579.

———. 1972. Coexistence of Species. In *Challenging Biological Problems,* ed. J. A. Behnke. New York: A.I.B.S., Oxford University Press.

Maienschein, Jane. 1983. Experimental Biology in Transition: Harrison's Embryology, 1895–1910. *Studies in History of Biology* 6:107–127.

Marcus, Jane, ed. 1982. *The Young Rebecca: Writings of Rebecca West, 1911–1917.* New York: Viking Press.

Mark, Joan. 1984. Personal Views of an Enigma. *Science* 225:1014–1015.

Mattox, Brenda. 2002. *Rosalind Franklin: The Dark Lady of DNA.* New York: HarperCollins.

May, Robert M. 1975. Patterns of Species Abundance and Diversity. In *Ecology and Evolution of Communities,* ed. Martin L. Cody and Jared M. Diamond, 81–120. Cambridge: Belknap Press.

May, Robert M., and Robert H. MacArthur. 1972. Niche Overlap as a Function of Environmental Variability. *Proceedings of the National Academy of Sciences* 69:1109–1113.

McIntosh, Robert P. 1985. *The Background of Ecology; Concept and Theory.* New York: Cambridge University Press.

Mead, Margaret. 1949. *Male and Female: A Study of the Sexes in a Changing Environment.* New York: William Morrow.

Mills, Eric L. 1989. *Biological Oceanography: An Early History, 1870–1960.* Ithaca: Cornell University Press.

Murphy, Robert Cushman. 1950. Review of G. Evelyn Hutchinson, *Biogeochemistry of Vertebrate Excretion. Ecology* 321:567–569.

Naumann, E., and A. Thienemann. 1922. Vorschlag zur Grundung einer internationalen Vereinigung für theoretische und angewandte Limnologie. *Archiv für Hydrobiologie* 13:585–605.

Odum, Eugene P. 1953. *Fundamentals of Ecology.* Philadelphia: W. B. Saunders.

Odum, Howard T. 1957. Trophic Structure and Productivity of Silver Springs, Florida. *Ecological Monographs* 25:55–112.

Odum, Howard T., and Eugene P. Odum. 1955. Trophic Structure and Productivity of a Windward Coral Reef Community on Eniwetok Atoll. *Ecological Monographs* 25:291–320.

Olson, Jerry S. 1964. Advances in Radiation Ecology. *Nuclear Safety* 6:78–81.
———. 1965. Equations for Cesium Transfer in a Liriodendron Forest. *Health Physics* 11:1385–1392.
———. 1966. Progress in Radiation Ecology: Radionuclide Movement in Major Environments. *Nuclear Safety* 8:53–58.
Osborn, Fairfield. 1948. *Our Plundered Planet*. Boston: Little, Brown.
Pankhurst, Sylvia. 1932. *The Home Front*. London: Hutchinson.
Patrick, Ruth. 1967. The Effect of the Invasion Rate, Species Pool, and Size of Area on the Structure of the Diatom Community. *Proceedings of the National Academy of Sciences* 58:1335–1342.
Pickford, Grace E. 1973. Introductory Remarks. *American Zoologist* 13:711–717.
Pickford, Grace E., and James W. Atz. 1957. *The Physiology of the Pituitary Gland of Fishes*. New York: New York Zoological Society.
Platil, G. P., and M. L. Rosenzweig. 1979. *Quantitative Ecology and Related Econometrics*. Fairland, Md.: International Cooperative Publishing House.
Quammen, David. 1996. *The Song of the Dodo. Island Biogeography in an Age of Extinctions*. New York: Scribner.
Raine, Kathleen. 1975. *The Land Unknown*. New York: George Braziller.
Richmond, Marsha L. 1997. A Lab of One's Own: The Balfour Biological Laboratory for Women at Cambridge University, 1884–1914. *Isis* 88:4422–4455.
Riley, Gordon. 1971. Introduction. In Yvette H. Edmondson, *Limnology and Oceanography* 16, no. 2:177–178.
Ripley, S. Dillon. 1975. *The Paradox of the Human Condition*. New Delhi: Tata McGraw.
Rollyson, Carl. 1996. *Rebecca West: A Life*. New York: Scribner.
Roughgarden, Jonathan. 1983. Competition and Theory in Community Ecology. *American Naturalist* 122:583–601.
Rowbotham, Sheila. 1987. *Friends of Alice Wheeldon*. New York: Monthly Review Press.
Schaffner, Perdita. 1988. Rebecca West Bio Charts Rare Life. *New Directions for Women*, March/April.
Schoener, Thomas W. 1982. The Controversy over Interspecific Competition. *American Scientist* 70:586–595.
Simberloff, Daniel S. 1982. The Status of Competition Theory in Ecology. *Annales Zoologica Fennici* 19:241–254.
Simberloff, Daniel S., and William Boeklen. 1981. Santa Rosalia Reconsidered: Size Ratios and Competition. *Evolution* 35:1206–1228.
Sitwell, Edith. 1925. The Little Ghost Who Died for Love. In *Troy Park*. New York: Alfred A. Knopf.
Slack, Nancy G. 1977. *Species Diversity and Community Structure in Bryophytes*. Albany: University of the State of New York, State Education Department. Bulletin 428.

————. 1990. Bryophytes and Ecological Niche Theory. *Botanical Journal of the Linnean Society* 104:187–213.

————. 1995. A Continuum of Couples in Ecology and Botany. In *Creative Couples in the Sciences,* ed. H. M. Pycior, N. G. Slack, and P. G. Abir-Am, 235–253. New Brunswick, N.J.: Rutgers University Press.

————. 2003. Are Research Schools Necessary? Contrasting Models of Twentieth-Century Research at Yale Led by Ross Granville Harrison, Grace E. Pickford, and G. Evelyn Hutchinson. *Journal of the History of Biology* 36:501–529.

Slack, Nancy G., Dale H. Vitt, and Diana G. Horton. 1980. Vegetation Gradients of Minerotrophically Rich Fens in Western Alberta. *Canadian Journal of Botany* 58:330–350.

Slobodkin, Lawrence B. 1961. Preliminary Ideas for a Predictive Theory of Ecology. *American Naturalist* 95:147–153.

————. 1962. *The Growth and Regulation of Animal Populations.* New York: Holt, Rinehart, and Winston.

————. 1977. On the Present Incompleteness of Mathematical Ecology. *American Scientist* 53:347–357.

Slobodkin, Lawrence B., and Nancy G. Slack. 1999. George Evelyn Hutchinson: Twentieth-Century Ecologist. *Endeavour* 23:24–30.

Snow, C. P. 1958. *The Search.* London: Macmillan.

Stopes, Marie C. 1918. *Married Love: A New Contribution to the Solution of Sex Difficulties.* London: A. C. Fifield.

Strong, Donald R., Jr., Daniel Simberloff, Lawrence Abele, and Anne B. Thistle, eds. 1984. *Ecological Issues and the Evidence.* Princeton: Princeton University Press.

Taylor, Peter L. 1988. Technological Optimism: H. T. Odum and the Partial Transformation of Ecological Metaphor after World War II. *Journal of the History of Biology* 21:213–244.

Thienemann, August. 1925. *Die Binnengewasser Mitteleuropas.* Stuttgart: E. Scheizerbart'sche Berglagsbuchhandlung.

Thomson, Keith S. 1991. *Living Fossil: The Story of the Coelacanth.* New York: W. W. Norton.

Twitty, Victor C. 1966. *Of Scientists and Salamanders.* San Francisco: W. H. Freeman.

Van de Velde, Theodor. 1928. *Ideal Marriage: Its Physiology and Technique.* London: Heinemann.

Vitt, Dale H., and Nancy G. Slack. 1975. An Analysis of the Vegetation of *Sphnagnum*-Dominated Kettle-Hole Bogs in Relation to Environmental Gradients. *Canadian Journal of Botany* 53:332–359.

Watson, Maxine. 1981. Patterns of Microhabitat Occupation of Six Closely Related Species of Mosses along a Complex Altitudinal Gradient. *Ecology* 62:1067–1078.

Welch, Paul S. 1935. *Limnology,* 1st ed. New York: McGraw-Hill.

West, Rebecca. 1941. *Black Lamb and Grey Falcon.* New York: Viking Press.

————. 1957. *The Court and the Castle.* New Haven: Yale University Press.

————. 1972. Evelyn Hutchinson. In E. S. Deevey, ed., Growth by Intussusception, 11–16.

————. 1977. *Rebecca West: A Celebration.* London: Macmillan.

————. 1982. *1900.* New York: Viking Press.

Wetzel, Robert G. 1975. *Limnology.* Philadelphia: W. B. Saunders.

Whittaker, Robert H., and Simon A. Levin, eds. 1975. *Niche Theory and Application.* Stroudsburg, Pa.: Hutchinson and Ross.

Wilson, Edward O., ed. 1988. *Biodiversity.* Washington, D.C.: National Academy Press.

————. 1992. *The Diversity of Life.* New York: W. W. Norton.

————. 1994. *Naturalist.* Washington, D.C.: Island Press.

Wilson, Edward O., and Evelyn G. Hutchinson [sic]. 1989. Robert Helmer MacArthur, April 17, 1930–November 1, 1972. *Biographical Memoir,* p. 319. Washington, D.C.: National Academy Press.

Winchester, Simon. 2007. *The Man Who Loved China.* New York: Harper Perennial.

Witcamp, Martin. 1966. Biological Concentrations of Cesium 137 and Strontium 90 in Arctic Food Chains. *Nuclear Safety* 8:58–60.

Wolfe, John. 1969. Radioecology: Retrospection and Future. In *Symposium on Radioecology,* ed. D. J. Nelson and F. C. Evans. Washington, D.C.: U.S. Atomic Energy Commission, Conference Number 670503.

Woodwell, G. M. and H. H. Smith, eds. 1969. *Diversity and Stability in Ecological Systems.* Upton, N.Y.: Brookhaven National Laboratory.

Worthington, E. B. 1983. *The Ecological Century: A Personal Approach.* Oxford: Clarendon Press.

Index

Page numbers in italics refer to photographs and other illustrations. GEH in index refers to G. Evelyn Hutchinson.

Abbott, W. L., 313
Abortion, 332
Adams, Charles C., 306–307
Adrian, E. D., 43, 45, 53, 56–57
Agassiz, Alexander, 66
Agassiz, Louis, 144
Agent Orange, xiii, 14, 263, 318–319
Aggression, 332
Albertus Magnus (author), 326–327
Albertus Magnus College, 204, 205, 210
Albrecht, W. A., 176
Aldabra, xiii, 14, 188, 271–272, 312–317, 368–369
Allee, W. C., 307
Allen, Winifred, 69–70, 401 (n. 13)
Aluminum, 175–177
American Association for the Advancement of Science (AAAS), 12–13, 145, 311
American Geographical Society, 104, 108
American Institute of Biological Sciences (AIBS), 309
American Museum of Natural History, 10, 175–177, 184, 236, 269, 406 (n. 15), 410 (n. 12)
American Scientist ("Marginalia"), xii, 9, 10–11, 51, 112, 252, 263, 297, 299, 323, 330–333, 340

American Society of Limnology and Oceanography (ASLO), 4, 76, 145, 158, 170, 309
American Society of Naturalists, 279
Anderson, Edgar, 59
Anderson, Myrdene, 231, 343, 344, 344
Andersson, K. J., 185
Andrews, Henry Marwell, 70, 251, 253, 258, 260, 261, 265, 267
Angell, J. R., 102–104, 108–111, 404 (n. 19)
Animal behavior, 343–344, 358–359
Animal diversity, 9, 157, 270, 279–288
Anthropology, 54, 101, 132, 240. *See also* Mead, Margaret
Aquatic ecology, 2. *See also* Limnology
Aquatic insects. *See* Water bugs
Archeology, 18, 179, 181, 185–186, 193, 194, 330
Aris, Ben, 229, 230, 231, 274, 343, 345, 359, 365–366, 369, 415 (n. 66)
Aris, Yemaiel Oved: and Alzheimer's disease suffered by Margaret Hutchinson, 221–223, 230, 334; as ballet dancer, 228, 230; childhood of, as GEH's foster child, 10, 29, 200, 223–232; and GEH during her adulthood, 229, 231–232, 250, 262, 343, 345, 354, 359; and GEH's old age, 10, 200,

Aris, Yemaiel Oved (continued)
  365–366, 369; marriage of, 230; parents
  of, 223–225, 227, 229, 414–415 (nn. 55–56);
  photographs of, 226; and Rebecca West's
  death, 250, 274
Art, 9, 41, 69, 70, 112, 210, 217–220, 230, 265,
  324–329, 325, 343–345, 354
Aruba, 189, 358
Aschenbach, Anna, 153, 334, 346, 363, 364,
  365
Ashmole, M. J. and N. P., 287–288
Atkins, W. R. G., 81
Atomic Energy Commission (AEC), 165–
  170, 373
Atz, James, 208, 413 (n. 14)
Auden, W. H., 33, 37, 39
Auer, V., 128
Auerbach, Stanley, 165–167
Austin, Ramona, 341
Australopithecus, 77, 82

Baas Becking, L. G. M., 151
Baitsell, G. A., 87, 89, 90
Baldwin, Raymond, 297
Balfour, Francis M. (Frank), 53, 64
Balfour Laboratory, Cambridge, 50, 53, 201,
  399–400 (n. 19)
Ball, J. N., 207, 210, 211
Ball, Mary, 336
Ball, S. C., 90, 107, 110, 404 (n. 19)
Barcroft, Joseph, 53
Barrington, Joyce, 46, 58, 203–204
Barrow, James, 207, 209, 210
Bateson, Gregory, xii, 10, 30, 54, 56, 58,
  235–240, 244, 247–249, 352, 357
Bateson, John, 30, 236
Bateson, Martin, 236
Bateson, Mary Catherine, 238, 239, 247–
  248, 357
Bateson, William, 30, 54, 235, 240, 357
Bats, 185, 193–194
Baumert, A., 379–380
Beament, Sir James, 338
Beamish, Tony, 314, 316, 317
Beaverbrook, Lord (Max), 251, 258
Beermaker, George, 184
Behavioral ecology, xi, xiv, 320
Benedict, Ruth, 237–238, 248, 260
Benitez, Carlos H., 179, 184
Bennett, Arnold, 257
Bennett, Charles E., 308–309, 423 (n. 35)
Benoit, R. J., 276

Benson, Keith, 170
Benthos, 154–155
Benton, Dorothy, 99
Berenson, Bernard, 264
Bernal, J. D., 46, 219, 320, 322–323, 424
  (n. 1)
Bertrand, Didier, 177
Bey, Mira, 30
Bidder, Anna, 28, 46, 47, 50, 66, 72, 200,
  202, 210–212, 249, 399 (n. 13)
Bidder, George P., 49–51, 52, 59, 66, 67
Bidder, Marion Greenwood, 50
Bingham, Harry Payne, 125
Bingham Oceanographic Laboratory, 125,
  127, 175, 206–207
Biochemistry, 51–53, 55
Biodiversity, 279–289, 294–295, 303–305,
  319, 321, 353, 367, 374, 378, 420 (n. 15)
Biogeochemistry, x, xi, 2, 8, 10, 139, 148–149,
  152, 171–198, 276, 372, 373
Biogeochemistry of Vertebrate Excretion,
  The (GEH), 177–195
Biogeography, 54, 103, 105, 293
Biosphere, 171, 240, 295, 310–312
Birds, 73, 137, 177–186, 190–195, 231, 260,
  269, 285–288, 302, 312–313, 324–329, 375,
  381
Birge, E. A., 98, 102, 131, 145–151
Birth control, 23, 24, 302
Bishop, Allison. See Jolly, Alison
Black Lamb and Grey Falcon (West), 259,
  263
Bleek, Miss D. F., 75
Blosser, Gale, 184
Bocking, Stephen, 165–166, 410 (n. 14)
Boeklen, William, 377
Bohlen, Charles, 262
Bohr, Niels, xii, 14
Bonin, Gerhardt von, 237
Borradaile, L. A., 48, 84, 86
Boulton, David, 213
Bovary, Marcella, 205
Bowen, Jean, 363–364
Bowen, Vaughan T., 8, 159–161, 165, 169,
  176, 198, 276, 296, 331
Bowman, P. W., 128
Bradley, E. H., 198
Bragg, Sir William, 57
Brenchley, Winifred E., 36
Brewster, Kingman, 389, 390
Bridges, Calvin, 56
Briggs, Derek, 147

British Museum (Natural History), 74–75, 111, 144, 261, 319, 366

British National Health Service, 260, 261

Brookhaven National Laboratory, 165, 284, 420 (n. 15)

Brooks, John L., 8, 129, 152, 197, 389

Brooks, William K., 94

Brower, Jane van Zandt, 385

Brown, F. Balfour, 48

Brown, Harrison, 219, 220

Brown, W. L., x

Browne, Janet, 385

Burgess, Charlotte, 57

Burgess Shale, 67, 146–147

Burkenroad, Martin, 125

Burn, Carolyn, 13

Bush, George H. W., 355

Butler, E. A., 34–35, 48–49, 60–63, 72, 139–140, 202, 398 (n. 33), 399 (n. 12)

Cairns, John, 370

Cambridge Between Two Wars (T. E. B. Howarth), 42, 48, 398 (nn. 1, 4), 399 (nn. 8, 19), 400 (n. 22)

Cambridge University: biochemistry at, 51–53; Biological Tea Club at, 52, 56, 58–59, 62, 203, 212, 236; and Channel Islands student trip, 60–62; compared with Yale University, 90–91; degrees from, 43–44, 71, 202; Elizabethan Club at, 10; Emmanuel College of, 44, 46–49, 47, 53, 56, 57, 304; entrance examination for, 33; faculty of, 46–55; Fitzwilliam Museum of, 324, 327–328; food at, 46, 223; founding of, 43, 46; GEH at, xiii, 5, 10, 15, 24, 30, 42–62, 71, 139–140, 143, 161, 205, 236; GEH's graduation from, 42; GEH's honorary doctorate from, 337–338, 339; GEH's lack of earned doctorate from, 5, 71, 337, 387; and Hebrides collecting trip, 48, 49, 139–140; A. Hutchinson at, 3, 17–18, 43, 71, 322, 351, 396 (n. 5); organization of, 43; Pembroke College of, 17–18; science curriculum of, 51–56; and Stazione Zoologica (Naples), 65; students and student clubs at, 10, 21, 24, 28, 43–47, 55–60, 90–91, 249; Trinity College of, 53–54; women students at, 21, 44–46, 50, 53, 56, 58, 60–62, 90, 200, 201–202, 335, 360, 385, 399 (n. 13); Zoology Department of, 47–51, 80–81, 88, 161, 201–202

Cameron, Malcolm, 111–112, 405 (n. 27)

Cantlon, John, 308

Capote, Truman, 259

Carbon dioxide, 299, 300, 310–312, 372

Carlson, Linda, 8, 385

Carnegie Institution, 65, 198, 307

Carrington, Lord, 338

Carroll, Douglas, 185

Carson, Rachel, 170, 373–374

Carson, W. F., 379–380

Carter, H. Gilbert, 55

Cephalopods, 65–69, 85–86, 146, 205, 210. *See also* Octopus

Channel Islands, 60–62

Chapin, Edward L., 301

Chapin, James, 184

Chile, 177–181, 190

China, W. F., 144

Cieza de León, 179

Cladocerans, 99, 107, 129

Clapp, Cordelia, 66

Clear Mirror, The (GEH), xiii, 6, 109, 112–114, 252, 323, 351, 392

Clements, Frederic, 2, 174, 307, 371–373

Club mosses (*Lycopodium*), 176–177, 371

Cody, Martin, 290, 375

Coe, Wesley Roswell, 89, 116

Coelacanth (*Latimeria*), 197

Colchicine, 133–134

Cold Spring Harbor Symposium, 276, 420 (n. 15)

Cole, Sydney, 51

Colliton, Margaret, 363

Coming of Age in Samoa (Mead), 245, 247

Competitive exclusion theory, 57, 99, 279, 375–378, 380–382, 429 (n.19)

"Concluding Remarks" (GEH), 8–9, 207, 270, 276–277, 289–290, 292, 373–374

Connecticut Academy of Arts and Sciences, 110–112, 368

Connecticut Board of Fisheries and Game, 297

Connor, Edward, 377

Conservation movement, 297–298, 305–310, 358. *See also* Environmental issues

"Contribution to the Limnology of Arid Regions, A" (GEH), 148

Conway, Edward, 175–176

Cook, James, 178, 179

Cook, Robert, 151, 163

Cooke, G. Dennis (Denny), xviii, 383, 396 (n.19)

"Copepodology for the Ornithologist" (GEH), 9, 135, 380, 381–382
Copper, 123–125, 127, 129, 173
Corixidae (water boatmen), 34, 37, 59–60, 74, 75, 86, 139, 279, 283
Cornell University, 91, 390
Cottrell Award, 295–296
Court and the Castle, The (West), 265
Cousteau, Jacques, 313–314
Covich, Alan, 8
Cowgill, Ursula, 152, 155, 156, 197, 198, 267, 297, 331, 384
Cowles, Henry, 290
Cowper Reed, F. R., 28
Creative Sex (E. D. [Evaline] Hutchinson), 21, 22, 24, 397 (nn. 16, 21)
Crowley, Philip, 299–300
Culgaith, 36, 39, 41
Cultural diffusion, 57–58
Cyclomorphosis, 129, 152

Dalhousie University, 127, 142, 162–163, 271, 406 (n. 26)
Daphnia (water fleas), 78, 105, 129, 152, 161, 174, 302
Dart, R. C. A., 77
Darwin, Charles, xi, 3, 6, 63, 135, 178–183, 190, 302, 313, 315, 317, 347, 371
Darwin, Sir George, 25, 351
Daubenmire, R. F., 308
Davis, Margaret B., 130, 390–391
Davison, Emily, 256
Day, F. H., 35–36
Day, George Parmly, 101–102
Debenham, Frank, 54
Deevey, Dian, xviii, 406 (n. 27)
Deevey, Edward S., Jr.: and conservation issues, 298; death of, 359; dedication of GEH's Treatise of Limnology to, 154; friendship between GEH and, 249; Juday on, 149; photograph of, 131; research by, 7, 128–132, 143, 148, 155, 163–164, 276, 297, 331, 356; scientific career of, 130–131, 356, 406 (n. 26); second marriage of, 406 (n. 27); successor of, at Yale University, 390; and Rebecca West, 271; and World War II, 130–131, 356; as Yale faculty member, 164, 271, 406 (n. 26); as Yale graduate student, 7, 120, 124, 128–132, 137, 148, 169, 198, 406 (n. 26), 406–407 (n. 31)
Deevey, Georgiana, 179
Dendroica warblers, 283, 285–286, 302

De Terra, Hellmut, 100–104, 107, 108, 117
De Terra, Rhoda, 106, 108
Diamond, Jared, 288, 375, 377, 378
"A Direct Demonstration of the Phosphorus Cycle in a Small Lake" (GEH), 160
Dispersal-assembly theory, 378–379
Dodd, Rosemary (Hutchinson) and Reverend Peter, 339, 343, 345–346, 357, 364–365, 369
Dohrn, Anton, 63–66
Dohrn, Reinhard, 64, 67
Donaldson, Lauren, 166
Dreiser, Theodore, 257
Drosophila, 56, 58, 165
Du Bridge, Lee, 318
Duke Marine Laboratory, 304
Dulanto, Luis Gammara, 183
Dunbar, Maxwell (Max), 132–137, 402 (n. 27)

Ecological Monographs, 127, 148, 306
Ecological Society of America, 295, 306–308, 406 (n. 26)
Ecological Theater and the Evolutionary Plan, The (GEH), xiii, 6, 323
Ecologists Union, 306, 308–310
Ecology: at Cambridge University, 54–55; and Carson, Rachel, 373–374; competitive exclusion theory of, 57, 99, 279, 375–378, 380–382; controversies over mathematical ecology in 1970s, 375–378; current debate in, 378–380; decline of, at Yale University, 388–391; discipline building in, 306–307; and food chains, 168, 281; GEH's contributions to, ix–xv, 2, 157–158, 275, 321, 371–393; GEH's scientific philosophy about, 127; language of, 372; and mathematics, 2, 149–151, 157–158, 275–276, 285, 288–293, 372; niche-assembly theory versus dispersal-assembly theory, 378–379; 1:3 rule of, 283, 377; organisms named after GEH, 76; organizations for, 305–310, 423 (n. 35). See also Behavioral ecology; Ecosystem (systems) ecology; Hutchinson, G. Evelyn; Limnology; Niche theory; Population ecology; Radioecology and radioisotopes; Radiation ecology; Theoretical (mathematical) ecology; Yale University; and other scientists
Ecology (journal), 160, 174, 192, 194, 263, 306, 372, 381

Ecosystem (systems) ecology, xi, 2, 151, 165-170, 275-276, 290, 293, 372, 373

Eddington, A. S., 30, 375

Edison, Thomas, xii, 14

Edmondson, W. T. (Tommy): childhood and youth of, 115-117, 406 (n. 12); and cleanup of Lake Washington, 115, 312, 382; and environmental issues, 115, 300, 310, 312, 382; friendship between GEH and, 249, 340, 342; on GEH, 117, 118, 120-121, 124; and GEH's laboratory, 116-118, 406 (n. 15); and GEH's publications, 153, 331; and Harrison, 118; and Macy Conference, 237; photograph of, *119*; research by, 7, 121, 164, 356; as University of Washington faculty member, 312, 342; as University of Wisconsin student, 120, 146; as Yale student, 7, 119-121, 124, 146, 164, 406 (n. 15)

Edmondson, Yvette Hardman, 120, 152, 153, 340, 342, 365

Edwards, John, 153

Eichelberger, Marie, 248

Einstein, Albert, x, xii, 14, 33

Eisig, Hugo, 64

Ellis, Havelock, 22, 23

Ellsworth, Phoebe, 10, 231, 249, 331, 337, 341, 342, 345, 348, 354, 356-357, 385

El Niño, 181-183, 195

Elster, Hans-Joachim, 141, 142

Elton, Charles, 10, 82, 117, 134-135, 372

Embryology, 94-99, 119, 121-122, 399 (n. 18)

Enchanted Voyage, The (GEH), 260, 268, 323

Endocrinology, 55-56, 68-69, 206-208, 211

Entomology, 37, 39, 41, 48-49, 74-75. *See also* Water bugs

Environmental Fund, 301

Environmental issues, xii-xiii, 14, 167-168, 170, 240, 264, 294-319, 382

Epstein, Jacob, 223-224

Erdmann, Rhoda, 84

Eutrophication of lakes, xii-xiii, 298-301, 330

Ewer, R. F., 295

Extinction, 191, 194, 295, 303, 305, 310, 313, 316-317, 321, 353, 379, 380

Fadiman, Clifton, 219

Fantham, H. B., 5, 71, 72, 74, 82-83, 402 (n. 27)

Farrar, F. W., 32

Feminism, 19, 21-24, 254-258

Fermi, Enrico, xii, 14

Finch, F. E., 143

Finnerty, Patrick, 8, 222, 358

Fischer, Emil, 17, 18

Fish, 94, 110, 125, 141, 143, 144, 148, 162, 166, 183, 192-193, 195, 197, 206-209, 304-305

Fitzwilliam Museum, 324, 327-328

Flint, R. F., 130, 164, 406-407 (n. 31)

Florida, 231, 312

Florida State University, 376-378

Forbes, S. A., 117-118

Ford, Ford Maddox, 254-255, 257

Forel, Francois Alphonse, 141, 401 (n. 22)

Fossils, 6, 28, 37, 77, 91, 101, 103-105, 108-109, 128-129, 146-147, 406 (n. 28)

Foster, Michael, 53

Fountain Overflows, The (West), 266

Fox, H. Munro, 49

Frank, Lawrence K., 237, 248

Franklin, Benjamin, xii, 14, 306, 307, 322

Franklin, Rosalind, 322

Franklin Medal, xii, 12, 14

Frazer, James George, 30, 43, 54

Freeman, Derek, 247

Freshwater Biological Association (FBA), 143, 144

Freud, Sigmund (psychoanalysis), 22, 55, 59, 73, 146, 321

Frey, David, 145, 149

Freyer, J. C. F., 313

Friedman, Lulu and Bernard, 267

Frost, Winifred, 143

Fruton, Joseph, 18

Fugitive species, 381-382

Funk, C., 51

Furth, David, 157, 346, 367

Gadow, Hans, 48

Galapagos Islands, 181, 183-184, 312

Galathea expedition, 206, 206

Galston, Arthur, 318, 361

Garber, Marjorie, 10, 249, 273-274, 336, 342, 385

Gardiner, J. Stanley, 47-48, 66, 84-85

Gardner, Dame Helen, 338

Gause, G. F., 192, 279, 289, 372

Gaymer, Robert, 314

Genetics and genetically engineered organisms, 54, 56, 353

Geography and biogeography, 54, 103, 105, 293

Geology, 28, 101, 197–198, 397 (n. 8). *See also* U.S. Geological Survey (USGS)

Germany, 94–96

Ghandi, Mohandas, 30

Gilpin, Michael, 377

Girton College, 44, 45, 201

Gissing, George, 255

Glacier Bay, 133–134

Glaus, Karen. *See* Porter, Karen Glaus

Glendinning, Victoria, 252, 258, 274

Goethe, Johann Wolfgang von, 322

Goldenrods (*Solidago*), 379–380

Goldsby, Anne Twitty. *See* Hutchinson, Anne Twitty Goldsby

Goldsby, Richard, 233, 361

Goldschmidt, Richard, 8, 172

Goldsmith, H. H., 165

Goldsmith, Oliver, 260

Goldsmith, Tim, 390

Golley, Frank, 167, 169

Goodall, Jane, 267–268

Gordimer, Nadine, 266

Gorham, Eville, 168, 174–175

Gosse, P. H., 121

Gould, John, 329

Gould, Stephen Jay, xi, 11, 146–147, 249, 323, 382

Goulden, Clyde, 299, 330–331

Graduate students (Hutchinson's), xiv, xv, 4, 7–9, 115–137; listed, 420 (n.3)

Grave, Caswell, 121–122

Gray, H. M., 298

Gray, James, 47, 49

Greenwood, Marion, 50

Gresham's School, 3, 31–39, 35, 47, 55, 79, 90, 139

Griswold, A. Whitney, 298

Groeben, Christiane, 400 (n.5)

Gross, Alexandra Ellsworth "Sasha," 231, 331, 341

Gross, Sam, 231, 341, 356–357

Growth and Culture (Mead), 243–246

Guano, 177–195, 187

Guatemala, 178, 267, 297

Gul, Khan Sahib Afraz, 104, 108

"Gwendle." *See* Rendle, Gwendoline "Gwendle" (Sah Oved)

Haddon, A. C., 54

Haeckel, Ernst, 63

Hagen, Joel, 163

Haldane, J. B. S., 52, 56–57

Hamilton, Alex, 260

Hammick, Mr., 33, 398 (n.32)

Hancock, G. L. R., 75

Haraway, Donna, 8, 209, 320, 336, 385

Harper, John, 335

Harris, Rendel, 58

Harrison, Jane E., 30, 43

Harrison, Ross Granville: education of, 94, 403 (n.8); and GEH, xiii, 6, 83, 86, 88–89, 92, 102, 106–110, 122, 147, 352; and graduate students at Yale, 7–8, 92, 94–95, 99, 118, 122–123; and National Academy of Sciences, 160; and Needham, 52; and North India Expedition, 6, 101–102, 106–110, 404 (n. 19); photograph of, 93; research by, 94, 399 (n. 18); at Stazione Zoologica, 66; teaching methods of, 94–96; as Yale faculty member and zoology chair, 6, 88–89

Hartman, Willard, 8, 19, 209, 210, 212, 214–215, 270, 346, 363, 388

Harvard, John, 46, 304

Harvard University, 66, 196, 249, 269, 304, 390

Harvey, Ethel Browne, 69, 74

Harvey, H. W., 79, 125

Harvey, Newton, 69

Haskins, Caryl, 333

Hasler, Arthur, 131, 149, 163, 166

Hastings, Anne, 46

Hayden, Ada, 307, 308

Hayes, F. Ronald, 162–163, 170

Hebrides, 48, 49, 139–140, 365

Hemiptera-Heteroptera. *See* Water bugs

Hill, Robin, 56, 57, 60

Hiram College, 207, 208–210, 212

History of Science, 15, 18, 51, 302, 303, 320, 324–327, 336

Hogben, Lancelot, 69, 77–78, 80, 83–86, 122, 140, 202, 351

Holloway, Jeremy, 303–304

"Homage to Santa Rosalia" (GEH), 9, 157, 270, 279–285, 287, 374, 376

Homosexuality, 31–33, 246–247, 323, 347–348, 352

Hopkins, F. Gowland, 30, 45, 51, 53, 54, 56–57

Horie, Shoju, 368

Horstmann, Dorothy, 273

Howarth, T. E. B., 42, 48, 52, 53

Hubbell, Stephen, 378–380, 390

Hubbs, Carl, 308

Hudson, C. T., 121

Humboldt, Alexander von, 6, 178, 180, 322

Hutchins, Robert, 219

Hutchinson, Andrew, 229, 343, 360, 365, 369

Hutchinson, Anne Twitty Goldsby, 10, 11, 189, 199–200, 231–234, 345–349, 346, 350, 354–364

Hutchinson, Arthur, 3, 15, 17–19, 28, 30–33, 30, 39, 43, 71, 155, 235, 322, 351, 396 (nn. 5–6), 397 (n. 8), 415 (n. 66)

Hutchinson, Dorothea: career of, 21, 24, 215, 335, 359, 360; childhood of, 25–26, 28–29, 396 (n. 5), 397 (n. 8); correspondence between GEH and, 11, 212, 335–337, 348, 358–360, 387; death of, 360; education of, 21, 45, 202, 215, 335, 360; eightieth birthday party for, 356; father of, 18, 397 (n. 8); on GEH's desire for children, 215; and GEH's eightieth birthday party, 340; and GEH's honors and awards, 337, 338, 341, 342, 349–350; and GEH's last months in England, 365–367; and GEH's wives, 72, 210, 212, 214–216, 221–222, 229, 232–233, 246, 345, 346, 360; on Gresham's School, 31; on Gwendoline Rendle, 223; mother of, 19–21, 24; old age of, 360; photographs of, 20, 339; and religion, 341–342; travel to Spain by, 351

Hutchinson, Evaline Demesey Shipley, 16, 16, 19–25, 20, 30, 257–258, 387, 399 (n. 13)

Hutchinson, Francis and Joyce, 343, 345, 355, 363–370

Hutchinson, G. Evelyn: and alcoholism of wife Anne, 345–346, 361–364; birth of, xi, 15, 25; childhood and youth of, 3, 5, 15–18, 25–41, 138–139, 272–273, 321, 351; and children, 200, 211, 215, 334, 357; contributions of, ix–xv, 2, 157–158, 275, 321, 371–393; correspondence of, 9–11, 34–36, 48–49, 58, 61, 70, 80, 105, 203–204, 216–217, 227–228, 235, 246–247, 250, 259–274, 386, 398 (n. 33), 399 (n. 12), 401 (n. 13), 418 (n. 27); death of and memorial service, 153, 292, 369–370, 378, 416 (n. 13); divorce of, from Grace Pickford, 9, 148, 199, 203–204, 211–213; education of, 3, 5, 26–27, 31–39, 42–62, 71, 139, 236; environmental work by, xii–xiii, 14, 240, 264, 294–319; family background and parents of, 3, 15, 17–25, 69, 360, 387, 397 (n. 11); finances of, 85, 87, 333; foreign language

abilities of, 70, 85, 142, 156; foster child of, 10, 29, 200, 221–232, 268, 338; friendships and social life of, xii, 10–11, 200, 235–274, 334, 338, 340–345, 346, 348–349, 354, 356–358, 365, 367–370, 385; health problems of, 223, 232, 268, 270, 348, 355, 357–358, 361, 365–367; honorary degrees, honors, and awards for, xii–xiii, 5, 10, 12, 14, 97, 169, 216, 231, 234, 270, 294–296, 296, 312, 319, 337–338, 339, 341, 342, 345, 349–354, 352, 355, 392; intellectual family tree of, 3, 4, 382, 391–392; in Italy, 5, 42, 50, 63–70, 85–86, 156, 216–220, 260, 263–264; job offer for, from U.S. Geological Survey, 3, 198; lack of earned doctorate for, 5, 71, 337, 387; last years of, 235, 334–370, 415 (n. 1); and Mead's writings, 235, 241–247; name of, 8, 67, 400 (n. 6); personality and character of, x, 12–13, 61, 84–85, 124, 222, 239, 383, 386–387; photographic memory of, 61, 330–331; photographs of, 16, 29, 35, 38, 40, 106, 119, 136, 157, 189, 296, 344, 346, 350; and politics, 14, 169, 260, 273, 355, 359; as polymath, ix, xii, 1, 9, 133, 137, 320–333, 347, 371, 400 (n. 24); portrait of, 231, 343; and religion, 213, 239, 341–342, 345, 416 (n. 13); retirement of, 270–273, 390; *Washington Post* interview of, 347–348; wives of, xii, 5, 9–10, 24, 62, 72–73, 199–234, 345–349, 354–364; writings by, xi, xiii, 6, 8–11, 51, 110–114, 138, 146–148, 151–157, 160, 173–195, 204, 237, 252, 271–272, 276–287, 293, 303, 323, 328–333, 373–376, 378, 380–382, 391–393; writing style of, 112–114, 392. *See also* Cambridge University; Ecology; Limnology; North India Expedition (Yale University); South Africa; Yale University; *and titles of specific writings*

Hutchinson, Hannah, 211, 272, 339, 345, 354

Hutchinson, Leslie, 21, 24, 25, 28–29, 31, 211, 216, 272–273

Hutchinson, Margaret Seal: and Alzheimer's disease, 154, 199, 220–221, 230, 274, 334–337, 340–341; birth date of, 215; childhood of, 213–214; church membership of, 213; death of, 213, 221–222, 274, 334, 341, 369; foster child of, 10, 29, 200, 221–231, 268, 338; marriage of, to GEH, 9–10, 212–231; music and art interests of, 9, 199, 212, 213, 215–218, 221, 222, 336;

Hutchinson, Margaret Seal (continued)
  personality of, 215, 230; photograph of,
  214; relationship between Grace Pickford
  and, 212; travels of, 13, 213, 215–217; and
  Rebecca West, 263–265, 268, 274
Hutchinson, Paul and Marie, 343, 363, 369
Hutchinson, Philip, 339, 343
Hutchinson, T. J., 180
Hutchinson Medal, 158
Huxley, Julian, 315
Huxley, Thomas, 1, 6, 64
Hyde, Ida, 66
Hydrobiology, 79, 80–81, 139, 140, 156. See
  also Limnology
Hydrography, 401 (n. 22). See also
  Limnology

India. See North India Expedition (Yale
  University)
Innes, Christina, 26, 30, 236
Insects. See Entomology; Water bugs
Institute for Intercultural Studies (ISS),
  237–238
International Biological Program (IBP),
  373
International Botanical Congress, 378
Introduction to Population Ecology, An
  (GEH), 274, 288, 293, 302–303, 359, 374
Isotopes. See Radioecology and radio-
  isotopes
Israel, 178, 297
Italy, 5, 12, 19, 42, 50, 63–70, 85–86, 154,
  156–158, 216–220, 260, 263–264. See also
  Stazione Zoologica (Naples)
Itinerant Ivory Tower, The (GEH), 262–263,
  323

Jackson, Margaret, 24
Jaeger, Bob, 288
Jaffe, Ruth, 179
James, Henry, 257
Japan, 350–354
Jefferson, Thomas, 307
Jenkin, Penelope, 46, 48, 58, 60–61, 73, 126,
  139, 143, 200, 210, 249
Jenkin, V. M., 143
Jennings, O. E., 308
Johnson, Lyndon, 318
Johnson, Roswell, 374
Jolly, Alison, 249, 267, 310, 340, 367–368,
  385
Journal of Marine Research, 127, 133

Juday, Chauncey, 98, 102, 120, 127, 131,
  145–151

Kabaservice, Geoffrey, 389
Kammerer, Paul, 57, 236
Kashmir, 100, 102, 103, 104, 106, 108, 109,
  110, 148
Keilin, David, 52
Kelvin, Lord (William Thomson), 44
Kendeigh, S. C., 308
Kendrew, John, 219
Kennedy, I. M., 255
Keynes, John Maynard, 43, 45
Kindly Fruits of the Earth, The (GEH), 48,
  62, 201, 212, 294, 335, 366
Kingsland, Sharon, 290, 292, 329, 385
Kirby, Harold, 83, 88
Klopfer, Peter, xiv, 8
Knight, Gwenllian, 213
Koestler, Arthur, 57
Kofoid, Charles, 65, 143, 144, 400 (n. 2)
Kohn, Alan J., 8, 153, 283
Kohn, Marion, 4, 76, 153
Kubie, Lawrence S., 237
Kubler, George A., 179, 182
Kyoto Award, xii, 10, 25, 97, 234, 349–355,
  352, 392, 397 (n.24)

Lack, David, 10, 249, 264, 279, 303
Ladakh. See North India Expedition (Yale
  University)
Lake ecology, xii-xiii, 2, 6. See also Eutro-
  phication of lakes; Limnology
Lake Huleh, 155, 156
Lake Malawi, 304–305
Lake Tanganyika, 99, 161, 173–174
Lake Washington, 115, 121, 312, 382
Lake Windermere, 81–82, 143–144
Lang, Anton, 318
Lansbury, George, 257
Leavy, Stanley and Margaret, 361–364
Leconte, Joseph, 144
Leeuwenhoek, Antony van, 142
Leigh, Egbert, 8, 275, 285, 310
Leopold, Aldo, 308
Letter to a Grandfather (West), 253
Levin, Simon, 374, 378, 380–382
Levins, Richard, 287
Lewin, Roger, 375–377
Lewis, G. E., 104
Libby, Willard, 130, 164
Likens, Gene, 149, 166

"Limnological Studies in Indian Tibet" (GEH), 112, 148

Limnology: arid lakes of Nevada, 148, 203, 373; Birge-Juday era of, 145–151; definition of, 5, 79, 401 (n. 22); early history of, 141–144; equipment for, 80, 86, 91–92, 102, 105, 124, 146; GEH's contributions to, x, xi, 2, 5, 157–158, 321, 371–393; GEH's early interest in, 78–80, 138–141; GEH's early teaching and research on, at Yale University, 98–99; GEH's infamous remarks on, 99; hypothesis-testing and mathematical approach to, 149–151, 157–158, 373; in Italy, 154, 156–158; light and dark bottle experiments, 126–127; North American studies in, 144–155; and pollen analysis, 128–130, 390, 406 (n. 28); radioisotopes in, 159–162, 296. See also Biogeochemistry; Ecology; Hutchinson, G. Evelyn; North India Expedition (Yale University); South Africa; Treatise on Limnology (GEH); Yale University; and other scientists

Limnology and Oceanography, 120

Lindeman, Raymond, 2, 142, 146, 149–150, 150, 163, 166, 169, 174, 275, 290, 372

Linnaeus, Carl, x

Linsley Pond, 2, 7, 117, 123–130, 148, 152, 153, 159–162, 164, 171, 174, 296, 330, 373

Livingstone, Daniel, 162–163, 169, 234, 331, 392–393

Lloyd George, David, 21

Lloyd Jones, Audrey, 28

Lo Bianco, Salvatore, 65

"Long Meg Reconsidered" (GEH), 330

Lotka, A. J., 276, 372

Lovejoy, Thomas, 8, 11–12, 249, 303–305, 310, 340, 347, 354, 358, 369

Low, Eva, 176

Lull, Richard Swann, 90, 404 (n. 19)

Lundquist, G., 128

Lynge, B., 191

Macan, T. T., 41, 82, 143, 144

MacArthur, Betsy, 291

MacArthur, Robert, 8, 168, 192, 249, 264, 273, 275, 281–282, 285–293, 286, 302, 332, 358, 373–378, 382–383

MacClintock, Dorcas, 343

Macgregor, Frances Cooke, 244, 246

MacInnes, M. P., 27

MacMillan, Margaret, 256

Macy, Michele, 208

Macy Conferences, 10, 236–237, 352

Madson, Dora, 254

Male and Female (Mead), 241–243, 247

Mall, F. P., 94

Mandrake root, 58

Mangione, Laura, 268

Mangold, Otto, 95, 96

Manton, Sydnie, 46, 73

Marcus, George E., 247

Marcuse, Sybil, 221

Margalef, Ramon, 154

"Marginalia" column. See American Scientist ("Marginalia")

Margulis, Lynn, 342, 385

Markert, Clement, 205, 209, 271, 388–391

Marsh, C. D., 145

Marsh, O. C., 89

Marshall, Alfred, 43

Marshall, George, 262

Martz, Edwine, 222

Mason, Victoria, 153

Mathematics, 2, 149–151, 157–158, 275–276, 285, 288–293, 348, 372, 374–375

May, Robert, 287, 290, 292, 358, 377, 386

Mayr, Ernst, 176, 249, 410 (n. 12)

McCarthy, Joseph, 262, 263

McConnell, Robert Earl, 177

McIntosh, Robert, 2, 165, 174, 275, 276, 289, 371–372

McKaye, Kenneth, 304

McNamara, Robert, 317

McNaught, Donald, 300

Mead, Margaret, xii, 10, 30, 176, 200, 223, 229, 235–248, 242, 352, 357, 385, 410 (n. 12)

Meaning of Treason, The (West), 241, 259

Medieval animal art, 324–329, 424 (n. 11)

Merian, Maria Sibylla, 41

Merriman, Dan, 206–208

Meselson, Matthew, 318

Methuen, P. A., 78

Meyer, Paul, 64

Midwife toad (Alytes obstetricans), 57, 236

Miller, R. M., 288

Mills, Eric, 2, 3, 121, 123, 126, 142

Mineralogy, 17–18, 30, 43, 155, 396 (n. 6)

Mitchell, Henry, 347–348

Mochi, Ugo, 343

Moffett, George Monroe, 177

Molecular biology, 388–391

Monti, Rina, 154

Moore, Emmeline, 130
Moore, Jon, 153
Moreau, A. E., 303
Morgan, Thomas Hunt, 56, 94
Morris, Simon Conway, 147
Morse, Philip, 165
Mortimer, Clifford, 143
Mosely, M. E., 111, 405 (n. 25)
Muller, Otto Friedrich, 142
Murphy, Robert Cushman, 179, 194–195
Murray, John, 143
Murrow, Edward R., 267
Music, xii, 1, 9, 47, 96, 133, 137, 199, 210, 212–222, 336, 348, 362
Myers, F.J., 406 (n.15)

Nabis boops, 35, 37
Naples Table Association for Promoting Laboratory Research by Women, 66
Nash, Philleo, 238
National Academy of Sciences (NAS), 8, 145–146, 160, 290, 295–296, 300, 305, 306, 315, 317–319, 341, 355–356, 358, 368, 374–375
National Research Council (NRC), 197–198
National Science Foundation, 11, 207, 306, 307, 389
"Natural History of Lakes, The" (GEH), xiii
Nature Conservancy, 305–306, 308–310, 423 (n. 35)
Naumann, Einar, 78, 141, 401 (n. 22)
Needham, Dorothy, 51, 53
Needham, John, 107
Needham, Joseph, 22, 48, 51–53, 55, 57, 99, 236, 399 (n. 17)
Neumann, John von, 237, 352
Nevada, 148, 203, 212, 373
New Guinea, 240, 269, 288
Newnham College, 5, 44, 201, 202, 210, 335
Niche theory, x, xi, 8–9, 75, 154, 156, 168–169, 275–279, 278, 280, 285–288, 292, 373–374, 378–380, 420 (n. 23)
Nicholas, John S., 88, 89, 92, 97–98, 111, 133, 204, 387–388
Nichols, George E., 129, 406 (n. 31)
Niering, William, 308
1900 (West), 259
Nixon, Richard, 14, 169, 273, 318, 355
North India Expedition (Yale University), xiii, xiv, 6, 9, 60, 72, 80, 100–114, 118, 138, 147–148, 161, 168, 201, 203, 269, 373, 404 (n. 19), 405 (nn. 25, 27); photo, 106

Notonectidae (backswimmers), 34, 37, 59, 74, 75, 86, 139
Nuclear war, 332
Nuclear wastes, 166–167, 264, 317–318

Oceanography, 2, 121–127, 142, 145, 176, 206, 265, 400 (n. 24); biological, 2
Octopus, 5, 63, 66–69, 140, 205, 210. See also Cephalopods
Odum, Eugene, 12, 166, 167, 174, 281
Odum, Howard T., 8, 142, 167, 174, 176, 276, 290, 373, 383
Ohlhorst, Sharon, 156
Olsen, Yngve, 125
Olson, Jerry, 167–168, 170
Omer-Cooper, Joseph, 58, 201, 203–204, 249
"On Living in the Biosphere" (GEH), 240, 310–312
Onslow, Huia, 52
Onychophora, 72, 73, 146
Osborn, Fairfield, 297–298, 311
Osborn Zoological Laboratory, 89, 94, 101, 104, 108
Oved, Moishe, 223, 224, 229, 414 (n. 55)
Oved, Sah. See Rendle, Gwendoline "Gwendle" (Sah Oved)
Oved, Yemaiel. See Aris, Yemaiel Oved
Overpopulation, 301–303, 332
Oxford University, 43, 45, 65, 71, 90–91, 132, 358

Pacific atolls, 186–190, 187
Paleolimnology, 7, 105, 128–130, 143, 148, 297, 300, 406 (n. 28)
Paleontology, 101, 104, 193, 194. See also Fossils
Panama, 379
Panayotova, Stella, 328
Pankhurst, Christabel, 254
Pankhurst, Sylvia, 21
"Paradox of the Plankton, The" (GEH), 9, 157, 380–381
Parapsychology, 59
Park, Orlando, 308
Park, Thomas, 149, 150–151, 289
Parker, Pamela, 343, 345
Parr, Albert Eide, 98, 125, 127, 133, 175, 177, 195, 410 (n. 12)
Patrick, Ruth, 124, 152–153, 223, 232, 297, 303, 331, 340, 368, 370, 385
Patten, Bernard, 167
Pauling, Linus, 318

Payne, Cecilia, 46, 172–173, 200, 249
Peabody Museum (Yale), 6, 89–90, 98, 101, 107–109, 125, 133, 147, 205, 208, 269, 305, 358, 388, 404 (n. 19)
Pearce, Jim, 28, 58
Pearsall, W. F., 143
Pearse, A. S., 307
Peat moss (*Sphagnum*), xiii, 49, 140, 152, 155
Peebles, Florence, 66
Pelikan, Jaroslav, 358
Pennak, Robert, 149
Pennington (Tutin), Winifred, 143
Percy Sladen Expedition, 81, 313
Peru, 177–183, 190, 195
Perutz, Max, 338
Pesta, Otto, 109
Petrunkevitch, Alexander, 89, 91–92, 99, 107, 147, 202, 205
Phosphorus, 2, 124, 160–163, 175–177, 181, 183, 188–190, 194, 294, 296, 299
Pickford, Grace: awards and honors for, 209–210; as Cambridge student, 46, 55–56, 58, 60–62, 139, 200–202, 205, 236; collaboration between GEH and, 80, 201–203, 205, 211–212; correspondence of, 9, 70, 80, 105, 203–204, 210, 211, 401 (n. 13); death of and memorial service for, 209, 349; dedication of GEH's *Treatise on Limnology* to, 154; divorce of, from GEH, 9, 148, 199, 203–204, 211–213; engagement of, to GEH, 62, 72, 202; at Hiram College, 207, 208–210, 212; marriage of, to GEH, 5, 72–73, 199, 201–204, 211–212, 401 (n. 13), 412 (n. 6); personality of, 210–211; photograph of, 206; physical appearance and clothing of, 206, 210–211; relationship between Margaret Hutchinson and, 212; research equipment for, 80, 91–92; scientific career of, 55, 68–69, 86, 91, 147, 197, 199, 201, 204–212, 368, 413 (n. 14); in South Africa, 5, 72–75, 77–80, 86, 98, 110, 117, 140–141, 147, 202, 212, 412 (n. 6); as Yale graduate student, 6, 91–92, 202; and Yale instructorship for GEH, 86, 87–88, 202; and Yale North India Expedition, 80, 105, 110, 147–148, 201, 203; as Yale research associate and faculty member, 201, 204–212, 388; and Yemaiel Oved, 227; youth and family background of, 201
Pickford, William, 201
Pickford Medal, 209–210
Pisan, Christine de, 329

Planck, Max, x, 14
Plankton, 9, 78, 125, 129, 133, 134, 145, 152, 157, 175, 185, 186, 380–381
Platil, G. P., 275
Plymouth Laboratory, 77, 81, 125, 126
Poland, 217–218
Politics, 14, 169, 260, 273, 355, 359
Pollard, E. C., 164
Pollock, David, 28
Polytrichaceae (mosses), 280, 287, 289
Popper, Karl, 333, 376–378, 429 (n. 19)
Population. *See* Overpopulation
Population ecology, xi, 2, 8, 52, 157, 161, 168, 191–192, 274, 276, 288–290, 293, 373, 374
Porter, Karen Glaus, 8, 13, 249, 331, 340, 356–357, 385
Pottle, Fred, 355
Potto, cover, 1, 395 (n. 2)
Potts, F. A., 48
Pound, Ezra, 254–255, 342
President's National Medal of Science, 14, 169, 273, 355
Press, Michelle, 340
Prichard, Matthew, 70
Priestly, R. E., 54
Princeton University, 216, 270, 273, 358
Provasoli, Luigi and Rose, 222
Psalters. *See* Medieval animal art
Psychoanalysis. *See* Freud, Sigmund (psychoanalysis)
Punnett, R. C., 52, 54

"Quantitative Radiochemical Study of the Phosphorus Cycle in Linsley Pond, A" (GEH), 160

Rachewiltz, Princess Mary de, 342
Rachootin, Stan, 157, 346
Radiation ecology, xi, 2, 159, 165, 166, 167, 168, 293, 320, 373
Radioecology and radioisotopes, xi, 2, 130, 159–170, 293, 296, 317–318, 373
Raine, Kathleen, 42–43, 45–46, 398 (n. 5)
Rainger, Ron, 170
Ramage, Hugh, 173
Ramsey, Frank, 56
Rappaport, Ray, 119
Rats, 60–61, 397 (n. 11)
Raven, Charles E., 22, 46
Reagan, Ronald, 355, 359
Religion, 9, 70, 112, 213, 239, 341–342, 345, 416 (n. 13)

Remington, Charles, 210, 212, 346, 363
Rendle, Gwendoline "Gwendle" (Sah Oved) 29, 223–225, 227, 229, 414–415 (nn. 55–56, 58–63)
Research schools, 7–8, 94, 96
Richard, Oscar, 127, 129
Richards, Paul, 338
Richmond, Fred, 301–302
Ridgeway, Sir William, 44
Riley, Gordon A., 7, 120, 121–130, 122, 142, 146, 148, 150–151, 154, 198, 275, 331, 356, 373
Ripley, S. Dillon, 112, 264, 269–270, 296, 305, 340, 368–369
Ripley, Sidney, 270
Rivers, W. H. R., 59
Rockefeller Foundation, 66, 130, 164, 219, 307
Rodhe, W., 141
Rollyson, Carl, 266
Rosen, Walter, 305
Rosenberg, Marie, 143
Rosenzweig, Michael L., 275, 289, 421 (n. 31)
Ross, Harold, 250, 253, 259, 261
Rothschild, Baron Lionel, 178, 317
Rotifers, 7, 116–118, 120–121, 123, 129, 406 (n. 12)
Roughgarden, Jonathan, 376, 377
Roughton, F. J. W., 53
Royal Society, 17, 36, 46, 51, 204, 271, 303, 315–317
Rudnick, Dorothea, 346
Ruff, Willie, 1, 362
Runcie, Robert, 338
Rushton, W. A. H., 47
Russell, Bertrand, 21, 22, 30, 56
Rutherford, E. R., 30

Saint Faith's School, 27, 28, 30, 54, 58, 236
Sanders, Howard, 340
Sanger, Margaret, 23
Saunders, J. T., 48, 54, 61, 79–82, 139, 143
Saurat, Denis, 272
Save the Children Fund (SCF), 360
Saybrook College, Yale, 10, 231, 343, 362
Schlesinger, Arthur, Jr., 262
Schobel, Emil, 64
Schoener, Thomas, 377
Schuchert, Charles, 90
Schuurman, Johanna F. M., 74, 77–80, 86, 110, 117, 147

Science: government-funded big science, 169, 373; Popperian view of, 376–378; relation of art and, 7, 262–263, 271; social significance of, 331–333. See also Ecology; Limnology; Women scientists; and specific scientists
Scott, Hugh, 48
Scudo, F. M., 289
Seal, Margaret. See Hutchinson, Margaret Seal
Sears, Paul, 128, 308
Semonides of Amorgos, 44–45
Setlow, Jane, 175, 179, 197
Sexuality, 21–24, 57, 332. See also Homosexuality
Seychelles Expedition (Yale University), 314
Seymour, Charles, 204
Shannon, Claude, 236
Shapiro, Joseph, 8, 331
Shaw, George Bernard, 255, 257
Shelford, Victor, 2, 174, 308, 371–373
Shipley, Sir Arthur, 19, 33, 54, 216, 270, 397 (n. 11)
Shipley, Emily, 19–21
SIL (International Association of Limnology) 12, 141, 145, 156, 158
Simberloff, Daniel, 332–333, 376–377
Singer, Maxine, 358
Singer, Stephanie, 346
Sitwell, Edith, 312
Skinner, Brian, 196
Skinner, Catherine, 195–197, 198
Slack, Nancy G., xiii, xv, 152, 155, 170, 201, 300, 327–328, 360–362, 365, 368, 383, 392, 430 (n. 42)
Slobodkin, Larry, xv, 3, 8, 135, 135, 249, 275, 281–282, 289, 312, 322, 341, 382–384
Smith, Cyril, 331
Smith, F. E., 8, 222, 289
Smith, Grafton Eliot, 57–58
Smith, Sydney, 337
Smith, Walter Campbell, 17, 396 (n. 6)
Smithsonian Institution, xv, 65, 67, 112, 269–270, 305, 358, 368
Snow, C. P., 1, 322–323, 424 (n. 1)
Sodestrom, Tomas, 169
South Africa: apartheid in, 266–267; GEH in, xiv, 5, 24, 42, 71–80, 82–83, 412 (n. 6); GEH's research on South African pans, 77–80, 82, 98, 110, 117, 140–141, 147, 171, 202, 373; guano in, 185–186; Grace Pick-

ford in, 5, 72–75, 77–80, 86, 98, 110, 117, 140–141, 147, 201, 202, 212, 412 (n. 6); human mummies in, 185–186; Pickford's research on earthworms of, 86, 91, 202–203, 205; Sharpeville massacre in, 267; Rebecca West in, 266–267
Spemann, Hans, 95
Spiders, 91–92, 107, 202, 205, 206, 207
Spiesel, Christina Olson, 231, 343, *344*, 346, 363
Spittle bug (*Philaenus spumarius*), 37, 39
Stanford University, 342, 376
Stanley, J., 313
Starfish (*Asturias forbs*), 197
State University of New York at Albany, 300–301
Stazione Zoologica (Naples; *also as* Naples Zoological), 5, 8, 63–69, *64, 68*, 85–86, 264, 400 (n. 2), 400 (n. 5)
Stella, Emilia, 154
Stevens, Nettie, 66
Stevenson, Marjorie, 53
Stewart, G. B. S. "Bill," 25
Stone circles, 330, 331
Stopes, Marie, 23, 397 (n. 19)
Strange Necessity, The (West), 252, 263, 271, 329
Strangeways, S. T. P., 49
Strickler, Rudi, 345
Stroud Laboratory, 232
Suzman, Helen, 267
Systems ecology. *See* Ecosystem (systems) ecology

Tansley, Arthur, 54–55
Taylor, Dennis, 209, 210
Taylor, Peter, 166
Terborgh, John, 377
Territoriality, 260, 302–303
Theoretical (mathematical) ecology, 157, 276, 285, 373–378
Thienemann, August, 61, 78–79, 82, 98, 112, 140, 141, 401 (n. 22)
Thom, Alexander, 330
Thompson, D'Arcy, 59, 323
Thomson, E. F., 307
Thomson, J. J., 30
Thomson, Keith, 197
Thomson, William (Lord Kelvin), 44
Thursby-Pelham, Dorothy Elizabeth, 28
Tolstoy, Leo, 257
Tomlinson, J. N., 184

Tonolli, Vittorio and Livia Pirocchi, 154, 156
Tortoises, 188, *189*, 271, 312–313, 316
Treatise on Limnology, A (GEH), xiii, 5, 9, 13, 118, 151–155, 197, 220, 299, 301, 305, 334, 340, 345, 347, 348, 357, 365, 368, 373, 378, 381
"Trophic-Dynamic Aspect of Ecology," 2, 149, 163, 408 (n. 35)
Truman, Harry S., 260
Twain, Mark, 307
Twitty, T. W., 94–96, 403 (n. 12)
Twitty, Tracey and Michael, 357, 361, 364
Twombly, Saran, 12, 153, 304, 334, 342, 385
Tyler Prize, 295, 312

Ultraviolet (UV) light, 105–106
United Nations, 165, 265
University of Colorado, 342
University of Michigan, 149, 249, 385
University of Tulsa, 261, 418 (n. 27)
University of Washington, 115, 312, 342
University of Wisconsin, 120, 131, 143, 145, 149, 163
University of Witwatersrand, 71, 73–77, 82–83, 88, 202, 216
U.S. Geological Survey (USGS), 3, 196, 198

Vallentyne, J. R., 300
Van de Velde, Theodor, 23, 24
Van Dyne, George, 167
"Variations on a Theme by MacArthur" (GEH), 375
Vernadsky, V. I., 91, 171–173, *172*
Verwoerd, Hendrik, 267
Vienna Academy of Sciences, 342
Vietnam War. *See* Agent Orange
Vinogradov, A. P., 172
Vitt, Dale H., xiii, 152
Vogt, William, 311
Volterra, Vito, 57, 192, 276, 289, 372
Vorticellid protozoans, 132–133

Walcott, Charles D., 147
Walker, James, 195, 196, 198
Wallace, Alfred Russel, 6
Wallace, Henry, 260
Wallach, Anne, 174, 297
Wangersky, Peter, 8
Washington Post, 347–348
Water bugs, xii, xiv, 3, 6, 28, 34–35, 37, 41, 49, 58–60, 74–75, 81, 86, 114, 117, 138–140, 155, 206, 279, 283, 321

Watson, Maxine, 8, 280, 287, 289, 385
Weese, A. O., 307, 308
Weil, Simone, 262–263
Weismann, August, 91, 92, 202
Welch, Paul, 69, 142, 145, 149, 150
Wells, Anthony, 251, 256, 258, 261, 265
Wells, H. G., 251, 255–258, 261, 265, 419–420 (n. 68)
Wendt, Herbert, 265
West, Rebecca: birth date of, 250, 253; as Dame of the British Empire (DBE), 250, 259, 266; death of, 230, 250, 274, 334; education of, 254; family background of, 254; farm of, 260–261, 268; friendship and correspondence between GEH and, xii, 10–11, 200, 235, 250–253, 259–274, 315, 317–319, 337, 338, 387, 418 (n. 27); friendships and lovers of, 250–251, 256–258, 261; GEH as literary executor of, 11, 250, 274; on GEH's understanding of women, 272; health problems of, 254, 268, 269, 274, 335, 340; honors for, 259, 266; journalism career of, 11, 250, 253–259, 261–262, 266–267; and Margaret Seal Hutchinson, 216; marriage of, 70, 251, 258, 260; photographs of, 251, 266; as pseudonym, 254; son of, 251, 256, 258, 261, 265; tribute to GEH on his retirement, 271–273; and H. G. Wells, 251, 255–258, 261, 265, 419–420 (n. 68); writings by, 241, 252–259, 261–262, 264, 265–266, 329; and Yemaiel Aris, 337, 338
Westing, Arthur, 318
Wetzel, Robert G., 153, 155
"What Is Science For?" (GEH), 331–333, 359, 376
Wheeldon, Alice, 21
Wheeler, Bernice, 165
Wheldale, Muriel, 52
Whewell, William, 52
Whitehead, A. N., 22, 30
Whittaker, Robert H., 380–381
Wicken Fen, 56, 80, 85, 139, 157, 157, 205, 320–321, 324, 338, 359, 367
Wiener, Norbert, 236, 352
Wilcox, Mary, 66
Wilde, Oscar, 255
Wilhelmi, A. E., 207
Wilkinson, J. A., 77
Williams College, 65–66
Wilson, E. B., 94, 146

Wilson, Edward O., ix-x, xi, 26–27, 279, 290, 292, 374
Witkamp, Martin, 168
Wittgenstein, Ludwig, 56
Wittington, Harry, 147
Wolcott, Alexander, 261
Wollack, Anne, 148, 174
Women scientists, 8, 13, 42, 46, 56, 58, 66, 74, 195–197, 200–212, 267–268, 340, 385. See also specific scientists, such as Pickford, Grace
Women students: at Cambridge University, 21, 44–46, 50, 53, 56, 58, 60–62, 90, 200, 201–202, 385, 399 (n. 13); at Harvard University, 196; at University College, London, 399 (n. 13); at Yale University, 90, 99, 179, 200, 385
Women's suffrage. See Feminism
Woodruff, Lorande L., 84, 86–90, 92, 116
Woolf, Virginia, 45
World War I, 17, 21–22, 28–31, 42, 46, 54, 70, 223–224, 236, 258
World War II, 29, 130–131, 134, 156, 206–207, 238–239, 269, 354, 356, 360
World Wildlife Fund, 11, 303–304, 306, 328
Worthington, E. B., 143, 144
Wren, Christopher, 46
Wright, Helena, 24
Wright, Margaret, 99, 205–206, 210–212
Wulff, Janie, 334, 346

Yale University: Art Gallery of, 344–345; biogeochemistry at, 195–197, 276; Botany Department of, 128, 129, 318; compared with Cambridge University, 90–91; Conservation Program of, 310; decline of ecology at, 388–391; Ecology and Evolutionary Biology Department of, 391; fellowship application by GEH to, 83–87; food and tea for GEH and his visitors at, 223, 363, 398 (n. 7); Forestry School of, 298, 390, 391; GEH's archives at, xv, 11, 385, 398 (n. 33), 399 (n. 12); GEH's first years at, 89–99; GEH's laboratory at, 61, 118, 120, 124, 136; GEH's office at, 124, 363; GEH's retirement from, 270–273, 390; GEH's students at, 3, 4, 7–9, 11–13, 115–137, 155, 169, 198, 249, 275–276, 291–292, 331, 382–386, 388; GEH's teaching methods and courses at, xiii, 96–97, 119–120, 123, 173, 180, 195–197, 202–203,

291–293, 367–368, 396 (n. 6); Geochronometric Laboratory of, 130, 164;
Gesell Clinic at, 244; Harrison's graduate students at, 7–8, 92, 94–95, 99, 118,
122–123; Harrison's teaching methods
at, 94–96; "Hutchinson research group"
at, 8, 164, 169, 198, 382–385; instructorship in zoology for GEH at, 6, 83, 86–88,
87, 97; length of GEH's career at, 88, 89;
Markert as Biology Department chair at,
205, 209, 271, 388–391; Medical School
of, 195, 196, 207, 361; molecular biology
at, 388–391; Osborn Zoological Laboratory of, 89, 101, 104, 108; promotion and
tenure for GEH at, 92, 97, 111; recommendations written by GEH for students
at, 292, 354, 386, 389; Seychelles Expedition of, 314; social position of undergraduates at, 91; West's correspondence
in Beinecke Library of, 261, 419–420
(n. 68); Rebecca West's lectures at, 265–
266; women faculty at, 200, 201, 273–274,
388; women students at, 6, 90, 99, 179,
200, 385; Zoology Department organization and faculty of, 88–99, 111, 135, 387–
388. See also Hutchinson, G. Evelyn;
Linsley Pond; North India Expedition
(Yale University); Peabody Museum
(Yale); and other faculty and students
Yale University Press, 243–244, 350
Young, R. B., 77, 82

Ziegler, J. R., 289
Zinn, Donald, 8
"Zoological Iconography in the West after
A.D. 1200" (GEH), 328–329
Zoological Society of London, 231, 345